American Institute of Chemical Engineers

# Path Forward to a Hydrogen Economy

Topical Conference at the
2007 AIChE Annual Meeting

November 4-9, 2007
Salt Lake City, Utah, USA

**Printed from e-media with permission by:**

Curran Associates, Inc.
57 Morehouse Lane
Red Hook, NY 12571
**www.proceedings.com**

**ISBN: 978-1-60423-843-3**

**Some format issues inherent in the e-media version may also appear in this print version.**

ISBN: 978-1-60423-843-3



American Institute of Chemical Engineers
Proceedings
Three Park Avenue
New York, NY 10016-5991
Phone: 212-591-8100

www.aiche.org

American Institute of Chemical Engineers

Path Forward to a Hydrogen Economy

# TABLE OF CONTENTS

## Results of the EC funded project HYTHEC on massive scale hydrogen production via thermochemical cycles

**Alain LE DUIGOU[1] – Jean-Marc BORGARD[1] – Bruno LAROUSSE[1] – Denis DOIZI[1] – F. WERKOFF[1]**

**Ray ALLEN[2], Bruce C. EWAN[2], Geoff H PRIESTMAN[2], Robin DEVONSHIRE[2], Rachael ELDER[2], Manu MINOCHA[2]**

**Giovanni CERRI[3], Coriolano SALVINI[3], Claudio CORGNALE[3], Ambra GIOVANNELLI[3] -**

**Martin ROEB[4], Nathalie MONNERIE[4], Mark SCHMITZ[4], Adam NOGLIK[4], Christian SATTLER[4]**

**Daniel de LORENZO MANZANO[5], Alfredo ORDEN MARTINEZ[5], Jorge CEDILLO ROJAS[5]**

**Stephane DECHELOTTE[6], Olivier BAUDOUIN[6]**

[1]Commissariat à l'Energie Atomique (CEA) – CEN Saclay 91191 Gif Sur Yvette Cedex, FRANCE / alain.le-duigou@cea.fr

[2]University of Sheffield (USFD) – Firth Court, Western Bank S102TN, Sheffield, UK r.w.k.allen@shef.ac.uk

[3]Università degli studi – Roma tre (DIMI – I) – Via Della Vasca Navale 79, 00146 ROMA, ITALIA / cerri@uniroma3.it

[4] Deutsches Zentrum für Luft und Raumfahrt E. V. (DLR) - Linder Hohe, 51147 Cologne, GERMANY / Martin.roeb@dlr.de

[5] Empresarios Agrupados (EA) – Magallanes 3, 28015 Madrid SPAIN / ddm@empre.es

[6] PROSIM SA – Stratège Bâtiment A, BP 27210 F-31672 Labège Cedex, FRANCE / olivier.baudouin@prosim.net

**ABSTRACT**

HYTHEC - HYdrogen THErmochemical Cycles – is a EC funded project to investigate the potential for massive hydrogen production by the Sulfur-iodine (S_I) thermo-chemical cycle and the Hybrid-Sulfur (H_S) cycle. These cycles have in common the $H_2SO_4$ decomposition reaction. The approach for accomplishing the project objective is multi-faceted. It includes the thermochemical steps modelling, flow-sheeting, experimental approach to the decomposition of $H_2SO_4$ in a solar furnace at high temperatures, and of the partial pressures above a mixture of HI, $H_2O$ and $I_2$, modelling and experimental exploratory research into alternative, low energy separation techniques relevant to the S_I process such as membranes, and the industrial scale-up, safety and cost analysis of $H_2$ production by coupling S_I cycle to a nuclear reactor and the Hybrid-Sulfur cycle to a solar and/or a nuclear source.

A first analysis round of both cycles has been completed : S_I flow-sheet and industrial scale-up including a first economic evaluation, experimental results on $H_2$ production step (liquid vapour equilibria and membranes) and on $H_2SO_4$ solar decomposition, and solar coupled H_S cycle modelling (flow-sheets, solar components sizing and costs). The second experimental step is ongoing, such as industrial scale-up and costs assessments of both cycles, that take into account modelling from the first experimental round. HYTHEC collaborations have been engaged with other European projects (RAPHAEL, HYSAFE, EXTREMAT). The HYTHEC project has been recognized and endorsed by the Steering Committee of the International Partnership for the Hydrogen Economy (IPHE).

**Keywords:** hydrogen, thermochemical, nuclear, solar, Sulphur Iodine, Hybrid Sulphur

## 1. INTRODUCTION

Today, Hydrogen is mainly produced from fossil resources. In the long term, given the prospect of an increasing energy demand (+20% by 2020 & expected to double by 2030, with a possible threefold increase by 2050), a lack of fossil resources and limitations on the release of greenhouse gases, only water and biomass are viable, long term candidate raw materials for hydrogen production. The two processes that have the greatest likelihood of successful massive hydrogen production from water are electrolysis and thermo-chemical cycles. Thermo-chemical cycles are processes where water is decomposed into hydrogen and oxygen via chemical reactions using intermediate elements which are recycled. As the heat can be used directly, these cycles have the potential of a better efficiency than alkaline electrolysis. The required energy can be either provided by nuclear energy or by solar energy, and, since the production requires a continuous supply of heat, hybrid solutions including solar and nuclear energy input are conceivable and desirable.

## 2. HYTHEC: THE SEARCH FOR AN EFFICIENT HYDROGEN PRODUCTION ROUTE FROM RENEWABLES

The Sulfur-Iodine (S_I) cycle (Figure 1) has been extensively studied by the General Atomics Company [1]. Japan has recently built a small pilot plant of this process [2]. Thus, the S_I cycle seems to be the best known, internationally leading candidate, as a promising thermo-chemical option. Beyond that, the Hybrid-Sulfur (WH) process (Figure 1) which offers a combination of electrolysis and thermo-chemical reactions is also the focus of much international attention. The objective of HYTHEC is to investigate, and compare, the effective potential of those two leading candidates, which have in common the high temperature process step: the $H_2SO_4$ decomposition reaction. These thermo-chemical options can only be considered viable if they meet two major criteria: demonstrated large scale technical feasibility, and competitive cost.

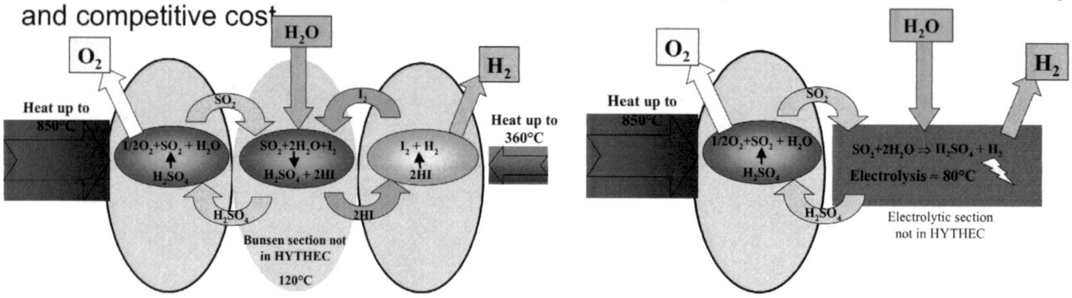


**Figure 1: The Sulphur-Iodine and Hybrid-Sulphur (WH) cycles**

## 3. BASIC KNOWLEDGE AND NEEDS FOR S_I AND WH THERMO-CHEMICAL CYCLES

The S_I cycle :

In terms of the chemical reaction stages involved, the S_I process may be summarized as follows:

R1 - $9 I_2 + SO_2 + 16 H_2O \rightarrow (2HI + 10H_2O + 8I_2) + (H_2SO_4 + 4H_2O)$ [120°C]

R2 - $2 HI \rightarrow H_2 + I_2$ [220-330°C]

R3 - $H_2SO_4 \rightarrow SO_2 + H_2O + ½ O_2$ [850°C]

These sum to:

R4. $H_2O \rightarrow H_2 + ½ O_2$

The first reaction, called the Bunsen-reaction, proceeds exothermically in the liquid phase and produces two immiscible aqueous acid phases whose compositions are aqueous sulfuric acid and a mixture of hydrogen iodide, iodine and water named $HI_x$. HI and $H_2SO_4$ are then decomposed according to reactions R2 and R3. Only high temperature heat sources, such as nuclear VHTR or solar energy devices, may be relevant for the reaction R3. Reaction R3 is the major endothermic reaction releasing water, oxygen and sulfur dioxide. It takes place in the vapor phase in a catalytic reactor at about 850°C. The concentration by distillation of the two acids HI and $H_2SO_4$ involves significant energy consumption, which has a direct influence on the efficiency of the cycle. Among all options available for the $HI_x$ section, HYTHEC is focusing on the reactive distillation concept as proposed by Knoche et al. [3]. Their approach allows this step to be done in one reactor so it seems to have the highest efficiency potential. An improved version has already been proposed in [4]. However necessary vapor-liquid equilibrium data and energy integration calculations are not available so that it is difficult to develop a reliable conceptual design. On the other hand, we propose to investigate the use of membranes in the distillation section of $HI_x$ in order to get the maximum HI concentration in the vapor phase, and therefore to improve the overall thermal efficiency of the process.

A solar furnace gives the opportunity to study the chemical reaction in an original manner, both at nuclear VHTR reactor temperatures (HTR: 850 – 900°C) and at higher ones (1100 – 1200°C). The kinetics could thus be improved, and the possibility exists to directly split a concentrated $H_2SO_4$ mixture into $SO_2$ + ½ $O_2$, even without the use of a catalyst as needed at lower temperatures.

The Hybrid-Sulphur (WH) Cycle :

This cycle is a two-step thermo-chemical cycle for decomposing water into hydrogen and oxygen. The reaction scheme is as follows:

R9 $SO_2 (g) + 2 H_2O (l) = H_2 (g) + H_2SO_4$ (aqueous) [electrolysis, 25-100 °C]

R10 $H_2SO_4 (g) = H_2O (l) + SO_2 (g) + ½ O_2 (g)$ [thermo-chemical, 850°C]

Due to the fact that the Westinghouse cycle has the decomposition of the sulphuric acid in common with the S_I cycle, it is worthwhile studying this cycle in the same project, even if the study is limited to literature review and engineering calculations to minimize cost.

For both S_I and WH cycles, industrial scale-up studies are of great importance, for the assessments of the safety aspects of the process, the feasibility of the main components at industrial scale, and $H_2$ production costs. Specific R&D needs may arise too. Beyond that, the combination of electrolysis and thermo-chemical process steps in the Westinghouse process offers the opportunity for a combi`ned use of solar and nuclear heat and this will be considered as well.

## 4 – THE HYTHEC PROJECT

### 4.1. Assessment of the S_I thermo-chemical cycle; technical and industrial viability

A critical analysis of previous flowsheets published in the literature has been undertaken with optimization of heat transfer using the process flowsheeting code ProSimPlus. A minimum temperature difference of 10°C between hot and cold streams has been assumed for internal heat recovery. The $H_2SO_4$ section is handled with a multiple flash distillation, followed by $H_2SO_4$ evaporation and $SO_3$ decomposition. $SO_3$ is recombined in the late stages of the distillation process to optimize the global efficiency. The $HI_x$ section is handled by a reactive distillation process as proposed in [4]. To enhance the heat recovery a water heat pump has been adopted to transfer the heat rejected by the column condenser to the reboiler. The flash at the exit of the column is needed to recover some of the latent heat of water for efficiency optimization. The flow-sheets of the $H_2SO_4$ and $HI_x$ sections are given in Figure 2. The Bunsen section has been assumed according to General Atomics as reported in [1]. As a result to produce 1mol of hydrogen the total heat required from the external source is of 763 kJ. Realist efficiency of S_I cycle (considering HHV) is ~ 35%.

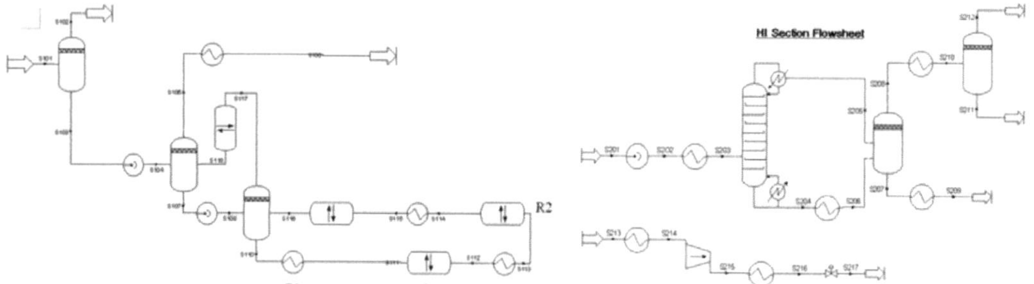


**Figure 2: The Sulphur-Iodine and Hybrid-Sulphur (WH) cycles**

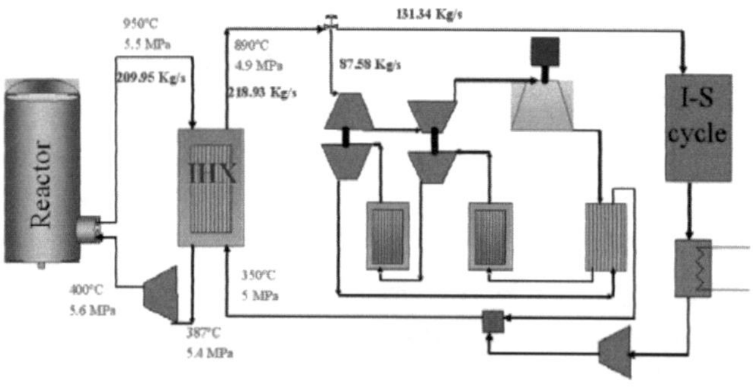


| | |
|---|---|
| Nuclear Power reactor [MW] | 600 |
| H2 production | 100% |
| IHX Inlet / Outlet Helium temperature [°C] | 350/890 |
| IHX Inlet / Outlet S_I temperature [°C] | 890/422 |
| Helium mass flow rate [kg/s] | 218.93 |
| Helium pressure [MPa] | 5 |

**Figure 3: S_I Cycle coupling to a Nuclear Reactor, "self-sustaining concept"**

The coupling of the S_I cycle to a Nuclear Reactor has been studied. Due to the high temperatures needed for the $H_2SO_4$ decomposition, the best connection option is with a VHTR (Very High Temperature Reactor). Figure 3 shows a connection scheme as an example, between HYTHEC and the European Project RAPHAEL. This scheme represents a self-sustainable plant concept, in which, in addition to the heat supply to the S_I cycle, the electrical demand of the internal consumers is provided by the nuclear reactor. Then, the high temperature flow coming out of the reactor is passed to an Intermediate Heat eXchanger (IHX), which provides heat to a secondary loop that interacts with the S_I cycle components. The flow goes partially to the S_I Cycle and another part goes to a Brayton cycle to generate electricity equal to that required for the S_I cycle consumption. This leads to a daily $H_2$ production of 110 tons/day.

In addition to this static analysis, a dynamic model is being used to evaluate the interactions of the chemical plant's instability towards the reactor's behaviour during transient or hazardous events.

An extensive safety assessment of the S_I cycle has also been performed. It includes a chemical plant hazards identification and description, safety considerations regarding the chemical plant, a preliminary Failure Mode Effects and Criticality Analysis (FMECA), and Chemical plant – Nuclear power plant interaction safety features analysis. Events from the VHTR affecting the $H_2$ production plant and vice versa have been addressed. That includes a preliminary calculation of Tritium production in the reactor core and its transmission to the $H_2$ storage devices as a consequence of a series of leaks. In addition, an estimate of the safety distance between both plants, in order to protect the nuclear reactor facilities from an eventual fire or explosion of $H_2$, have been performed with calculations based on the US and German Guides.

In order to select suitable plant design solutions the following main aspects have been taken into consideration:

- definition of the process in terms of required unit operations and related process quantities (pressures, temperatures, compositions, heat and work requirements and so on);
- coupling between the nuclear thermal source and the S_I process;
- set up of the industrial plant lay out in order to find the most effective arrangement to perform the required operations;
- selection of appropriate technology and materials for each plant component;
- sizing of components and hydrogen production cost evaluation.

To perform the component sizing and cost accounting activities, $H_2SO_4$ and $HI_x$ section flow-sheets shown in Figure 2 and VHTR data reported in Figure 3 have been taken as reference. With these assumptions, only one design degree of freedom (DOF) has to be specified to completely define the VHTR-S_I cycle matching. The hydrogen production level has been assumed as DOF.

A $H_2$ production level of 633 mol/s with an overall plant efficiency of some 35% has been taken as reference. Plant components have been sized and costs evaluated by using standard chemical engineering methods and by means of "ad hoc" developed databases. The $SO_3$ catalytic decomposition reactor has been sized as a tubular one. The $H_2SO_4$ concentration endothermic flashes have been sized as kettle boiler units. The $HI_x$ section heat recovery devices have been assumed and sized as shell and tube heat exchangers. Suitable acid resistant materials have been selected, taking ceramic materials into particular consideration. Costs of installed equipment have been evaluated by adopting a factorial method. The investment cost of the $HI_x$ section has been found to be approximately eight times greater than that of the $H_2SO_4$ section, and roughly 80% of $HI_x$ section overall cost is made up of heat recovery devices. A hydrogen production cost of approximately 5.2 €/kg$H_2$ has been estimated. The influence of investment costs on overall production costs is of some 40%.

In order to find the best compromise between investment and operating costs a parametric analysis has been carried out by varying the hydrogen production level keeping the overall available thermal power from the nuclear reactor at the reference value (600 MW). For different $H_2$ productions detailed plant sizing and cost accounting have been performed. Results are reported in Figure 4. The reduction of $H_2$ production levels has led to a decrease of investment costs mainly due to the reduction of the size of the $HI_x$ section recovery heat exchangers. The minimum cost (4.2 €/kgH2) is achieved at 540 mol/s and does not correspond to the maximum achievable efficiency. In order to obtain further reduction of costs, technology improvements concerning the HIx section recovery heat exchangers have been introduced. To enhance heat transfer coefficients bare tubes have been replaced by corrugated ones. In Figure 5 results achieved by adopting HPHE (High Performance Heat Exchangers) are reported and compared with those already given in Figure 4 (Bare Tubes).

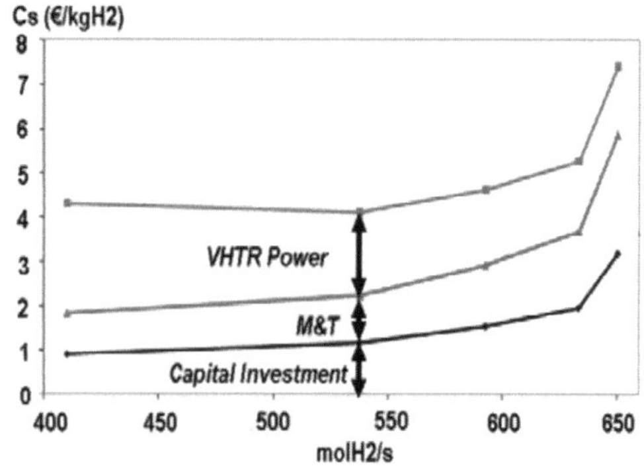


Fig.4: H2 specific costs vs. H2 production

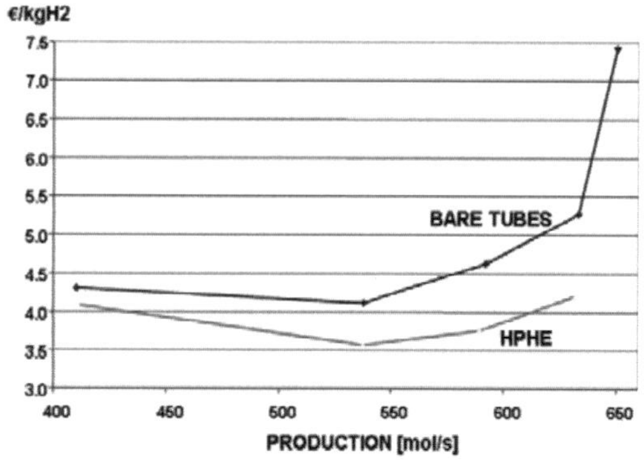


Fig. 5. Hydrogen production costs: HPHE and Bare Tubes comparison

A reduction of costs is observed all along the production range taken into consideration and the minimum is still achieved at 540 mol/s. A value of some 3.6 €/kg$H_2$ is found, which represents a noticeable 40% cost reduction if compared to the reference case. Finally the optimisation of $HI_x$ section heat recovery network at 540 mol/s by adopting the pinch technique has led to a production cost of some 3.4 €/kg$H_2$. To achieve further hydrogen production costs reductions future activities are going to be focused on the possibility to use less expensive materials, the adoption of compact heat exchangers and the improvement of plant flow-sheet to reduce thermal as well as electrical demand.

### 4.2. The S_I cycle H2 production section / HIX system liquid-vapor equilibria measurements

The $HI_x$ section is the most important section for the efficiency of the sulphur-iodine cycle. To design and optimize the reactive distillation column we have chosen, knowledge of the total and partial pressures of the liquid vapour $HI_x$ mixtures is required up to 320°C and 100 bar. We have developed a progressive methodology around three experimental devices which contain these corrosive and concentrated mixtures:

- I1 is an experimental device devoted to the measurement of the total pressure up to 130 bar and 315°C. It is composed of a microautoclave made of tantalum and placed in a thermoregulated oven. A pressure gauge equipped with a tantalum

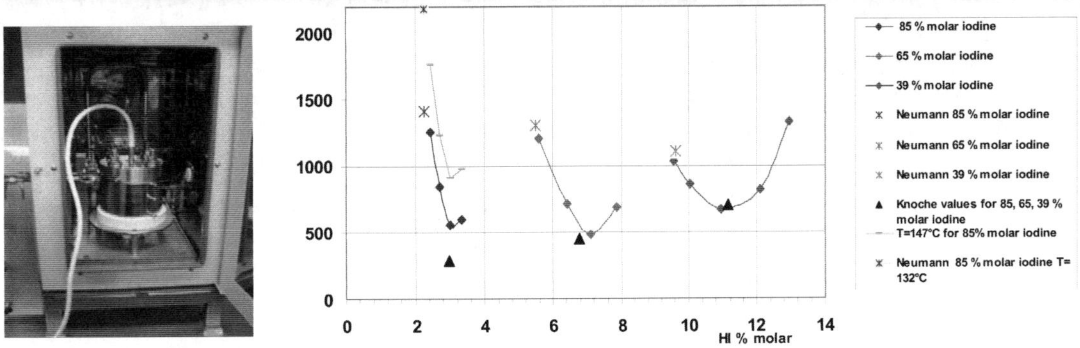


**Figure 7: I2 device / total pressures for ternary systems**

membrane enables the pressure measurements for different $HI_x$ compositions. First experiments have been completed with water up to 30 bar and 240°C. Four HI-H2O binary media close to azeotropic compositions have been tested, and HI-I2-H2O ternary system containing 39% of iodine.

- I2 is an experimental device devoted to the qualification of the analytical diagnostics for partial pressure measurements up to 130°C and 2 bar (Figure 6). It is composed of a glass cell equipped with a total pressure gauge and placed in a thermoregulated oven. This cell is equipped with two optical pathlengths because optical "online" diagnostics have been chosen for partial pressures measurements in order not to alter the vapour composition and prevent tedious manipulations. UV-Visible spectrometry has been chosen to measure iodine concentration and FTIR spectrometry for HI and $H_2O$ concentration measurements. Total and partial pressure measurements have been conducted in the experimental device I2. Pure samples, binary HI-$H_2O$ and ternary HI-$H_2O$-$I_2$ mixtures have been studied; the results have been compared with the literature and with Prophy code. Up to the azeotrope, a good agreement is observed, beyond the azeotrope, the total pressure measured is higher than the pressure calculated. Ternary systems with an iodine composition close to the Bunsen exit (39 % molar) have also been studied and exhibit the same behaviour beyond the azeotrope (Fig.7).

- I3 is an experimental device devoted to the partial pressure measurements of the $HI_x$ mixtures in the process domain. It is composed of the I1 device equipped with a vapour chamber. The same "online optical techniques" are used. First experiments have been conducted on pure water to validate the choice of FTIR spectrometry. Experiments are in progress on iodine to validate the choice of UV-Visible spectrometry

A campaign of experiments exploring different iodine compositions is currently under way. The objective of the work is to collect unknown data, especially beyond the azeotrope, which will be included in Prophy code for the scaling of HI reactive distillation column.

### 4.3. The S_I cycle H2 production section / The use of membranes to improve the S_I cycle $HI_x$ section efficiency

Applying a liquid phase membrane separation unit to the $HI_x$ processing section has two potential advantages, an increase in efficiency and less extreme operating conditions. A membrane separation process could be used to separate the $H_2O$ and HI, thus having the effect of dehydrating the $HI_x$ by removing some water from the process stream. The membrane separation unit has been considered in the following system positions: on the column feed, at the column reboiler and on a column sidestream.

Ideally the membrane would allow HI through and reject water and iodine as there is much less HI than water and iodine in the feed stream. In the majority of membranes, however, it is likely that $H_2O$ will have a higher flux than HI. The amount of water removed from the $HI_x$ stream depends on the fraction of the feed that is distilled and the flux ratio of the components. A distillation column can separate the distillate to give the azeotrope and pure water [5] . As iodine is present there may be added complications due to membrane fouling and a high HI vapor pressure.

The three key factors that need to be considered when selecting the membrane unit are the solution as a highly corrosive, strongly acidic, aqueous system, the process temperature about 398K, the selectivity of the membrane that must be selective to either HI or $H_2O$.

Both PTFE and ceramic membranes fulfill these criteria [5]. Recent work in America shows that Nafion-117 membranes can be used in a pervaporation process to remove water from aqueous HI [6]. The results are very promising for the application of liquid phase membrane separations to the SI process.

In order to investigate the effect of dewatering, simulations were carried out using the process flowsheeting code ProSimPlus, using a modified version of the process flowsheet which includes the membrane separation unit (figure 8). The membrane is modeled as a splitter with zero enthalpy drop. This means that the retentate exits at a lower temperature than it would in practice. This is accounted for in the heat integration and so will not significantly affect the efficiencies calculated.

The base efficiency of using a 50 bar column with no membrane separation unit is 0.379. As shown in figure 8, with 9% feed dewatering this value is increased to 0.412, a significant increase. A maximum in efficiency occurs as the power required by the pump and that required surplus to heat integration decreases, whereas the additional power required by the heat pump increases. Using a 55 bar column the maximum efficiency obtained is 0.418, again at 9% feed dewatering. As the column pressure decreases the efficiencies obtainable also decrease. With approximately 7% feed dewatering a 30 bar column has the same efficiency as the 50 bar column with no membrane. A decision as to the best column pressure would depend on a detailed sizing and cost evaluation. With 1% HI removal along with 9% dewatering the maximum efficiency decreases to 0.401. Although this is lower than with no HI removal it is still considerably better than the efficiency obtained with no membrane unit.

Placing a membrane separation unit at the column reboiler or on a sidestream increases the efficiency slightly although it never exceeds 0.382. The simulations therefore suggest there is little benefit in applying a membrane unit in this way.

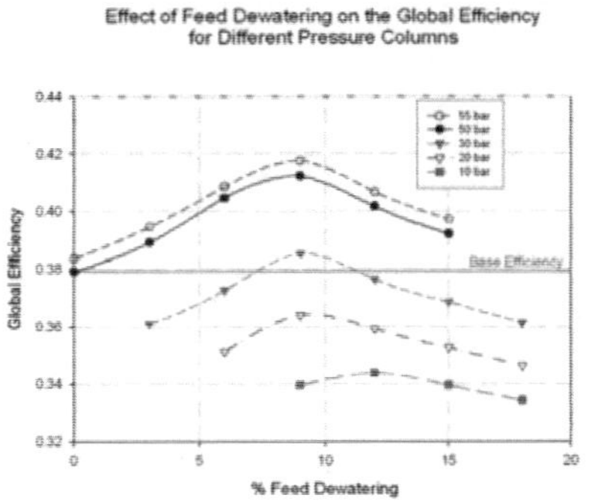


Figure 8 – theoretical effect of feed dewatering

Figure 9 – selective membranes experimental device

### 4. 4. Assessment of the WH thermo-chemical cycle, for a solar and/or nuclear driven process

The co-application of the electrolysis and the thermo-chemical step in the Westinghouse process offers the opportunity for a combined use of heat and power. The required electrical and thermal energy can be either provided by concentrated sunlight or by a very high temperature nuclear reactor. This opens a wide variety of operational strategies. Amongst different configurations, two "solar-only" cases were selected. The first one foresees a directly irradiated solar receiver-reactor which provides heat to generate electricity for the electrochemical step and at the same time splits sulphuric acid at 1200°C. The second one comprises an indirectly irradiated solar heat-exchanger-reactor, which catalytically splits sulphuric acid at 850°C and a receiver providing heat for the production of solar electricity. The overall flow-sheets of the both solar Westinghouse processes are given in Figure 10.

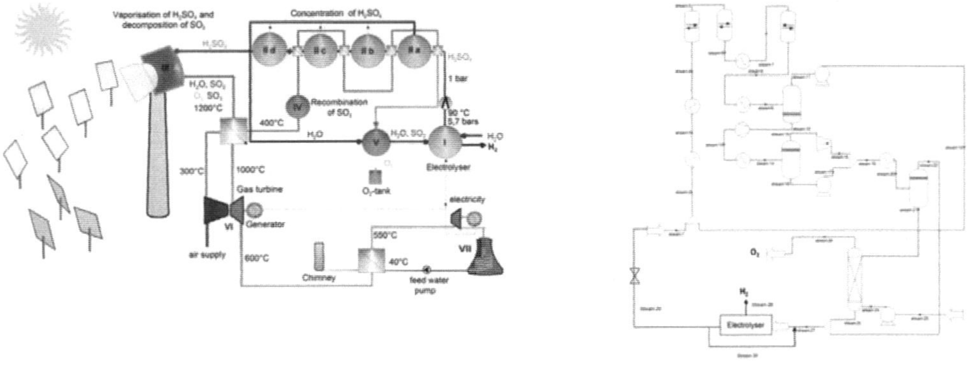


**Figure 10: Overall flow-sheet of the solar-only plant for the Westinghouse process**

The calculations of heat balances and of the solar field size and shape were made at the region close to Lake Nasser in Egypt for a plant with an annual average thermal power of 50 MW entering the chemical part of the plant. As the solar energy fluctuates strongly, solar plants have to harvest more energy during hours of sunshine to compensate outages at night or during cloud passages to reach the same annual level of constant nuclear energy input. The considered 50 MWth solar plant requires a maximum thermal capacity of 140 MWth. The decomposition of $H_2SO_4$ is powered by concentrated solar energy of a central receiver system (CRS), consisting of sun-tracking mirrors (heliostats), through a receiver at the top of a tower. Time design point is June, 21st, noon. The solar power plant requires 2677 heliostats of each 121.34 $m^2$. The tower height is calculated at 200 m.

The production varies with the season so that e.g. the peak hydrogen production is in June with a maximum output of 406 g/s. The calculations of the chemical part of the process were performed with ProSimplus for a steady state at the design point of the plant. Preliminary sizing of the main components has been done. For the high temperature heat exchangers, tubes are made of Incoloy, whilst for the other components, tubes are made of SiC.

Hybrid solutions including solar and nuclear energy input are also conceivable and analyzed with regards to their technical and economic feasibility. Different operation and plant concepts are generated, including the solar and nuclear supply of heat for the thermo-chemical step and of nuclear and solar power for the electrolysis step. The studied flow-sheets were selected regarding the technical feasibility, the feasibility of a constructive interaction of nuclear and solar energy supply, the connection between the solar and the nuclear heat circuit, the operation time and temperature, safety aspects, and finally the expected plant efficiency and economics. As a result one hybrid case has been selected, which foresees a catalytic splitting of $H_2SO_4$ at 850 °C by indirect heating.

The characteristics, composition, temperature, pressure, energy balance, of each stream and unit operation of the flow sheet have been calculated. The hybrid powered process is running with two $H_2SO_4$ splitting units: one is powered by solar energy and the other by nuclear energy. The required absolutely constant and undisturbed operation on the nuclear side of the process can only be ensured by widely separating process units depending on intermittent solar supply from the part powered by nuclear heat. There is only one electrolyser unit which is powered constantly with nuclear electricity. The solar thermal splitting process is carried out at 850 °C to allow the hybridisation with nuclear heat. $SO_2$ and $H_2SO_4$ storages are applied to ensure a constant supply to the electrolyser unit and thereby to ensure a constant production of hydrogen. The calculations are made for an average annual thermal power of 50 MW. The component sizing for the hybrid plant is in progress. It will enable an economic comparison between solar-only plants and plants powered by nuclear and solar energy simultaneously.

### 4. 5. Development of a dedicated solar receiver-reactor for the decomposition of sulphuric acid

The decomposition of sulphuric acid is a highly heat consuming and at the same time technically challenging step in the sulphur-iodine cycle as well as in the Westinghouse cycle. One important task of this project is the investigation of sulphuric acid decomposition by concentrated sunlight in the DLR solar furnace in Cologne with the development of a technology and a process to allow the coupling of concentrated solar radiation into the $H_2SO_4$ splitting process. A significant enhancement of the splitting of $SO_3$ under the influence of concentrated solar radiation was showed when using a receiver-reactor which allows for direct absorption of sunlight by the reactants itself, giving the chance to operate the process without a catalyst or to accelerate the reaction and thereby to increase the potential throughput [7].So the goals of the current activity are the development of a direct absorbing receiver-reactor, the porous absorber reactor, and of a corresponding process, the verification of their feasibility and viability with respect to the splitting of sulphuric acid, and their assessment in comparison to other reactor and process concepts.

The porous structure ([8]) has to be adapted to the requirements of the two-step splitting reaction of concentrated $H_2SO_4$, which are connected to the characteristics of corrosive liquid and gases, to the performance of a change of phase, to the kinetics of the reactions, and to the temperature levels. The central innovation of the concept is the execution of the evaporation of sulphuric acid in a solar heated porous structure. The liquid sulphuric acid is fed onto a solar heated foam with a large inner surface to initiate a rapid and homogenous evaporation of the acid at the inner surface of the porous structure. The necessary heat is supplied by the sun and transferred to the liquid by the porous structure. During and directly after the evaporation, the sulphuric acid is dissociated. The resulting $SO_3$ molecules have to be split into $SO_2$ and $O_2$ to provide the necessary feed for the ensuing step of the S_I cycle and Westinghouse cycle respectively. For the execution of this splitting step at 1000-1200 °C another porous structure is foreseen: a honeycomb structure stable and robust enough to stand the required temperature level and capable of being coated by catalyst materials if necessary.

Due to the temperature level and the corrosive potential of the reactants only SiSiC seems feasible for the honeycomb structure. Different options are conceivable as construction materials for the foam vaporiser with SiSiC as the favourite option. The design of the receiver-reactor comprises as main elements the modular arrangement of ceramic components (SiSiC) as heart of the reactor which contains the reactive zone; a window; a housing; a support for the central ceramic part of the reactor; a flange junction between reactor and off-gas line; a suitable inlet enabling the feed of liquid sulphuric acid into the reactor. Figure 11 shows the design principle of a porous absorber reactor to fulfil those requirements.

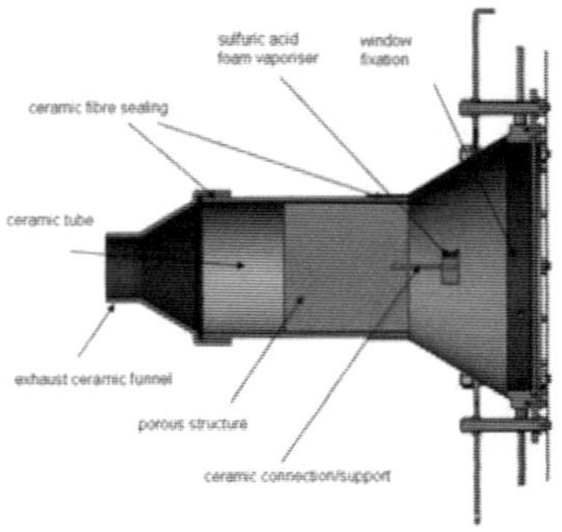


Figure 11: Design of the porous absorber reactor

The reactor concept has been investigated with a finite element model using ANSYS as well as with computational fluid dynamics (CFD) to assist in enhancing and optimising the geometry and arrangement of components. The effect of the diameter and of the distance from the honeycomb structure of the foam vaporiser, the effect of the position of the focal plane in the axial direction of the reactor and the effect of the solar flux power have

**Figure 11: Design of the porous absorber reactor**

The reactor thermal behaviour assures sufficient temperature distribution and heat transfer. Structural deformations and stresses are acceptable, the quartz window holder design is appropriate and no leak tightness hazards are foreseen. The simulations predict that the use of porous structures turns out a good solution to achieve a uniform temperature field in the reactor body. Moreover they seem suitable to allow a homogeneous vaporization and an adequate residence time in order to achieve a sufficient dissociation rate.

After being heated by heat supplied by a blower-heater and achieving a certain temperature level the foam was fed in a first time with water droplets. The droplets impinged on the foam at a depression at the topside of the foam and penetrated into the foam. Three different regimes of operation were observed by thermographic inspection of the foam. In regime A, a comparably high temperature of the foam is combined with a comparably low flow of liquid. This leads to an excess of the so called Leidenfrost temperature of the liquid: straight after a brief contact with the foam a thin film of vapour is built surrounding the individual droplets. The film decreases the heat transfer to the droplet significantly and thereby delays the vaporisation process. The droplets cross the surface with little friction and exit the foam without being vaporised. If the foam is charged with higher volume flows of liquid at constant heat supply the regime changes (regime B). Due to the higher heat demand for the evaporation, the temperature significantly decreases in a zone adjacent to the impingement of the droplets (<200 °C). The surface temperature falls below the Leidenfrost temperature of the liquid. The foam is wetted and the liquid is completely vaporised without any liquid exiting the foam. Application of even higher volume flows of liquid combined with lower maximum temperatures lead to another regime of operation (regime C) which is not desirable: the foam is more than completely wetted and the heat supply is too low to vaporise the complete amount of liquid.

The decomposition rate of $H_2SO_4$ in a laboratory-scale reactor heated by an electric furnace was measured for several temperatures and sulphuric acid concentrations. A significant kinetic barrier to the decomposition of $H_2SO_4$ was observed even for large residence times (>40s) in the high temperature reactor. Satisfactory reaction yields (>90%) were obtained only at temperatures higher by about 300-350 K compared to thermodynamic predictions.

Three experimental campaigns were performed in the solar furnace in Cologne. Firstly the feasibility of a solar decomposition of sulfuric acid in a receiver-reactor containing volumetric absorbers was investigated and proven. Then the reactor (Figure 12) was qualified at different operating points. Finally the receiver-reactor and strategy of operation was iteratively optimized with respect to chemical conversion and reactor efficiency.

Several test series were performed with variation of the absorber temperature, the mass flow and the dilution rate. Partial pressure of $SO_3$, residence time, absorber temperature, and the kind of catalyst applied were identified and quantified as parameters with the most relevant influence on chemical conversion and reactor efficiency. The operation behaviour observed and the detailed knowledge of dependencies of different operation parameters assist in evaluating the potential of scaling up the described technology.

The concept of a volumetric receiver-reactor for the decomposition of sulfuric acid is feasible as has been proven. High conversions almost up to the maximum achievable conversion determined by thermodynamic equilibrium can be achieved with a platinum catalyst. A less expensive catalyst, iron oxide has been tested as well exhibiting activities less than platinum but still sensible when used as coating on a SiC honeycomb. An operation without any catalyst and applying only homogeneous decomposition appears not profitable as the low measured conversions show.

**Figure 12: Reactor in the solar furnace**

The evaporation of sulfuric acid on ceramic foams is well suited for a direct application of solar energy. A modified design of the reactor with a separated chamber for the evaporation enables a larger vaporizer and therewith a higher flow rate and a separate control of the solar input to the vaporizer. Higher mass flows of the reactants improve the reactor efficiency because energy losses are almost independent of the mass flow. The mass flow in the performed experiments was limited by the capacity of the foam vaporizer which could not be enlarged because of shadowing of the honeycomb absorber. As the reradiation is the main reason for energy losses in the reactor a cavity design can also raise the efficiency.

## 5 - CONCLUSION

HYTHEC is a European collaboration involving a restricted number of partners (6, in 5 countries) to give a initial evaluation, both from technical and economic points of view, of a potentially interesting route for future $H_2$ production, via $CO_2$ free Thermo-chemical Cycles. It has concentrated mainly on the Sulphur-Iodine cycle, and to a lower extent the Westinghouse cycle as an alternative "hybrid" solution.

On the basis of an S_I cycle reference flow-sheet, and a "self-sustaining" coupling scheme with a VHTR (600 MWth), scale-up to industrial levels and a cost estimation, performed using standard chemical engineering methods based on suitable materials and devices to support a high temperature aggressive environment, we showed that the minimum cost (4.2 [€/kgH2]) does not correspond to the maximum achievable efficiency, due to the importance of heat exchange recovery. Further cost improvements have been obtained by introducing high performance heat exchangers (3.6 [€/kgH2]) and by optimizing the plant lay out (3.4 [€/kgH2]). To achieve further hydrogen production cost reduction, future activities will be focused on the possibility of using less expensive materials, adopting compact heat exchangers and flowsheet improvements to reduce thermal as well as electrical demand.
HIx section measurements improved the fundamental knowledge of the complex liquid vapour equilibrium system, and led to a new preliminary thermodynamic model which, when considered in a new flow-sheet, gives promising trends toward lower $H_2$ production costs. A first distillation and pervaporation membrane database and characteristics have been given. The best use and place of membranes is feed dewatering (up to 4% efficiency improvement) upstream the HIx section. A first round of stability testing, fluxes and selectivity properties of three membranes has been carried out as well as the first ever Raman spectra measurements of the gas and liquid phases of HIx solutions. Flux and selectivity experiments are ongoing.

The concept of a volumetric receiver-reactor for the decomposition of sulfuric acid has been proven to be feasible. High conversions almost up to the maximum achievable conversion determined by thermodynamic equilibrium can be achieved with a platinum catalyst. The evaporation of sulfuric acid on ceramic foams is well suited for a direct application of solar energy. As the reradiation is the main reason for energy losses in the reactor a cavity design can also raise the efficiency.

A 100% solar WH process analysis has been completed, including thermal part loading and solar components sizing, and a full cost evaluated. Nuclear and hybrid concepts evaluations are ongoing, including safety aspects.

This project is funded by the European Community - Sixth Framework Program Priority [6.1] - Sustainable Energy Systems, Medium to Long Term – (contract number : 502704).

## REFERENCES

1. G.E.Besenbruch and coll – GA-A18257 (1982)

2. H. Nakajima, K. Ikenoya, K. Onuki et al. – "Closed-Cycle Continuous Hydrogen Production Test by Thermochemical S_I Process", Kagaku Kogaku Ronbunshu, 24 (1998) 352 (in Japanese)

3. M.Roth, K.F. Knoche – Thermo-chemical water splitting through direct HI decomposition from $H_2O$ -HI- $I_2$ solutions - Int. J. Hydrogen Energy, 14, N°8, 545-549 (1989).

4. Stephen Goldstein, Jean-Marc Borgard and Xavier Vitart - Upper bound and best estimate of the efficiency of the iodine sulphur cycle - International Journal of Hydrogen Energy, Volume 30, Issue 6, May 2005, Pages 619-626

5. Elder, R. H., Priestman, G. H., Ewan, B. C. and Allen, R. W. K. (2005) The Separation of $HI_x$ in the Sulphur-Iodine Thermochemical Cycle for Sustainable Hydrogen Production, *Trans IChemE, Part B, Process Safety and Environmental Protection,* **83,** 343-350.

6. Orme, C. J., Jones, M. G. and Stewart, F. F. (2005) Pervaporation of water from Aqueous HI using Nafion-117 Membranes for the Sulfur-Iodine Thermochemical Water Splitting Process, *J Membrane Sci,* **252,** 245-252.

7. E. Lüpfert, K.-H. Funken, in: M. Becker, M. Böhmer (eds.): Solar Thermal Concentrating Technologies, Proc. of the 8$^{th}$ Int. Symposium, Oct. 6-11, 1996, Köln, Germany, Vol. 3, C.F. Müller-Verlag, Heidelberg.

8. T. Fend, B. Hoffschmidt, R. Pitz-Paal, O. Reutter, P. Rietbrock, Porous Materials as Open Volumetric Receivers: Experimental Determination of Thermophysical and Heat Transfer Properties. Proceedings of the 11th SolarPACES International Symposium on Solar Power and Chemical Energy Technologies, pp. 379 - 387 (2002).

# 89b Status of Sulfur Iodine Cycle Assessment at Cea

**Philippe Carles**[1], Xavier Vitart[1], and Pascal Yvon[2]. (1) DEN/DANS/DPC, Commissariat à l'Energie Atomique, GIF SUR YVETTE Cedex, 91191, France, (2) DEN/DDIN, Commissariat à l'Energie Atomique, GIF SUR YVETTE Cedex, 91191, France

French Commissariat à l'Energie Atomique has been studying the sulfur-iodine cycle for hydrogen production using nuclear heat. The program comprises thermodynamic measurement to assess the efficiency of the cycle, corrosion studies and technological developments, but also large scale component design, plant design, reactor coupling, and cost evaluations. The presentation will provide an overview of the program, highlight recent results, and focus on how they impact the projected hydrogen production cost.

## Life Cycle Assessment of the Sulfur-Iodine Cycle

**William C. Lattin**, Department of Environmental Sciences, University of Idaho, Idaho Falls, ID 83402, and
Vivek P. Utgikar, Department of Chemical Engineering, University of Idaho, Idaho Falls, ID 83402

Several thermochemical processes have been proposed for large-scale production of hydrogen from nuclear reactors, including the sulfur-iodine cycle (SI), UT-3 (University of Tokyo), and Ispra Mark 9. The sulfur-iodine cycle combined with a new-generation nuclear reactor as the source of energy for the process is favored for implementation and deployment. The decision to use this process was based partially on a study performed by General Atomics, University of Kentucky, and Sandia National Laboratories. That study ranked numerous thermochemical processes using qualitative parameters, such as the number of chemical reactions involved, the number of chemical separations necessary, and the number of reports and studies published. However, according to the International Nuclear Societies[1], the study "neglects some industrial-scale issues, like heat exchangers and the size of equipment, and . . . it tends to over-penalize the cycles lacking relevant thermodynamic data." The decision to use a specific thermochemical cycle should be based on objective criteria, as well as technical feasibility and life-cycle environmental impacts.

Life-cycle assessments (LCAs) have been conducted for several different processes for hydrogen production. However, because each uses different methodologies and assumptions, comparison of results is difficult. For example, some studies include storage, distribution, and use of the hydrogen product in the LCA, as well as production. Life-cycle assessments should be performed which examine each process objectively, consistently, and equitably.

The goal of this life-cycle assessment is to evaluate the environmental impacts of producing hydrogen using the sulfur-iodine thermochemical cycle and a nuclear reactor heat source. The LCA will identify significant environmental aspects and assess their impacts. The assessment can be used "stand-alone" or may be compared with similar life-cycle assessments.

One area that has significant effect on the outcome of an LCA is the definition of system boundaries. This LCA defines the boundaries of the hydrogen production system as the nuclear reactor and the hydrogen plant. The analysis of the nuclear reactor begins with mining, milling, conversion and enrichment of uranium ore; fabrication and transportation of nuclear fuel; construction, operation and decommissioning of the nuclear power plant; and, nuclear waste disposal. The boundary of the hydrogen plant begins with construction of the

---

[1] International Nuclear Societies Council (INSC), *Nuclear Production of Hydrogen: Technologies and perspectives for Global Deployment.* November 2004. p.56.

physical plant and acquisition of raw materials for the thermochemical process (i.e., iodine and sulfuric acid). Also included is the interface between the two plants, the heat transfer loop consisting of the intermediate heat exchanger (IHX) and heat transfer fluid (helium gas). Some studies do not include disposal of radioactive waste and spent nuclear fuel in the analysis. However, operation of one nuclear reactor for 20 years results in over 35 ton of heavy metal for disposal or reprocessing, not a trivial amount when the large number of nuclear reactors required to support a hydrogen economy is considered. This LCA does not include storage and distribution of hydrogen product in the analysis since they are independent of the method of production. End-use of hydrogen is also excluded from the study since a single purpose is assumed for the product (e.g., transportation using fuel cells or internal combustion engines) and is not reflective of the current usage of hydrogen.

Once the system boundaries are established, inputs to the system (i.e., inventory) must be defined. Major input to the system being studied include: materials of construction for the reactor and hydrogen plant, such as concrete, structural steel, stainless steel, and other materials; nuclear fuel; feed material for the processes (helium, iodine, sulfuric acid, and water); fossil fuels and electricity necessary for mining and construction; and, energy and materials needed to operate the facility (replacement nuclear fuel, electricity, fossil fuels and electricity, make-up helium gas, process materials, and water).

Each inventory item is associated with environmental aspects. For example, fossil fuel usage for generating electricity and mining uranium ore results in emissions of airborne pollutants. These emissions can further be broken down as greenhouse gas emissions ($CO_2$, $CH_4$, and $N_2O$), criteria pollutants (volatile organic compounds, CO, $NO_x$, $SO_x$, and PM10). Plant construction and operation affect resource usage, liquid effluents, waste heat, land use, solid and radioactive waste.

The S-I cycle requires a heat source capable of operating at over 1000°C reactor vessel outlet temperature (peak fuel temperatures of 1200°C). Since current-generation light-water reactors operate nominally at 600° C, new-generation high-temperature gas-cooled reactors must be used. For the system being considered, one or more 600 MW(t) reactor modules are coupled to a hydrogen production plant. All 600 MW of thermal energy are transferred to the process through an intermediate heat exchanger (IHX) to a secondary helium loop. Design of an IHX capable of operating at the required temperature (1000°C) poses an engineering challenge, requiring advanced design and materials.

Another significant difference among thermochemical processes involves chemicals used in the processes. For example, the UT-3 process uses bromine, calcium, and iron. Ispra Mark 9 uses iron and chlorine, whereas the S-I process uses sulfuric acid and iodine. Since energy is required to circulate the material, processes using liquid or gas are preferred over solid materials.

This LCA addresses the method of production for each chemical, the energy required, and its relative abundance. In order to produce one ton of hydrogen requires circulation of 500-10,000 tons of process material. Iron, chlorine, and calcium are relatively abundant and easily produced in quantities required. Bromine and iodine, on the other hand, are less abundant and require energy-intensive separation techniques. The S-I cycle requires close to 10,000 tons of material. Assuming nominal process losses of iodine, a significant percentage of the current world's production of iodine is involved. Some chemicals (e.g., iron and sulfuric acid) may be recycled from other industrial processes and may be a positive influence on the LCA.

Ultimately, the results of a life-cycle analysis must be examined in the context intended. If the purpose of hydrogen generation is to replace hydrocarbon-based fuels in automobiles and enhance energy independence, then higher greenhouse gas emissions may be acceptable from the process. On the other hand, if reducing GHG emissions is the ultimate goal, various LCAs can be compared and the appropriate technology selected.

Results of various energy analyses and life-cycle analyses show that nuclear-based production of hydrogen results in significantly lower fossil energy use and greenhouse gas (GHG) emissions than the current method of production via steam methane reforming using natural gas (11200g $CO_2$-eq/kg-$H_2$). High-temperature electrolysis results in 2000g $CO_2$-eq/kg-$H_2$ and 4.8g $SO_2$-eq/kg-$H_2$. Utgikar, *et al.*, estimate 2515g $CO_2$-eq/kg-$H_2$ and 11.2g $SO_2$-eq/kg-$H_2$ using Ispra Mark 9 (1250g $CO_2$-eg/kg-$H_2$ is due to nuclear reactor construction and operation). Wu, *et al.,* performed an LCA for generic thermochemical processes resulting in 25-30g $CO_2$-eq/km, which includes end-use in hydrogen fuel-cell vehicles. Assuming 95 km/kg-$H_2$ results in ~2700g $CO_2$-eq/kg-$H_2$. Each thermochemical cycle studied results in one-fifth to one-sixth the GHG emissions from steam methane reforming.

Comparing these results, it is evident that relatively little difference in GHG emissions exists between thermochemical hydrogen production processes using nuclear heat. Further work needs to be done to standardize life-cycle analyses for the various available nuclear options to enable valid comparisons. However, the environmental LCA provides only one input to the decision-making process. Other factors to be considered may include technical feasibility, socio-economic impacts, resource usage, politics and national strategy. Sensitivity analyses should be performed to determine which factors have the greatest effect on results.

# Relative Economic Incentives for Hydrogen from Nuclear, Renewable, and Fossil Energy Sources

Charles W. Forsberg
Oak Ridge National Laboratory
Oak Ridge, TN 37831; Tel: (865) 574-6783
E-mail: forsbergcw@ornl.gov

Maximilian B. Gorensek
Savannah River National Laboratory
Aiken, SC 29808; Tel: (803) 725-1314
E-mail: maximilian.gorensek@srnl.doe.gov

**Abstract**

The value of hydrogen from different sources depends upon the specific market for which it is intended. Each hydrogen production technology has its own distinct characteristics. For example, steam reforming of natural gas produces only hydrogen. In contrast, nuclear and solar hydrogen production facilities produce hydrogen together with oxygen as a by-product or co-product. For a user who needs both oxygen and hydrogen, the value of hydrogen from nuclear and solar plants is higher than that from a fossil plant because "free" oxygen is produced as a by-product. Six factors that impact the relative economics of fossil, nuclear, and solar hydrogen production to the customer are identified: oxygen by-product, avoidance of carbon dioxide emissions, hydrogen transport costs, storage costs, availability of low-cost heat, and institutional factors. These factors imply that different hydrogen production technologies will be competitive in different markets and that the first markets for nuclear and solar hydrogen will be those markets in which they have a unique competitive advantage. These secondary economic factors are described and quantified in terms of dollars per kilogram of hydrogen.

## 1. Introduction

Different methods of hydrogen production have different characteristics. Table 1 shows the characteristics of hydrogen produced by nuclear energy, steam reforming of coal, and decentralized solar energy. The economically preferred choice depends upon what the customer requires. There are several important characteristics.

- *Hydrogen.* If this is the only customer requirement, the economic comparison is based on dollars per kilogram of hydrogen.
- *Oxygen.* If the customer needs oxygen, the oxygen is a "free" by-product of nuclear and solar hydrogen production. For such customers, the cost of an oxygen plant must be added to the cost of hydrogen if the hydrogen is produced from fossil fuels.
- *Greenhouse gases.* If restrictions are imposed on release of greenhouse gases to the atmosphere, the customer will add to the cost of hydrogen from fossil fuels the costs of disposing of the carbon dioxide or the taxes that will have to be paid.
- *Transportation.* Transportation costs for hydrogen strongly impact the relative costs of hydrogen from different sources. A customer who needs large quantities of hydrogen will add hydrogen transport costs to his centralized facility if the hydrogen is produced by dispersed renewable technologies. However, if the hydrogen is needed for decentralized applications, the cost of hydrogen transport will be added to that for centralized methods of hydrogen production.

- *Storage*. The only proven low-cost method of hydrogen storage is in large underground facilities. If hydrogen storage is required by the customer, the cost of transporting hydrogen to the storage facility will be added. This cost is high for decentralized methods of hydrogen production.
- *Heat.* If the customer needs heat, the cost of providing that heat will be considered. Nuclear systems provide low-cost heat relative to natural gas; thus, large users of hydrogen and heat have incentives to use nuclear-hydrogen production facilities and simultaneously buy low-cost heat.
- *Institutional.* Associated with every hydrogen production technology are specific institutional constraints that have economic implications.

Table 1.
Characteristics of different methods of hydrogen production

| Production Method | $O_2$ | $CO_2$ Free | Large-Scale Use | Bulk Storage | Low-Cost Heat | Institutional |
|---|---|---|---|---|---|---|
| Nuclear | X | X | X | X | X | ? |
| Steam Reforming of Coal | | | X | X | | ? |
| Decentralized Solar | X | X | | | | ? |

Different customers have different requirements. Table 2 lists several major customers and some of their requirements. In this example, the *X*'s in the table indicate where a match exists between nuclear-hydrogen production characteristics and various markets.[1] Such tables provide a perspective on which markets are likely to be preferred for specific hydrogen production technologies. Similar tables can be constructed for any hydrogen production system to identify those markets where a specific hydrogen production technology will have a competitive advantage.

Table 2.
Characteristics and value (*X*) of nuclear hydrogen relative to other methods of hydrogen production for specific applications

| User | $O_2$ | $CO_2$ Free | Large-Scale User | Bulk Storage | Low-Cost Heat | Institutional |
|---|---|---|---|---|---|---|
| Small Local User | | X | | | | |
| Pipeline | | X | X | | | |
| Chemical/Iron Industries | | X | X | | X | |
| Refinery | ? | X | X | ? | X | |
| Liquid Fuels Production | X | X | X | ? | X | |
| Peak Electric Production | X | X | X | X | X | X |

These markets have different characteristics.

- *Pipelines.* In parts of the United States and Europe there are hydrogen pipelines that deliver hydrogen to multiple customers from multiple producers. As these systems grow, they will develop the characteristics of the electrical grid. In these systems hydrogen is a commodity market.
- *Chemical/Iron indusries.* Several types of chemical plants use massive quantities of hydrogen and could consume all of the hydrogen from a nuclear-hydrogen plant. These include the production of ammonia and the production of iron from iron ore. The characteristic of most of these markets is that there is a single customer with a single need: hydrogen. In such markets, centralized methods of hydrogen production are favored.
- *Refineries.* Refineries use massive quantities of hydrogen to remove sulfur from crude oil and convert heavy oils to liquid fuels such as gasoline, diesel, and jet fuel. The facilities also use massive quantities of heat and thus are simultaneously potential markets for high-temperature heat. Refineries use over 7% of the total energy demand of the United States and thus are a mega-market by themselves.
- *Liquid fuels production.* Plants using coal or biomass to produce liquid fuels require both hydrogen and oxygen.
- *Peak electric production.* This market requires hydrogen, oxygen, and heat. It is part of the same industry that produces base-load electricity from existing nuclear power plants; thus, it is a market where nuclear energy is accepted.

The following sections examine each of the characteristics that impact the relative value of hydrogen in specific markets.

## 2. Oxygen as a by-product

Oxygen is a by-product of hydrogen production methods in which water is the feed input. This is the case for most nuclear and solar methods of hydrogen production. If there is a demand for oxygen, it becomes a second product for a hydrogen production plant. The cost of oxygen is a strong function of the price of electricity, because typical large cryogenic separation plants that separate oxygen from air consume ~230 kWh/ton $O_2$. At \$0.04/kWh, the energy costs total \$9.20/ton of oxygen.[2] For the analysis herein, it is assumed that oxygen costs \$20/ton. On a mass basis, 8 times as much oxygen is produced as hydrogen. Consequently (assuming there is a demand), the value of the oxygen per kilogram of hydrogen produced is

$$(\$20/\text{ton } O_2)\ (\text{ton}/2000\text{ lb})\ (2.2\text{ lb/kg})\ (8\text{ kg } O_2/\text{kg } H_2) = \$0.176/\text{kg } H_2.$$

## 3. Avoidance of carbon dioxide releases

Hydrogen today is produced primarily by steam reforming of fossil fuels, especially natural gas. The overall chemical reaction is

$$2H_2O + C \rightarrow 2H_2 + CO_2.$$

Concerns about global climate change may lead to taxes on the release of carbon dioxide to the atmosphere or to requirements for geological sequestration of carbon dioxide. A wide range of estimates have been proposed for possible future taxes on carbon dioxide releases to the atmosphere and the cost of carbon dioxide sequestration. Low-end estimates indicate taxes or sequestration costs of ~$25/ton of carbon dioxide.[3]

The implications of adding a carbon dioxide tax or disposal cost of $25/ton to the cost of hydrogen were calculated by assuming that carbon is used for hydrogen production and that the efficiency of converting carbon and water to hydrogen and carbon dioxide is about 60%. Typical coal-to-hydrogen plants have efficiencies near 60%.

Input assumptions:

- Free energy of formation of water: 54.63 kcal/gmol
- Free energy of formation of carbon dioxide: 94.26 kcal/gmol
- Energy efficiency in conversion of carbon to hydrogen: 60%

Energy release in the production of 1 ton of $CO_2$ from burning carbon is

$(2000 \text{ lb } CO_2)(1000 \text{ g}/2.2 \text{ lb})(12 \text{ g C}/44 \text{ g } CO_2)(1 \text{ gmol C}/12 \text{ g C})(94.26 \text{ kcal/gmol C})$
$= 1.948 \times 10^6$ kcal.

Energy in the form of hydrogen with 60% conversion efficiency from the production of one ton of $CO_2$ is

$(1.948 \times 10^6 \text{ kcal})(0.6) = 1.169 \times 10^6$ kcal.

Hydrogen produced from using carbon and releasing one ton of $CO_2$ to the atmosphere

$(1.169 \times 10^6 \text{ kcal})(\text{gmol } H_2/54.63 \text{ kcal})(2 \text{ g } H_2/1 \text{ gmol})(\text{kg}/1000\text{g}) = 42.8$ kg of $H_2$.

If there is a $25/ton disposal cost for carbon dioxide, the cost impact on hydrogen if the hydrogen is produced from carbon is

$(\$25.00/\text{ton } CO_2)(1 \text{ ton } CO_2/42.8 \text{ kg } H_2) = \$0.58/\text{kg } H_2$.

Constraints on greenhouse gases can significantly alter the economics of hydrogen production from fossil fuels versus hydrogen production via other methods.

### 4. Hydrogen transport

Hydrogen transportation by pipeline is expensive. Over any distance such transport can add $1.00/kg or more. If there are centralized users of hydrogen, strong incentives exist to use centralized production of hydrogen to meet those needs to avoid the transport costs. The exact benefit or cost is dependent upon the specific case.

A number of commercial hydrogen pipelines that service a number of chemical users are currently in operation in the United States. Their lengths range from less than a mile to several hundred miles. Refineries are the largest users of imported hydrogen, at about 50 million stand cubic feet per day (MMSCFD), while chemical plants typically use less than 10 MMSCFD (with the exception of those producing ammonia and methanol).

The cost[4] of hydrogen transport by pipelines of different sizes carrying different quantities of gas is shown in Table 3. The investment and operating cost of pipeline compressors is also shown. Hydrogen has the lowest possible molecular weight, so reciprocating compressors are needed regardless of the size of the pipeline.

Table 3.
Cost of hydrogen transport by pPipeline[4]

| $H_2$ Gas Capacity, MMSCFD | Line Size, In | Pipeline Investment, $M/100 miles | Compression Investment, $M/100 miles | Compression Power Cost, $/day | Pipeline cost/100 miles, $/MMBtu HHV |
|---|---|---|---|---|---|
| 740 | 20 | 149,000 | 33,000 | 20,300 | 0.50 |
| 600 | 18 | 138,000 | 30,000 | 18,400 | 0.57 |
| 490 | 16 | 127,000 | 29,800 | 18,300 | 0.65 |
| 370 | 14 | 116,000 | 25,200 | 15,500 | 0.77 |
| 270 | 12 | 105,000 | 21,400 | 13,200 | 0.94 |
| 240 | 10 | 94,000 | 41,500 | 25,600 | 1.28 |
| 110 | 8 | 83,000 | 12,000 | 7,000 | 1.65 |
| 70 | 6 | 72,000 | 13,600 | 8,300 | 2.43 |
| 32 | 4 | 62,000 | 12,500 | 7,200 | 4.62 |

Specific constraints are associated with fossil and solar hydrogen production. Because of hydrogen transport costs, constraints on greenhouse gas releases may significantly alter the economics of hydrogen production from fossil fuels versus hydrogen production via other methods. Multiple projects are under way to develop methods to produce hydrogen from fossil fuels with underground sequestration of the carbon dioxide. However, only parts of the United States have geologies suitable for such sequestration. Hydrogen made from fossil fuels with carbon dioxide sequestration may have the lowest production costs in cases where fossil fuels and sequestration sites are located next to one another. Hydrogen made from fossil fuels with carbon sequestration will have higher delivered costs elsewhere because of the cost of transporting the hydrogen by pipeline.

Hydrogen from renewable energy sources faces similar geographical constraints. Unlike nuclear energy, the cost and availability of renewable energy sources are strongly dependent on location. Solar-based systems are most economic in the sunny southwest United States. Wind resources are most abundant on the upper Great Plains (North Dakota, Montana, South Dakota, Minnesota, etc.) and in coastal areas. There is a second factor: solar hydrogen systems are being developed for local hydrogen production and use. These small scale systems for small users are designed to avoid transport costs.

## 5. Hydrogen storage

Only one technology[5] currently exists for low-cost storage of hydrogen—storage underground in caverns or geological reservoirs. This is the same technology used to store natural gas. In the United States, natural gas is moved by pipeline throughout the year to northern markets from production in the south-central United States. The natural gas is stored in ~400 large facilities, with a total capacity to store a third of a year's production of natural gas. Underground storage is a low-cost technology with market prices for storage typically 10% of the value of the natural gas.

Commercial hydrogen storage facilities using this technology have been built in the United States and Europe. For hydrogen, the capital costs for such facilities are estimated to be $0.80–$1.60/kg, which is lower than the production costs for hydrogen. The hydrogen storage technology has been used commercially in several geologies but not yet demonstrated in all the geologies that have been used for natural gas.

Storage economics demand large facilities. Two factors drive the facility size. First are siting and site development costs (including understanding the local geology) that are almost independent of facility capacity. Second, gas storage requires compression of the gases, typically to pressures of ~1000 psi for storage. Gas equipment efficiencies and costs are strong functions of the size of the equipment. The practical implications is that if bulk hydrogen storage is required by the customer, strong incentives exist for large-scale centralized hydrogen production to avoid hydrogen collection costs from decentralized hydrogen production units.

## 6. Low-cost heat

Many industrial applications of hydrogen also require heat. The cost of nuclear heat from nuclear plants is far less than heat from natural gas (typically $6-11/MMBTU) and, in much of the United States, less than heat from coal. If hydrogen is being produced and the hydrogen consumer is across the fence, heat can be sold. How the heat is delivered depends upon the required temperature. For temperatures below 600°C, the heat will most likely be delivered as steam because this is the standard proven industrial technology to transfer heat to industrial facilities. At higher temperatures, helium or liquid salts may be used to transfer heat.

The price of nuclear plant heat can be estimated from the price of electricity. A nuclear power plant produces steam that can be sold or used to produce electricity. The utility will demand at least the same revenue from the sale of steam as from the sale of electricity. A rough estimate of the price of steam can be calculated from the wholesale price of electricity. The price depends upon the temperature and pressure of the steam because more electricity can be made from high-temperature, high-pressure steam than from low-temperature, low-pressure steam.

An example can clarify this. Because the technology is well defined, the sale of steam from a light-water reactor will be used. Once the parameters of a high-temperature reactor are defined, similar calculations can be done. Consider the sale of 150-psi steam used for general heating and many industrial processes—including drying of many products and production of fuel ethanol.

The price of electricity varies across the country; thus, this example uses the recent average market price for wholesale electricity in Minnesota, which is \$53.89/MWh(e).

The efficiency of nuclear power plants is ~33%; that is, if one less Btu of electricity is produced, 3 Btu of steam becomes available. Although nuclear reactors produce high-temperature steam, in this example the heat is delivered as 150-psi steam. In converting high-temperature steam to electricity, 40% of the electricity is obtained by the time the steam pressure is 150 psi, with the remaining 60% of the electricity produced in the low-pressure turbines. Using this information, a rough estimate can be made of the corresponding price of steam from a nuclear plant, given the price of electricity:

$$\$53.89/\text{MWh(electricity)} \cdot 0.33 \cdot 0.6 = \$10.67/\text{MWh(steam)} = \$3.13/\text{MMBtu}.$$

This cost represents less than half the price of natural gas in the United States. Therefore, if a customer needs both hydrogen and heat, there is an incentive to obtain the hydrogen from a nuclear plant if, at the same time, the customer can obtain lower-cost heat to meet his energy needs.

## 7. Institutional

All industries have institutional structures that include regulation, organizational structures, international trade agreements, and accepted business practices for that industry. These structures have major impacts on the choice of hydrogen production technology. Two examples provide a perspective on the extremes of these institutional structures.

- *Peak electric power production.*[6] Assuming that the technology and economics are favorable, the institutional structure for using nuclear hydrogen for peak electrical power is highly favorable. Nuclear power plants are built to produce base-load electricity. The production of peak electricity is an extension of that business by organizations that understand nuclear energy, operate nuclear power plants, and sell electricity. In such cases, the producer of hydrogen is the user of hydrogen; thus, the market is not an issue. Furthermore, the market for peak power is local. What happens in the rest of the world has only limited impacts on local peak power demand.

- *Ammonia production.* The primary cost of ammonia production is the cost of hydrogen; thus, economics are dependent upon hydrogen production costs. However, ammonia is transported by ship around the world and, in the United States, by pipeline from Louisiana as far north as Minnesota. The price of ammonia is set on a global basis. New ammonia plants are built where there is low-cost natural gas that has no local market. Over a period of a decade, the lowest cost natural gas in the world can strongly impact the cost of ammonia everywhere in the world. Governments can push ammonia production and drive down world prices to encourage local development. This is an institutional structure that potentially makes it risky to invest in capital-intensive technologies such as nuclear hydrogen for ammonia production.

## 8. Conclusions

User requirements ultimately determine the preferred method of hydrogen production. Because user requirements vary widely, it is likely that different hydrogen production techniques will be economic for different users. This also implies that for any new hydrogen production technology, there are specific markets in which that technology is most likely to become competitive relative to other production techniques.

## 9. References


1. Forsberg CW. *Competitive-Advantage Markets for Hydrogen from Nuclear Energy: Peak Electric Power, Liquid Fuels from Coal and Biomass, and Refineries*, Oak Ridge National Laboratory (in press).
2. Alstrom Power Inc., *Greenhouse Gas Emissions Control by Oxygen Firing in Circulating Fluidized Bed Boilers: Phase I – A Preliminary Systems Evaluation: Final Report,* National Energy Technology Laboratory, May 15, 2003
3. Massachusetts Institute of Technology, *The Future of Coal: Options for a Carbon-Constrained World*, Cambridge, Massachusetts, 2007.
4. Summers WA. "Interim Project Report, Task A.1, Nuclear Hydrogen Plant Definition," NERI Project 02-0160, *Centralized Hydrogen Production from Nuclear Power: Infrastructure Analysis and Test-Case Design Study*, WSRC-TR-2003-00484, Savannah River Technology Center, October 30, 2003.
5. Forsberg CW. "Is Hydrogen the Future of Nuclear Energy?" *International Topical Meeting on the Safety and Technology of Nuclear Hydrogen Production, Control, and Management*, American Nuclear Society, Boston, Massachusetts, June 24–28, 2007.
6. Forsberg CW. "Economics of Meeting Peak Electricity Demand Using Nuclear Hydrogen and Oxygen," *International Topical Meeting on the Safety and Technology of Nuclear Hydrogen Production, Control, and Management*, American Nuclear Society, Boston, Massachusetts, June 24–28, 2007.

# 166a Status of the Sulfur-Iodine Integrated Lab Scale Experiment

**Benjamin E. Russ**[1], Robert T. Buckingham[1], Lloyd Brown[1], Gottfried Besenbruch[1], Paul Pickard[2], and M. Helie[3]. (1) General Atomics, 3550 General Atomics Court, San Diego, CA 92121, (2) Sandia National Laboratories, P.O. Box 5800, Albuquerque, NM 87185, (3) DPC/SCCME, CEA Saclay, Gif sur Yvette, Cedex, 91191, France

The Sulfur-Iodine (S-I) thermochemical cycle is one of the baseline production methods that is being investigated under the Department of Energy's Nuclear Hydrogen Initiative (NHI). The current objective is to conduct an integrated lab scale (ILS) experiment in order to provide a technical basis for the assessment of the viability of the S-I cycle for nuclear hydrogen production.

The development of the S-I cycle ILS loop has been performed as an international effort involving the US Labs and Industry and the Commissariat à l'Énergie Atomique (CEA) of France. The S-I cycle consists of three reaction sections which are currently being developed by three different organizations. Sandia National Laboratory (SNL) is developing the high temperature sulfuric acid decomposition section, General Atomics (GA) is developing the HI decomposition (hydrogen generation) section, and the CEA is developing the primary Bunsen reaction section. Work has been ongoing in all three organizations for the past five years and has resulted in experiments on the initial stand-alone versions of the reaction sections. These three sections have been integrated into the ILS experiment at the General Atomics facility in San Diego. This paper will describe the status of the ILS system and experiments.

# Continuous Flow Operation of a Bunsen Reactor in the Sulfur-Iodine Thermochemical Water-Splitting Cycle

Alberto Giaconia, Salvatore Sau, Giampaolo Caputo, Claudio Felici, Pietro Tarquini

ENEA, "Casaccia" Research Center – via Anguillarese, 301, 00123 - Rome (Italy)

Bunsen reaction in the iodine-sulfur thermochemical cycle represents a key process step.

Although being less critical in terms of coupling with the high-temperature heat source (nuclear or solar), and despite the relatively low operating temperatures and pressures, other fundamental issues must be regarded for the development of the Bunsen section. It implies the direct contact of iodine with sulfur-containing recycling components, and is interfaced to both the sulfuric acid and hydrogen iodide decomposition sections. Therefore, the best match with the other sections is to be identified. Additionally, since iodine and water removal from the produced acids prior decomposition are operations responsible of the major heat losses, the two produced acid solutions must be as pure and concentrated as possible to minimize the energy consumption in the downstream units. Finally, side reaction occurrence must be avoided.

Our preliminary study [1] aimed to the characterization of the phase behaviour of the hypothetical products of the Bunsen. Results obtained were in good agreement with those previously published by Sakurai *et al.* [2], despite the different adopted temperatures and dilution.

More recently, a laboratory-scale glass-made Bunsen reactor with $SO_2$ feed flow rates the order of 10-50 L/h has been assembled and its performances are currently investigated. The main general objective is to assess the most favourable operative conditions to enhance downstream purification and concentration of the produced acid solutions. Particularly, the effect of temperature on one hand, and the water and iodine excess in the feed on the other hand, are currently investigated.

Both semi-continuous and continuous tests are carried out, and potential side products (sulfur or hydrogen sulfide) formation is determined at the end of each experiment.

When the S-I cycle is coupled with a solar source, semi-continuous operation of the Bunsen reactor can be considered: $SO_2$ can be fed at variable rate depending on the $H_2SO_4$ decomposition rate which, in turn, depends on solar radiation availability and intensity. Hence, $HI_x$ phase storage is required in the Bunsen section in order to allow continuous flow operation of the $HI_x$ section, whose management is more complicated because of the larger flow-rates involved (due to the excess water and $I_2$, and to the low HI single-pass decomposition conversion); hence, Bunsen section can consist of a series of reactors where storage of the formed phases occurs. Some preliminary considerations resulted in high costs because of the very large iodine inventory required.

Different viable reactor configurations can be conceived for practical continuous flow operation of a Bunsen reactor: mixed reactor separated or integrated with liquid-liquid phase separator, or a static column for $SO_2$ chemical adsorption on the liquid $HI_x$ phase.

Our first tests are currently being carried out with a packed column operating at 1 atm, without mechanical mixing device, and with iodine conveniently fed either as a pure melt (at temperatures higher than 113°C), or using a hydrogen iodide water solution as carrier. This system has been progressively changed and optimized to improve performances. The general objective is to obtain the basic processing data for subsequent larger scale design. Moreover, the minimum water and maximum process temperature allowed for the reaction to proceed (without side reaction occurrence) is determined.

Preliminary results show that it is possible to minimize water content working under $SO_2$ and $I_2$ saturation conditions, and the composition of the obtained streams (sulfuric acid and $HI_x$ phases), in good agreement with the results of the previous analysis [1].

Once optimized, this reactor type can find industrial application other than the sulfur-iodine cycle for desulfurization processes: sulfur containing components from crude oil, natural gas or coal, converted to $SO_2$ can be fed to the Bunsen reactor to be quantitatively adsorbed to produce concentrated sulfuric acid and pure hydrogen; the former is a very valuable and marketable commodity, the latter an important feedstock for several operations (e.g. crude oil refinery).

The effect of the presence of oxygen in the inlet $SO_2$ flow is assessed too.

The two produced liquid acid streams from the reactor are collected and distilled in order to remove impurities and $I_2$ from the lower phase.

As far as the sulfuric acid phase treatment is concerned, in the prospective of coupling the Bunsen reactor with a solar $H_2SO_4$ splitting system, we are studying the possibility to feed a solar reactor with a sulfate salt instead of $SO_3$ or $H_2SO_4$. As a matter of fact, solar reactors are appropriated to process solids instead of gases or liquids; moreover, solids are more easily storable and do not involve material resistance concerns, and according with our experimental results $Fe_2(SO_4)_3$ decomposes to $SO_2$ at temperatures not higher than pure $SO_3$ and without the use of catalysts or special materials components.

**References.**

[1] Giaconia A, *et al.* Int. J. Hydrogen Energy. 2007, 32, 531.

[2] Sakurai M, *et al.* Int. J. Hydrogen Energy. 2000, 25, 605.

# 166c Titania-Supported Platinum-Group Metal Catalysts For Sulfur-Based Thermochemical Water Splitting Cycles

**Lucia M. Petkovic**, Daniel M. Ginosar, and Kyle C. Burch. Chemical Sciences, INL, P.O. Box 1625, Idaho Falls, ID 83415-2208

Thermochemical cycles can be used to split water through a series of chemical reactions where the net result is the production of hydrogen and oxygen at much lower temperatures than direct thermal decomposition. The sulfur-based family of thermochemical cycles appears promising for producing hydrogen from water. These cycles employ a high-temperature sulfuric acid decomposition reaction step. The reaction produces oxygen and generates sulfur dioxide, which is used in other reaction steps of the cycles. The reaction takes place from 750 to 900°C, or higher, and is facilitated by heterogeneous catalysts. The high-temperature, harsh chemical reaction conditions present a significantly challenging environment for catalytic materials. In this work a number of titania (rutile)-supported platinum-group metal (i.e., Pt, Ir, Pd, Ru, and Rh) catalysts were investigated for the sulfuric acid decomposition reaction.

# 166d Supported Metal Particle Catalysts For Sulfuric Acid Decomposition Reaction

**Sergey N. Rashkeev**[1], Daniel M. Ginosar[2], Lucia M. Petkovic[2], and Helen H. Farrell[3]. (1) Center for Advanced Modeling & Simulation, Idaho National Laboratory, Idaho Falls, ID 83415, (2) Chemistry, Idaho National Laboratory, Idaho Falls, ID 83415, (3) Materials Sciences, Idaho National Laboratory, Idaho Falls, ID 83415

Production of hydrogen by splitting water in thermochemical water-splitting cycles, such as the sulfur-based group that employs the catalytic decomposition of sulfuric acid into SO2 and O2 is of considerable interest. Most of these processes occur at high temperatures ($T > 1{,}000$ K) and exposes catalysts to the extreme conditions such as steam, oxygen, and acid vapor that severely damage these catalysts within a short time. To develop an understanding of the factors that cause catalyst deactivation, we performed density-functional-theory (DFT)-based first-principles calculations and computer simulations for transition metal (TM) particles positioned on the two types of substrate (gamma-alumina and TiO2-rutile).

The catalytic activity of the considered systems is defined by several factors, namely:

(i) The efficiency of detaching oxygen atoms from the sulfur-containing species SOn ($n = 1,2,3$). The breaking of the S-O bonds may occur at both the substrate and the transition metal cluster. However, the bond-breaking at the substrate is endothermic (and takes about 1.5 eV per bond) while at low-coordinated metal atom of a cluster it is exothermic (with energy gain of about 0.5 eV per bond). This explains why the presence of transition metal clusters is necessary for catalytic activity;

(ii) The ability of the cluster to "clean" itself, i.e., to eliminate oxygen from its surface, in order to regain the catalytically active sites and to continue the process. We found that the clusters of Pd and Pt with the size > 2-3 nm are more efficient in this process (at $T > 1{,}000$ K) than the clusters of other TM's considered (Rh, Ir, Ru, and Os);

(iii) The ability of the cluster to keep its size to avoid sintering (that reduces the number of low-coordinated catalytically active sites at the surface of the cluster). We found that the sintering of Rh, Ir, Ru, and Os clusters is significantly suppressed in comparison with the sintering of Pd and Pt clusters of the same size (the individual atoms at the surface of Rh and Ir clusters have a tendency to have higher coordination number, i.e., the detachment of individual atoms from the surface is less likely).

Therefore, the activity of TM nanoparticles is mainly defined by the competing factors (ii) and (iii). At the present, we try to find (experimentally and theoretically) the most optimal combination of the structure, size, and composition of TM nanoparticles, for which the catalytic activity of sulfuric acid decomposition will be the highest.

# 166e Activated Carbon Catalysts For Hydrogen Production In The Sulfur-Iodine Thermochemical Water Splitting Cycle

**Daniel M. Ginosar**, Lucia M. Petkovic, and Kyle C. Burch. Chemical Sciences, INL, P.O. Box 1625, Idaho Falls, ID 83415-2208

Thermochemical cycles produce hydrogen through a series of chemical reactions where the net result is the production of hydrogen and oxygen from water at much lower temperatures than from direct thermal decomposition. All chemicals within the cycle are recycled and the heat to drive the reactions, which tend to be endothermic, must be provided by a primary energy source. When the primary energy driver is nuclear or solar heat, hydrogen can be generated without producing greenhouse gasses, and can provide independence from our dwindling supplies of fossil fuels.

Among the high number of thermochemical water-splitting cycles proposed in the literature, the Sulfur-Iodine (S-I) cycle has generated considerable interest. The S-I cycle consists of three simultaneous reactions; the decomposition of hydroiodic acid (HI) to produce hydrogen and generate iodine for recycle, the decomposition of sulfuric acid to produce oxygen and generate $SO_2$ for recycle, and a main reaction where water and the recycle chemicals react to regenerate HI and sulfuric acid.

Previous literature has reported that activated carbon is a desirable catalyst for the decomposition of hydroiodic acid. The Idaho National Laboratory is currently exploring the activity and properties of activated carbon catalysts for the HI decomposition reaction. In this presentation, the activity and stability of activated carbon catalysts will be presented, focusing primarily on commercially available activated carbons. Catalyst characterization, including surface area, temperature programmed desorption (TPD), Boehm's titration results, and contact pH of the activated carbons will be discussed.

# DECOMPOSITION OF HYDROGEN IODIDE IN THE S-I THERMOCHEMICAL CYCLE OVER Ni CATALYST SYSTEMS.

*Annarita Spadoni, Michela Lanchi, Claudio Felici, Paolo Favuzza, Raffaele Liberatore, Pietro Tarquini, ENEA Casaccia Research Centre, Rome, Italy*

*Alina Carmen Tito, Carlo Vittorio Mazzocchia, Polythechnique "Giulio Natta", Milan, Italy*

## Abstract

The Sulphur-Iodine thermochemical cycle for hydrogen production has been investigated by ENEA over the last five years, with a particular focus on chemical aspects. Regarding the hydrogen iodide decomposition, three γ-alumina supported nickel catalysts were produced and characterized, and then tested in terms of catalytic activity and stability by means of a tubular quartz reactor. In particular the relationships between catalytic activity and preparation procedure was investigated. From the experimental data acquired it can be concluded that two of the three catalysts tested demonstrated high catalytic activity, since hydrogen iodide conversion was almost coincident with theoretical equilibrium value. On the other side, for all the catalysts, a gradual but considerable deactivation phenomenon was observed at 500°C, while at a temperature higher than 650°C the catalytic activity was recovered.

## Introduction

The Sulphur-Iodine thermochemical cycle is one of the most promising routes currently under development for sustainable hydrogen production from water. It is based on three main reactions carried out at three different temperature levels. Among these reaction, the hydrogen iodide (HI) decomposition presents a rather low equilibrium conversion even at high operating temperatures (the equilibrium conversion at 750 K is about 0.22) and subsequently a considerable energy expense for the separation and recirculation of the unreacted species is required [1]. It is well known that the homogeneous gas phase decomposition of hydrogen iodide has a very low rate [2,3]. The use of a catalytic system is essential for the improvement of the whole cycle thermal efficiency. Moreover, the reduced temperature resulting from the use of a catalyst attenuates the corrosion problems associated with the presence of hydrogen iodide-iodine ($I_2$ ) mixtures in process equipment.

Heterogeneous catalytic reaction was extensively investigated by NCLI (National Chemical Laboratory for Industry, Japan) and GA (General Atomics, USA) [4]. Different catalytic systems were found effective, in particular the Pt supported catalysts showed very good performances with respect to activation energy values and lifetime [5,6,7]. Some authors proposed less costly catalysts: the cost savings resulting from the use of non precious metals may be an advantage in the design of a plant of large size. O'Keefe reported that nickel was for the most part a poor catalyst, although its particular deposition on an active carbon surface produced a catalyst that was notably good. Nickel alumina supported catalysts were also tested and demonstrated a remarkable catalytic activity. However, operating times were limited to the 100 hours range.

The aim of the present work is the analysis of the catalytic activity of three γ-alumina supported nickel catalysts obtained from different preparation methods, in particular focusing on the relationships between catalytic activity and preparation procedure. Two of them were prepared from γ-alumina by impregnation-calcination method and were obtained from the deposition of two different precursors, such as nickel acetylacetonate and nickel nitrate [8]. The third one is a nickel aluminate catalyst obtained by coprecipitation method, prepared in order to better understand the metal-support interactions in nickel acetylacetonate catalyst (Ni-acac), and, at the same time, to test the effectiveness of this catalyst in HI decomposition .

## Materials and methods

***Catalyst preparation***

Three different γ-alumina supported nickel catalysts with a nickel content from 0.83 to 11.93% were prepared by impregnation-calcination and co-precipitation methods.

For catalysts obtained by impregnation technique, Ni-acac and Ni-nitrate, neutral activated alumina was used (powdered γ-alumina of high purity from Aldrich, BET: 153.37 $m^2/g$; micropore volume:12.80 $mm^3/g$, average particle size: 150 mesh; pH=7.0, density: 0.397 $g/cm^2$; melting point: 2040°C).

Regarding the Ni-nitrate catalyst, it was prepared by conventional wetness impregnation adopting nickel nitrate salt ($Ni(NO_3)_2$_$6H_2O$) as precursor. In particular the alumina powder was impregnated with the nickel nitrate aqueous solution, stirred at room temperature for 5 hours, filtered by membrane and finally washed with water. The resulting solid was dried at 60°C for 12 hours in a $N_2$ atmosphere, calcined in air at 550°C for 10 hours (under 5°C/min ramp followed by holding isothermally for 2 hours), and then reduced in presence of a gaseous mixture of $H_2/N_2$ (1:1) at 550 °C (heating rate of 5°C/min and isotherm of 2 hours).

The Ni-acac catalyst was obtained by impregnation of powdered γ-alumina with an acetone solution of nickelacetylacetonate, stirred at 40°C for 5 hours. Then, the resulting solution was dried under vacuum in a rotovapor using a flask with internal breakfluxes, in order to have impregnation even at the first phase of the drying.

The co-precipitated Ni catalyst, designated as Ni-aluminate, was prepared from a mixed aqueous solution of nickel hydroxide $Ni(OH)_2$ and aluminium hydroxide $Al(OH)_3$. The precipitate was obtained by adding aqueous ammonia slowly to the solution to a final pH of 9,5. Then the resulting residue was filtered , washed with water, dried at 100°C in air and calcined in air at 900°C for 5 hours.

***Catalyst characterization***

BET surface area and pore volume ($N_2$ adsorption/desorption) were determined using a Kelvin 042 Sorptometer apparatus. X-ray diffraction (XRD) patterns were collected over fresh catalysts, before testing in the fixed-bed tubular reactor. ICP-OES analysis were also performed in order to determine Ni loading. A SEM analysis over Ni-acac was conducted.

***Catalytic tests***

Catalytic reaction experiments were carried out in a quartz vertical fixed-bed reactor (11.4 mm inside diameter.). The other apparatus parts were mostly made of pyrex. The reactor was positioned vertically in a thermostated oven and fed with 0.1 mL/min of HI aqueous solution at azeotropic concentration (7.5 M, Sigma Aldrich) by means of a microfeeder through a capillary glass inlet. The capillary inlet temperature was kept over 136°C in order to assure the acid solution flash vaporization (B. P. 127 °C). A flow of 50 mL/min of $N_2$ from the top of the reactor was also adopted to drive the vaporized HI throughout the apparatus. The reactor was filled with a catalyst system distributed in five different zones: quartz wool was placed in the top and the bottom layers, granular SiC (Carlo Erba) in the intermediate layers, while 0.25 g catalyst was mixed with 2.25 g of powder SiC (Sigma Aldrich, 230 mesh) in the central layer. The tests were carried out under a total pressure equal to 1.5 bar and at two temperature levels: 500°C and 650°C. All the outlet components except for $H_2$ and $N_2$ were absorbed in a first scrubber containing water. A second scrubber containing titrated thiosulfate ($Na_2S_2O_3$) solution was added to ensure instrumental safety from corrosion due to possible HI and $I_2$ residues in the gas stream. The $H_2$ concentration (% v/v) of the outlet gas was determined by the use of an electrochemical sensor (Orbisphere). The flow rate of the stream coming from the second scrubber ($H_2$+$N_2$) was measured by a massflow meter F, as indicated in figure 1. Every test started with the HI solution injection, after a reactor conditioning step consisting solely of $N_2$ flow for 1 hour at 500°C .

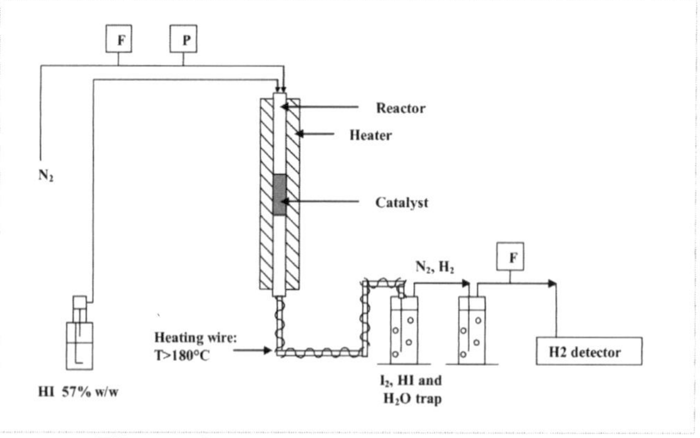


**Figure 1:** **Experimental apparatus layout**

## Results and discussion

***Catalyst characterization***

XRD patterns of fresh Ni-acac catalyst revealed that the phases detected were mainly nickel oxide (NiO) and nickel aluminate ($NiAl_2O_4$). Nickel crystal showed small intensity peak. This experimental evidence suggested that a phase transformation occurred on Ni-acac during calcination step with the result of $NiAl_2O_4$ formation. This compound is hardly reduced to metallic nickel at 500°C. Molina and Poncelet [9] evidenced in the calcination step strong interactions between NiO and alumina support producing the formation of nickel aluminate spinel which required temperatures higher than 500°C to be reduced.

ICP analysis detected a Ni loading of 11.93% (%w. dry basis). SEM analysis (figure 2) showed that there were large nickel particles and that their dispersion wasn't homogeneous. This could be attributed to the impregnation phase in the catalyst preparation. During this step, indeed, metal ions may not be easily diffused into the pores leading to a deposit on the outer surface and to their aggregation in large particles.

XRD analysis over Ni-aluminate fresh catalyst showed that the main phases of nickel crystal were $NiAl_2O_4$ and NiO, but no nickel peaks appeared. A γ-alumina peak was also reported, indicating that the coprecipitation and the thermal process were correctly executed.

During the preparation of Ni-nitrate several difficulties were encountered in the impregnation step: in particular, nickel ion was removed during washing stage of the solid. Catalyst obtained was found to contain only a few percentage of nickel (0.83 %), despite the expected concentration was around 20%.

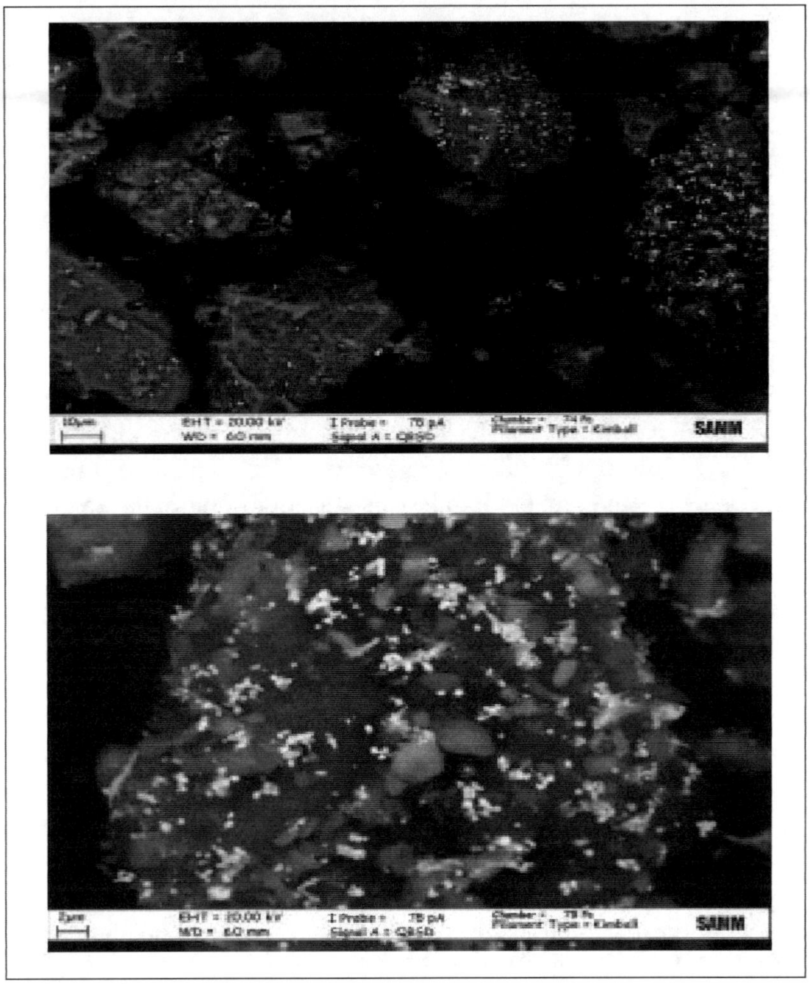


**Figure 2.** **SEM micrographs of Ni-acac at two different magnifications**

***Catalytic activity for HI decomposition***

The catalytic performance of nickel catalysts was evaluated operating a fixed-bed reactor in the temperature range of 500-650°C. The conversion of HI was measured after 1 hour from the start, when a steady state was reached. Table 1 reports HI conversions at 500°C and 650°C over the catalysts prepared. It is clearly deducible that both Ni-acac and Ni-aluminate catalysts demonstrated very high HI conversion at 500°C and 650°C. High conversions on $\gamma$-alumina supported nickel catalysts were also observed by other researchers. NCLI has shown that there was a rapid approach to theoretical equilibrium values, comparable to Platinum based catalysts for temperatures over 500°C. However, at lower temperatures this catalyst showed a lower activity than platinum based catalyst.

High catalytic activity of both Ni-acac and Ni-aluminate suggested that active sites for hydrogen iodide decomposition may not be ascribed only to nickel crystallites. As above mentioned, XRD characterization showed no nickel peaks existence in the calcined $NiAl_2O_4$ phase of Ni-aluminate catalyst. Hence, it is deducible that both the calcined $NiAl_2O_4$ phase of Ni-aluminate catalyst and the reduced $NiAl_2O_4$ phase in Ni-acac catalyst may be considered as the active site in the HI decomposition reaction.

**Table 1. HI conversion at 500°C and 650°C**

| Catalyst type | Conversion at 500°C | Conversion at 650°C |
|---|---|---|
| Ni-acac | 20.7 | 23.3 |
| Ni-nitrate | 12.0 | 21.7 |
| Ni-aluminate | 21.0 | 21.2 |

Moreover, conversion obtained on the Ni-acac catalyst at 650°C were slightly higher than in the case of Ni-aluminate. This result may be ascribed to :

(1) the synergy effect of the reduced $NiAl_2O_4$ and surface nickel particle;

(2) during Ni-aluminate catalyst preparation, $Ni^{2+}$ and $Al^{3+}$ are coprecipitated as hydroxide compounds and $Ni(OH)_2$ are coated by a large amount of $Al(OH)_3$. This configuration could prevent the contact of reactants to the active sites leading to lower conversions.

Regarding the behaviour of Ni-nitrate, the HI conversion was considerably lower than the theoretical equilibrium value at 500°C, but it significantly raised at 650°C. This could be attributed to the lower concentration of active sites, as the nickel loading was very poor compared to other catalysts.

***Deactivation of catalysts***

The catalysts stability experiments were carried out at 500°C. Assuming that the catalysts can be considered deactivated when the conversion is 25% lower than the initial value, the deactivation rates of the three catalysts demonstrated to be in the following order: Ni-aluminate~Ni-acac<Ni-nitrate.

Regarding the Ni-aluminate catalyst, in figure 3 the variation in HI conversion as a function of time is reported and it is evident a gradual decrease of the catalyst activity. The same behavior was experimentally found for the other two catalysts. O'Keefe [4] reported that a copper oxide catalyst was tested for the HI decomposition reaction. As expected, the $Cu_xO$ reacted with the HI to form what appeared to be $Cu_xI_2$, but this latter material was not a particularly active catalyst. In addition, the catalyst was seen to crumble off of the support alumina substrate and there was evidence of $Cu_xI_2$ material being transported downstream with the HI carrier gas flow. Moreover, previous NCLI studies over $\gamma$-alumina supported nickel catalysts evidenced solid deposits at the exit of the reactor, indicating the transport of the metal from the catalyst, as well as a significant decrease in catalyst surface area after the HI decomposition reaction. In this study, over the testing period black deposits were noted at the exit of the fixed-bed reactor.

Moreover, it was found that, keeping the deactivated catalysts at a temperature of 650°C for at least 1 hour, the HI conversion seemed to substantially increase. These observations suggested that the deactivation may be ascribed for all catalysts to $NiI_2$ formation during the experiment. Since $NiI_2$ thermal decomposition begins at about 500°C [10] and can be considered quantitative at about 650°C, as observed at ENEA labs, it can be deduced that catalyst thermal treatment produces a considerable $NiI_2$ decomposition and consequently the catalyst activity recovery.

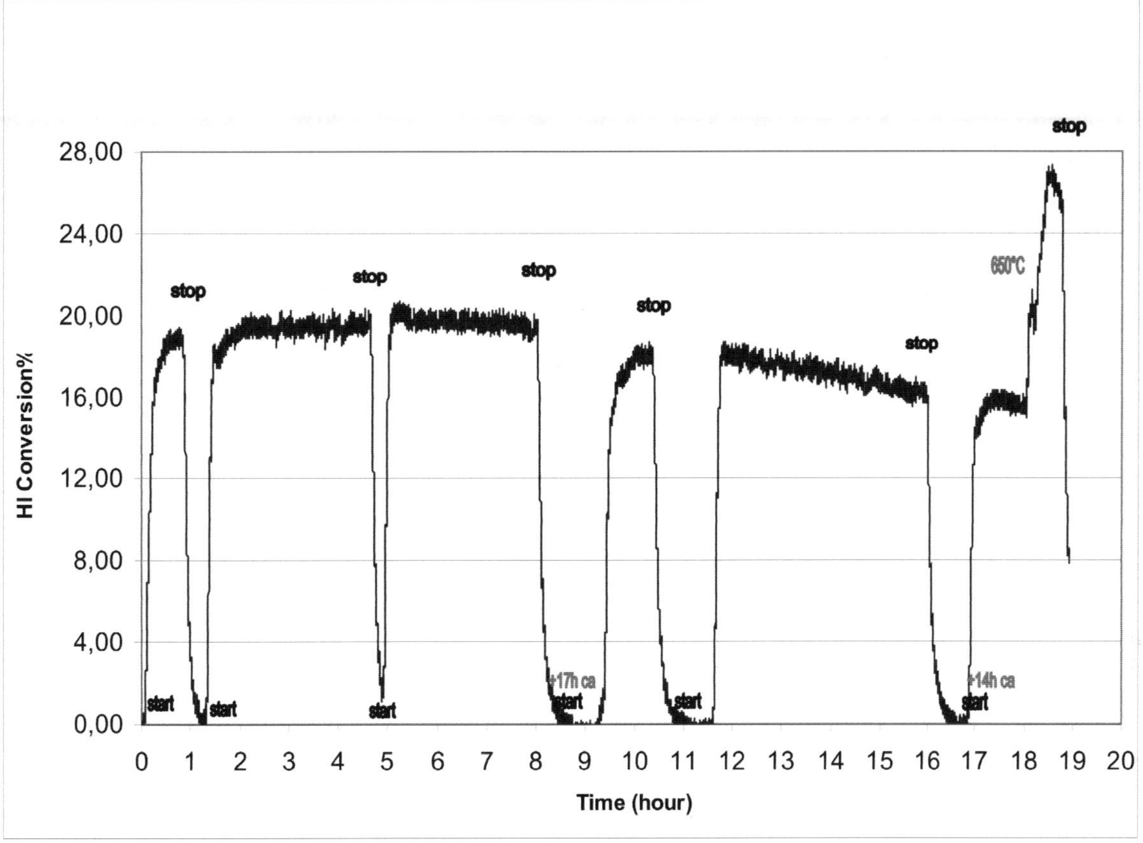


**Figure 3:** **HI conversion vs. time. Operating conditions: T=500°C-650°C; HI=0,1 mL/min. In correspondence of HI feed rate stops the reactor was kept at 500°C in a $N_2$ stream**

## Conclusion

The present experimental investigation has showed that both Ni-acac and Ni-aluminate catalysts have high catalytic activity at 500°C and 650°C, as HI conversion detected quickly approached the theoretical equilibrium value. These results suggest that catalytic activity in HI decomposition may not be ascribed only to nickel crystallites, and in particular it is deducible that the calcined $NiAl_2O_4$ phase of the Ni-aluminate catalyst and the reduced $NiAl_2O_4$ phase in Ni-acac catalyst may be considered as the active site in the HI decomposition reaction. In addition, the slightly higher activity of the Ni-acac compared to Ni-aluminate at 650°C could be attributed mostly to the synergy effect of the reduced $NiAl_2O_4$ and surface nickel particle. Catalysts stability experiments at 500°C showed that the deactivation rates of the three catalysts were in the following order: Ni-aluminate~ Ni-acac<$NiNO_3$. The experimentally demonstrated deactivation behavior was probably due to $NiI_2$ formation. In fact, since $NiI_2$ thermal decomposition occurs near 650°C, the catalysts deactivation has been found to be reversible over this temperature.

## References


1. R. H. Elder, G. H. Priestman, B. C. Ewan and R. W. K. Allen (2005), "The separation of $HI_x$ in the sulphur-iodine thermochemical cycle for sustainable hydrogen production", *Trans. IChemE, Part B, Process safety and Environmental Protection*, 83 (B4), pp. 343-350.
2. G. B. Kistiakowsky (1928), "Homogeneous gas reaction at high concentrations. 1. Decomposition of hydrogen iodide", *The Journal of the American Chemical Society*, vol. 50, pp. 2315-2332.
3. M. Bodenstein (1899), "Gasreaktionen in der chemischen Kinetik", Z. Phys. Chem., vol. 29, pp. 295.
4. D. R. O'Keefe, J.H. Norman, D. G. Williamson (1980), "Catalysis Research in Thermochemical Water-Splitting Processes", *Cat. Rev .Sci. Eng*, vol. 22, n°3, pp. 325-369.
5. C. N. Hinshelwood and R.E. Burk (1925), "The relation of homogeneous to catalysed reactions. The catalytic decomposition of hydrogen iodide on the surface of platinum", *J. Chem. Soc. Trans.*, vol. 127, pp. 2896-2900.
6. Y. Shindo, N. Ito, K. Haraya, T. Hakuta, H. Yoshitome (1984), "Kinetics of the catalytic decomposition of hydrogen iodide in the thermochemical hydrogen production", *International Journal of Hydrogen* Energy, vol. 9 n°8, pp. 695-700.
7. Y. Oosawa, T. Kumagai, S. Mizuta, W. Kondo, Y. Takemori, K. Fujii (1981), *Bull. chem. Soc. Japan,* vol. 54 n°3, pp. 742-748.
8. A. C. Tito, C. Mazzocchia, A. Spadoni, P. Favuzza, M. Lanchi, P. Tarquini (2007), "Synthesis of the catalyst $Ni/Al_2O_3$ for the production of $H_2$ from HI decomposition in the thermochemical water splitting", in *"2nd World Congress of young scientists on hydrogen energy systems"* , June, Turin, Italy.
9. R. Molina, G. Poncelet (1998), "α-Alumina supported nickel catalysts prepared from nickel acetylacetonate: a TPR study", *J.Catal.,* vol. 173, pp.257-267.
10. S. Sato, S. Shimizu, H. Hakajima, K. Onuki and Y. Ikezoe (1986), " Studies on the Nickel-Iodine-Sulfur process for hydrogen production-III", *Int. Journal of Hydrogen Energy,* vol. *11,* n°9, pp. 571-575.

# 166g Membrane Separations Technology for the Sulfur-Iodine Thermochemical Cycle

**Frederick F. Stewart** and Christopher J. Orme. Chemical Sciences, Idaho National Laboratory, P.O. Box 1625, Idaho Falls, ID 83415-2208

One of the more promising cycles for the thermochemical generation of hydrogen is the Sulfur-Iodine (S-I) process, where aqueous HI is thermochemically decomposed into H2 and I2 at approximately 350 degrees Celsius. Regeneration of HI is accomplished by the Bunsen reaction (reaction of SO2, water, and iodine to generate H2SO4 and HI). Furthermore, SO2 is regenerated from the decomposition of H2SO4 at 850 degrees Celsius yielding the SO2 as well as O2. Thus, the cycle actually consists of two concurrent oxidation-reduction loops. As HI is regenerated, co-produced H2SO4 must be separated so that each may be decomposed. Current flowsheets employ a large amount (~83 mol% of the entire mixture) of elemental I2 to cause the HI and the H2SO4 to separate into two phases. To aid in the isolation of HI, which is directly decomposed into hydrogen, water and iodine must be removed. Separation of iodine is facilitated by removal of water. Sulfuric acid concentration is also required to facilitate feed recycling to the sulfuric acid decomposer. Decomposition of the sulfuric acid is an equilibrium limited process that leaves a substantial portion of the acid requiring recycle. Distillation of water from sulfuric acid involves significant corrosion issues at the liquid-vapor interface. Thus, it is desirable to concentrate the acid without boiling. Recent efforts at the INL have concentrated on applying pervaporation through Nafion-117 and Nafion-112 membranes for the removal of water from HI/water and HI/Iodine/water feedstreams. In pervaporation, a feed is circulated at low pressure across the upstream side of the membrane, while a vacuum is applied downstream. Selected permeants sorb into the membrane, transport through it, and are vaporized from the backside. Thus, a concentration gradient is established, which provides the driving force for transport. In this work, membrane separations have been performed at temperatures as high as 134 degrees Celsius. Transmembrane fluxes of water are commercially competitive (~5000 g/m2h) and separation factors have been measured as high as 8000, depending on the membrane and the water content. In these experiments, the common trade off in membrane performance is observed in that as flux is increased, separation factor decreases. Nafion-112, a thinner membrane, exhibited much higher fluxes than the Nafion-117 with no significant loss in separation factor indicating that the permeability of iodine and HI through Nafion materials can be low depending on experimental conditions. Data for the sulfuric acid concentration suggests performance similar to the HI experiments. All membranes studied for the HI, HI/iodine and sulfuric acid feeds exhibited no degradation in membrane performance during use. Particular attention in paper will be paid to the performance dependence on the pressure gradient across the membrane.

# Paper 224a
# Generation of Hydrogen Using Electrolyzer with Sulfur Dioxide Depolarized Anode

John L. Steimke and Timothy J. Steeper, Savannah River National Laboratory, Aiken SC

## ABSTRACT

The HyS Process is a hybrid thermochemical cycle that may be used in conjunction with advanced nuclear reactors or centralized solar receivers to produce hydrogen by water-splitting. Like all other sulfur-based cycles, HyS utilizes the high temperature thermal decomposition of sulfuric acid to produce oxygen. The unique aspect of HyS is the generation of hydrogen in a water electrolyzer that is operated under conditions where dissolved sulfur dioxide depolarizes the anodic reaction, resulting in substantial voltage reduction. Sulfur dioxide is oxidized at the anode, producing sulfuric acid that is sent to the acid decomposition portion of the cycle. Hydrogen is generated at the cathode. The focus of this work was to conduct single cell electrolyzer tests to further improve cell performance.

A test facility for conducting $SO_2$-depolarized electrolyzer (SDE) testing had been previously built. The maximum cell current is 50 amperes, which is equivalent to a hydrogen production rate of approximately 20 liters per hour. Feed to the anode of the electrolyzer is a sulfuric acid solution saturated with sulfur dioxide. The partial pressure of sulfur dioxide may be varied in the range of 1 to 6 atm. Electrolyzer temperatures may be controlled in the range from ambient to 80°C. Hydrogen generated at the cathode of the cell is collected for the purpose of flow measurement and compositional analysis. The test facility proved to be easy to operate, versatile, and reliable.

A versatile electrolyzer cell was designed and built with wetted parts of carbon and Teflon. It was previously reported that the cell was tested with six Membrane Electrode Assemblies (MEA) during FY 2006. An MEA is a thin sandwich consisting of an anode, a Proton Exchange Membrane (PEM) and a cathode. During FY 2007 the cell was tested with twelve different MEA. A 100 Hour Longevity Test was conducted with MEA 12. The primary considerations were low cell voltage, MEA durability and low rate of sulfur formation at the cathode.

The primary results of FY 2007 testing follow.

1. The 100 Hour Test was completed successfully. The cell voltage increased about 50 mV during the test.

2. During the 100 Hour Test elemental sulfur formed at the cell cathode. For every mole of hydrogen generated about 0.001 mole of sulfur formed.
3. Increasing electrolyzer cell temperature generally decreased cell voltage, which is attributed to increased reaction rates and also to decreased viscosity of anolyte, allowing easier movement of reactants and product. There is significant room for improvement by increasing cell temperature beyond 80°C.
4. Hardware was added to provide a small water flow to the cathode which flushed out sulfur and also helped to hydrate the membrane. A Sulfur Collector was also added to separate product hydrogen from the sulfur and water mixture that exited the cell. This hardware functioned well.
5. The Sulfur Collector allowed visual approximation of the rate of sulfur formation. Periodically the contents were drained, filtered, dried and weighed. However, the sulfur particles were small enough that they passed through common filter paper and required weeks to settle.
6. The anode flow distributor was redesigned to function well with an anolyte flowrate that was lower by a factor of 7.5. This increased the conversion of sulfur dioxide by the same factor or to a range from 5% to 20%.
7. System upsets that generated a pressure difference across the MEA in the range from 20 psid to 40 psid damaged the catalyst layers on the MEA by flexing it.

## Overview of Test Facility

A test facility capable of testing sulfur dioxide depolarized electrolyzers at pressures up to six atm and temperatures up to 80°C was designed and constructed. The facility is capable of operating with electrolyzer currents up to 50 amperes or a current density of 1000 $mA/cm^2$ for a cell area of 50 $cm^2$. A current of 50 amperes is sufficient to generate approximately 20 liters per hour of hydrogen.

The test facility proved to be versatile and easy to use. One MEA was easily replaced in the cell with another. The data acquisition system computer controlled anolyte flowrate, even with changing hydraulic resistance. Computer interlocks were easily added. It was relatively easy to change liquids, flush both sides of the cell and isolate part of the piping. The accurate measurement of hydrogen generated helped to explain other observations and to quantify the amount of hydrogen that reacted with sulfur dioxide. In 2007 a water purge to the cathode was added to remove any sulfur formed and improve hydration of the membrane, which improves its ionic conductivity. A sulfur and water mixture was collected in a new Sulfur Collector where it separated from product hydrogen.

## Design of Electrolyzer Cell

The electrolyzer cell is of the MEA type. The MEA is a sandwich consisting of an anode where sulfur dioxide and water in the anolyte are oxidized to produce hydrogen ions and sulfuric acid, an electrically insulating Proton Exchange

Membrane (PEM) that allows the hydrogen ions and water to pass through, and a cathode where the hydrogen ions are reduced to hydrogen gas. The remainder of the electrolyzer serves to deliver the anolyte to the anode and remove the sulfuric acid reaction product, remove the moist hydrogen gas and any contaminants or by-products from the cathode, provide the current paths to the electrode, allow temperature control and contain the internal pressure. The cell flow paths are inter-digitated. Flow enters a header and is distributed to fingers, flows in a thin layer past the MEA to other fingers, is collected in the outlet header and is re-circulated. The cell has two heaters to maintain elevated temperature.

**Membrane Electrode Assemblies**

An electrolyzer cell was tested with six different MEA in sulfur dioxide depolarized operation. The electrolyzer has an active area ranging from 48 to 60 $cm^2$. The MEA tested in the electrolyzer cell during FY 2007 used Nafion® 115, Nafion® 117, Nafion® 212 or polyphenylene membranes which are 5, 7, 2 and 2 mils thick, respectively. Nafion® membranes are designed to pass water and hydrogen ions, but not gases. Unfortunately, Nafion® membranes also pass dissolved sulfur dioxide to some extent. The catalyst was either platinized carbon or platinum black. Catalyst loadings ranged from 0.75 $mg/cm^2$ to 1.5 $mg/cm^2$. The characteristics of the MEA are summarized in Table 1.

**Table 1 Summary of MEA Characteristics for FY 2007**

| MEA # | Membrane | Membrane thickness, mils | Anode flow field | Cathode flow field | Anode Pt loading, mg/cm2 | Cathode Pt loading, mg/cm2 | Active area, cm2 |
|---|---|---|---|---|---|---|---|
| 7 | Celtec-V | 4 | Carbon paper, 7 mil | Carbon cloth, 12 mils | 0.8 Pt-C | 0.8 Pt-C | 47 |
| 8 | Nafion-115 | 5 | Carbon paper, 7 mil | Carbon cloth, 12 mils | 0.78 Pt-C | 0.61 Pt-C | 60 |
| 9 | Nafion-117 Giner | 7 | Carbon paper, 7 mil | Carbon cloth, 12 mils | 4.0 Pt black | 4.0 Pt black | 60 |
| 10 | Nafion-117 Giner | 7 | Carbon paper, 7 mil | Carbon cloth, 12 mils | 1.0 Pt-C | 1.0 Pt-C | 60 |
| 11 | Nafion-115 | 5 | Carbon paper, 7 mil | Carbon cloth, 12 mils | 1.09 Pt-C | 0.72 Pt-C | 47.6 |
| 12 | Nafion-115 | 5 | Carbon paper, 7 mil | Carbon cloth, 12 mils | 1.01 Pt-C | 1.01 Pt-C | 60.8 |
| 13 | Nafion-115 | 5 | Carbon | Carbon | 1.02 | 0.59 | 60 |

| | | | | | | | |
|---|---|---|---|---|---|---|---|
| | | | paper, 7 mil | cloth, 12 mils | Pt-C | Pt-C | |
| 14 | Nafion-117 Giner | 7 | Carbon paper, 7 mil | Carbon cloth, 12 mils | 0.8 Pt-C | 0.8 Pt-C | 49 |
| 15 | polyphenylene SDAPP 2.2 Hickner | 2 | Carbon paper, 7 mil | Carbon cloth, 12 mils | 1.5 Pt black | 1.5 Pt black | 46.3 |
| 16 | polyphenylene SDAPP 2.2 Hickner | 2 | Carbon paper, 7 mil | Carbon cloth, 12 mils | 1.5 Pt-C | 1.5 Pt-C | 60.8 |
| 17 | Nafion-212 Lynntech | 2 | Carbon paper, 7 mil | Carbon cloth, 12 mils | 1.5 Pt black | 1.5 Pt black | 50. |
| 18 | Nafion-115 | 5 | Carbon paper, 7 mil | Carbon cloth, 12 mils | 0.75 Pt-C | 0.75 Pt-C | 61.6 |

**Test Facility**

Anolyte is saturated with sulfur dioxide in an absorber filled with Raschig Rings. An annulus surrounds the Absorber through which passes hot water. A backpressure regulator controls the Absorber pressure. Anolyte flows through a pump, a flowmeter, the cell and back to the Absorber. The first one controls the pressure in the Sulfur Dioxide Absorber.

The cathode side of the electrolyzer is supplied with a small flow of water. The effluent from the cathode, a three-phase mixture of hydrogen, water and elemental sulfur passes through the Sulfur Collector. Hydrogen flows on through another backpressure regulator and then to the Hydrogen Collector which measures flowrate. Care is taken to equalize the pressures on the anode and cathode sides of the cell.

**Summary of Testing and Results**

By far the longest test of an MEA was the 100 Hour Longevity Test of MEA 12. Figure 1 shows the cell voltage for that period for a temperature of 80°C, a sulfur dioxide pressure of 4 atm and a current density of 300 mA/cm$^2$. The cell voltage slowly increased by about 50 mV over the test. There were variations on cell voltage because of changes in current or acid concentration. The long test at steady conditions allowed careful measurements of hydrogen production. Hydrogen production was 100% of theoretical to within measurement uncertainty. Sulfur formed on the cathode and accounted for about 0.001 mole of sulfur for every mole of hydrogen.

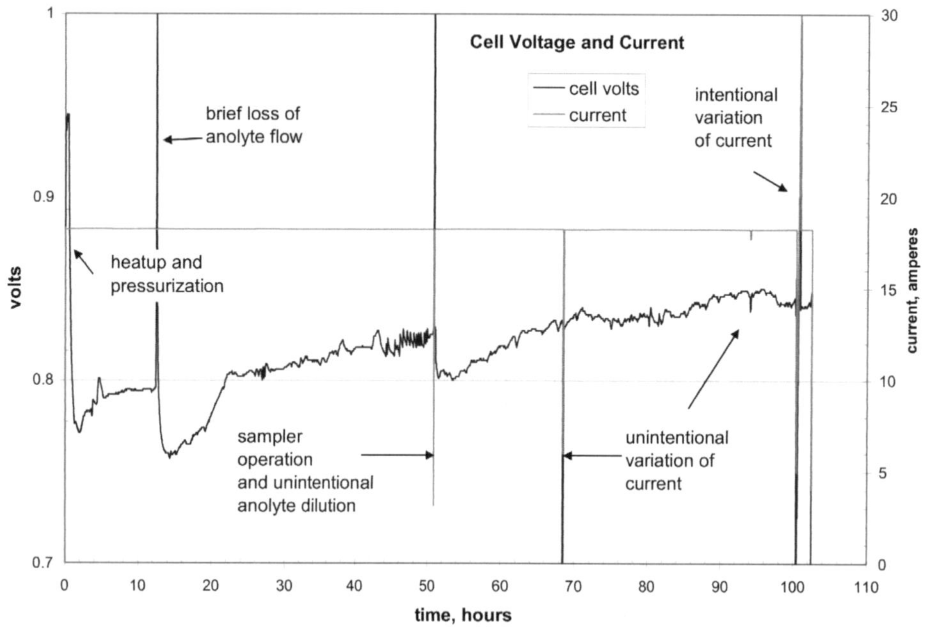


**Figure 1 Electrolyzer Cell Voltage During 100 Hour Test**

Electrolyzer testing always started at ambient conditions and then temperature and pressure were increased. As expected, the cell voltage decreased with increasing temperature as shown in Figures 2 and 3. Figure 2 shows the results for MEA 12 for which the final pressure was 4 atm. Note that initially there is a nearly linear decrease of voltage with temperature, but the effect diminishes at higher temperatures. Figure 3 shows the results for MEA 13 and for similar conditions except that the final pressure was 5 atm, so the concentration of sulfur dioxide was 25% higher. The linear trend of voltage decrease with increasing cell temperature continued up to 80°C. Therefore, the concentration of sulfur dioxide was insufficient at 4 atm and 80°C. Presumably, further increases in temperature would require further increases in pressure to prevent the effect shown in Figure 2.

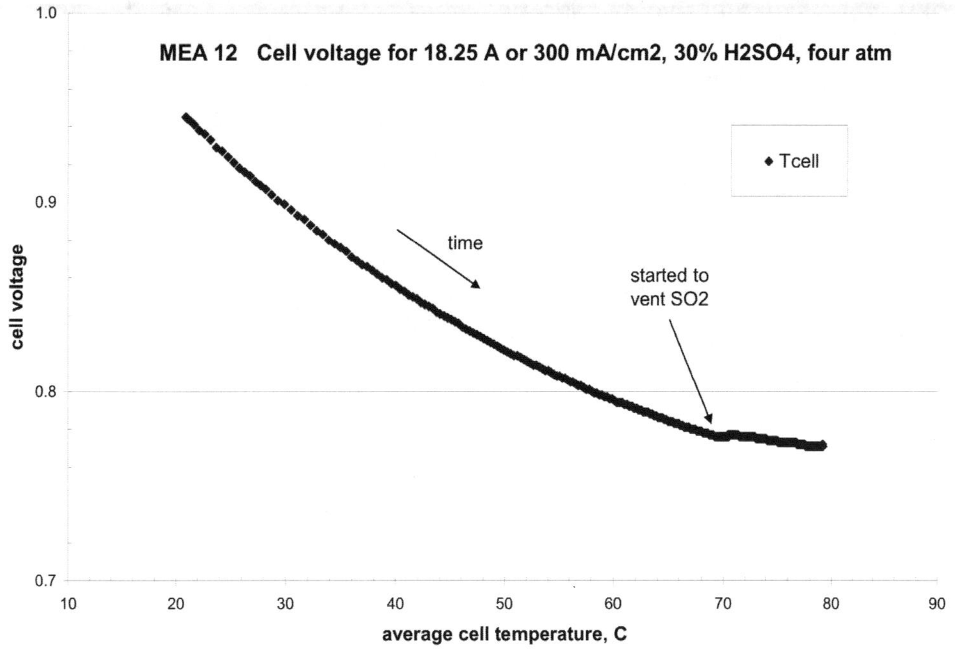


**Figure 2 Cell Voltage as Function of Temperature, Insufficient SO2**

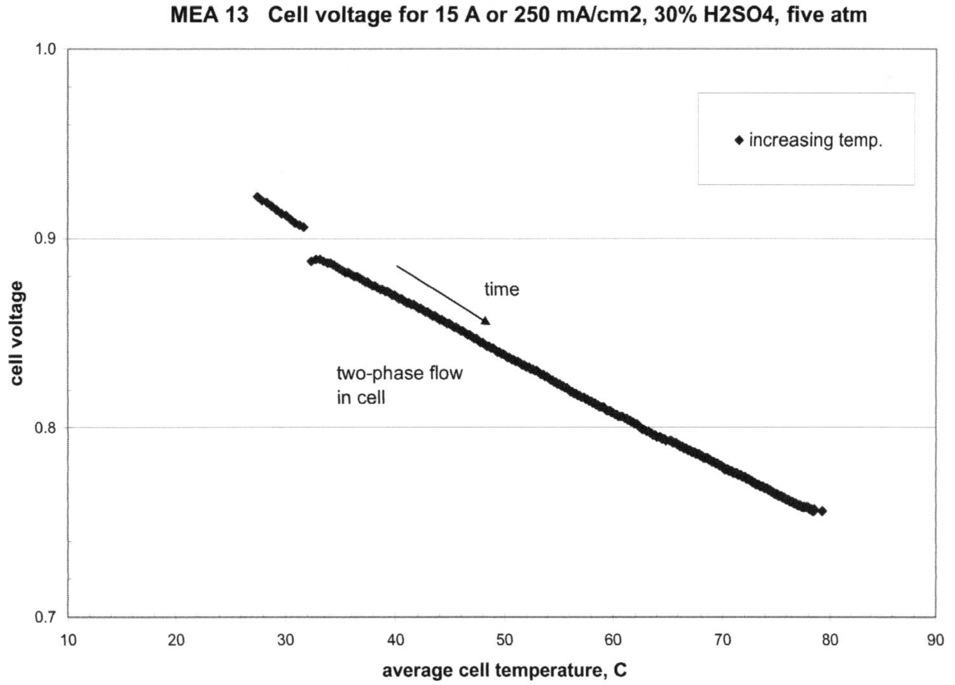


**Figure 3 Cell Voltage as Function of Temperature, Sufficient SO2**

### Tests of Other MEA

Three of the MEA, MEA 15, MEA 16 and MEA 17, used membranes that had a thickess of 0.002". The first two membranes were polyphenylene and the last was Nafion 212. It was hoped that these relatively thin membranes would have a low ionic conductivity. However, results were disappointing because MEA performance degraded quickly. The thin membranes probably flexed more than thicker membranes. Visual and microscopic observation showed loss of catalyst on both the anode and cathode sides.

Testing of MEA 11 showed the effect of increasing the anolyte concentration from 30 wt% sulfuric acid to 50 wt%. At 80°C and 500 mA/cm$^2$ the increase in acid concentration increased the cell voltage by 200 mV, which is significant.

## CONCLUSIONS

The primary conclusions resulting from this work are as follows.

1. SRNL successfully completed a 100 Hour Longevity Test of the $SO_2$ Depolarized Electrolyzer.
2. Pressure differences across the MEA such as caused by system upsets in the range from 20 psid to 40 psid can flex the membrane and crack off the anode and cathode layers. Thinner membranes are more susceptible to flexing.
3. The yield of hydrogen is 100% of theoretical to within measurement error.
4. Some hydrogen is consumed in the unwanted reaction with sulfur dioxide that has diffused through the membrane to the cathode to form. The moles of sulfur formed are about 0.1% of the moles of hydrogen generated.

## FUTURE WORK

Based on the progress shown in this study we recommend the following activities during FY08 for continued development of the $SO_2$ Depolarized Electrolyzer for hydrogen production.

1. Expand the range of test conditions beyond the present 80°C and 5 atm.
2. Test MEA with platinized carbon anodes and platinum black cathodes.
3. Test a three cell electrolyzer stack that was designed and built by Giner Inc.

## REFERENCE

Steimke, J. L. and T. J. Steeper, "Single Cell SO2-Depolarized Electrolyzer Longevity Test Results", WSRC-STI-2007-00300, Rev. 0, June 2007.

## Hydrogen generation by plasma-assisted electrolysis of $H_2O/SO_2$ gas mixture

W. M. Lee , I. G. Koo, M. S. Lee, J. H. Kim, M. Y. Choi, J. Sohn

*Division of Energy Systems Research, Ajou University, Suwon 443-749, Korea*

We demonstrated that the reaction between water and sulfur dioxide ($SO_2$) can be catalyzed plasma-chemically and hydrogen species, products of the reaction, can be pumped out electrochemically as hydrogen molecules ($H_2$) with the help of an electrochemical hydrogen pump [1]. The major byproduct of the reaction of $H_2O/SO_2$ mixture in the plasma reactor is sulfuric acid ($H_2SO_4$) which can be decomposed back to $SO_2$ gas using thermal energy source like nuclear heat [2].

Decomposing water into hydrogen and oxygen gas by brute thermal method is impractical because of the need for high thermal energy far above 3000°C, of the difficulty with selecting material withstanding such high temperature and of the need for the hydrogen separator. Most widely known method for decomposing pure water is electrolysis which alters the reaction path into two half-cell reactions, oxidation and reduction. The thermodynamic voltage of water electrolysis is 1.23 volt, but by adding to water some chemical species such as $Ti^{3+}$, Sb, Re, $SO_2$, Sb, As, Re, and Bi or organic fuels such as methanol the thermodynamic voltage can be reduced substantially. For example, at the anode of the electrolyzer the mixture of water and $SO_2$ are oxidized to generate hydrogen ion (proton) and $H_2SO_4$ at 0.18 V [2]. The anodic reaction of $H_2O/SO_2$ produces sulfuric acid ($H_2SO_4$) instead of oxygen gas. The protons, another product of the reaction, are reduced to hydrogen gas at the cathode after being transported through the electrolyte. The sulfuric acid can be decomposed to $SO_2$ and $O_2$ gas at temperature higher than 850°C if high thermal energy like nuclear heat is available and the regenerated $SO_2$ gas would be recycled. This scenario of hydrogen production is illustrated in Fig. 1 and is commonly called the hybrid model of nuclear hydrogen production since it combines the electrolysis and sulfuric acid decomposition [2, 3].

But the electrolysis of $H_2O/SO_2$ has some drawbacks such as kinetic sluggishness of the anodic reaction that leads to high anodic overpotential and cross-over of $SO_2$ across the electrolyte to the cathode site that may result in its poisoning. We have developed an alternative method based on microplasma to the electrolysis with the aim of solving these problems [1]. The essence of the method is to utilize non-thermal electrons generated in the plasma for activating the $H_2O/SO_2$ molecules and to pump out the hydrogen atoms from the reactor through an electrochemical hydrogen pump. Since the plasma reactor and the pump share a common electrode made of hydrogen-absorbing metal such as palladium (Pd) the cross-over of $SO_2$ is resolved naturally [4]. We named this device "plasma-assisted hydrogen pump assembly (PAHPA)" as illustrated schematically in Fig. 2. Microplasma generated in the reacting gas of the plasma reactor substitutes for the catalyst used for the oxidation reaction of $H_2O/SO_2$ mixture in conventional electrolyzer. Advances in microdischarge techniques in recent years now make it possible to generate microplasma in atmospheric pressure air or even water vapor [5] consuming very low electrical energy, often below 10 $W\cdot cm^{-3}$. Hydrogen atoms, products of the plasma-assisted reaction, are dissolved in the hydrogen absorbing metal like palladium (Pd). The dissolved hydrogen atoms are oxidized at the interface between Pd, now the anode of the pump, and the proton conducting membrane, the electrolyte of the pump and reduced into $H_2$ gas at the cathode of the pump. For low temperature operation below 180°C a polymer membrane like Nafion was used as the electrolyte of the pump while for high temperature operation between 600 and 850°C solid oxide membrane like $BaCe_{0.8}Y_{0.2}O_{3-\delta}$ (BCY)oxide was used as the proton-conducting electrolyte. The reactions occurring in the high temperature reactor and the pump are summarized as

| | | |
|---|---|---|
| $2\,H_2O + 2\,SO_2 + Pd \rightarrow H_2 + H\text{-}Pd\text{-}H + 2\,SO_3$ | in the plasma reactor | (1) |
| $H\text{-}Pd\text{-}H \rightarrow 2\,H^+ + 2\,e^- + Pd$ | at anode of the pump | (2) |
| $2\,H^+ + 2\,e^- \rightarrow H_2$ | at cathode of the pump | (3) |
| $H_2O + SO_2 + Pd \rightarrow H_2 + SO_3 + Pd$ | over-all | (4) |

The plasma reactor was constructed by sandwiching 1 mm thick porous alumina disk of 1 cm diameter (pore diameter of 100 μm and 30 volume % porosity) with two stainless steel (SUS) meshes of size 100 as the current collectors. One side of the disk was coated with palladium (Pd) film by plasma method [6]. BCY solid oxide (Praxair) and Nafion 115 were used as the electrolyte material for the high and lower temperature PAHPA, respectively. The MEA (membrane electrode assembly) for the high temperature hydrogen pump was constructed in the order of fabricating the BCY-embedded palladium (Pd) anode first, then coating one side of the anode with thin BCY film, and finally coating the bare side of the film with the platinum (Pt) cathode. For low temperature PAHPA, a catalyst coated (Pd at the anode and Pt at the cathode) Nafion 115 membrane (CCM) was attached as an electrochemical pump to the Pd-coated alumina disk. Input gas to the plasma reactor was prepared by mixing water vapor and $SO_2$ gas, the flow rate of each gas being 100 and 30 $ml \cdot min^{-1}$, respectively.

As the gas temperature in the plasma reactor was increased from 150 to 700°C, the power consumed for the discharge decreased from 0.45 to below 0.15 W, respectively, while the current was kept constant at around 1 mA. In actual experiments the electrical voltage at a given temperature was adjusted to a minimum value just to avoid extinction of the discharge and then it was increased gradually until the current reached the value around 1 mA. It was confirmed from the experiments that the electron number density at

700°C is almost three times higher than that at 150°C. In plasma-chemical terms the increase of the electron density implies that the $H_2O/SO_2$ molecules can be energized to higher degree as the temperature goes up, resulting in increase of the reaction rate.

Once the electrical discharge in the plasma reactor became stable, the electrical power of the pump was turned on to pump out the hydrogen atomic species absorbed at the bipolar Pd electrode. The i-V curve of the hydrogen pump is shown in Fig. 4. The electrical current due to hydrogen pumping was about 60 mA $cm^{-2}$ at 0.1 V for the low temperature (below 150°C) operation, but it did not change much as the voltage was increased to a value over 0.1 V. This is similar to diffusion limited current. The limiting current at 600°C is much lower than the value at 150°C possibly due to poor conduction of the BCY electrolyte at the temperature. But the limiting current density at 850°C was near 100 mA $cm^{-2}$, four times higher than the value at 600°C and even 30% higher than the value at 150°C. Since the conductivity of the BCY at 850°C is comparable with that of Nafion, the fact that the high temperature PAHPA registered a higher limiting current than the low temperature PAHPA can be attributed to the increase of the hydrogen species dissolved in the Pd electrode at higher temperature. The current-voltage curve of the conventional cell electrolyzing $H_2O/SO_2$ mixture is also shown in Fig. 4 for comparison.

The energy efficiency of PAPHA, defined as the ratio of the measured volume of the hydrogen gas generated to the total consumed electrical energy, was about 160 ml · $kJ^{-1}$. The energy needed to operate the PAHPA was mostly consumed at the plasma reactor (the energy to operate the pump is less than 5% that to operate the plasma reactor).This is equivalent to about 75% that for the electrolysis of $H_2O/SO_2$ mixture at 80°C. Thus the PAHPA system has the capacity to produce hydrogen gas without using platinum catalyst electrode needed in the conventional electrolyzer. Furthermore, the problem of $SO_2$ cross-over is no longer problem with PAHPA system. For comparison, the energy efficiency of pure water electrolyzer is about 70 ml · $kJ^{-1}$ if the working voltage is 1.8 V. There is a room for further improvement of the efficiency for the PAHPA system by further reducing the electrical energy needed to sustain the microplasma. The findings from this work provide an alternative scheme of the hybrid model of nuclear hydrogen production.

Acknowledgements

This study was supported by Korea Science Foundation through its National Nuclear Technology Program.

References


1. I. G. Koo and W. M. Lee (2007), "Hydrogen generation by plasma-assisted electrochemical pumping," *Electrochem. Commun.*, **9**, 2326.
2. P. W. Lu, S. Srinivasan (1979), "Advanced in water electrolysis technology with emphasis on use of the solid polymer electrolyte," *J Appl Electrochem.*,**9**, 269.
3. P. Sivasubramanian, R. P. Ramasamy, F. J. Freire (2007), C. E. Holland and J. W. Weidner, "Electrochemical hydrogen production from thermochemical cycles using a proton exchange membrane electrolyzer," *Int. J. Hyrdogen Energy*, **32**, 463.
4. J. H. Shim, S. M. Song, W. K. Her, I. G. Koo and W. M. Lee (2003), "Electrochemical acceration of hydrogen transfer through a methanol impermeable metallic barrier," *J. Electrochem. Soc.* **150**, 12.
5. I. G. Koo and W. M. Lee (2006), "Hollow-cathode based electrical discharge in atmospheric pressure water vapor at wide range of temperature," Jpn. J. Appl. Phys. **45**, 7888.
6. I. G. Koo, M. S. Lee, W. M. Lee (2006), "Low temperature plasma-chemical treatment of $PdCl_2$ film by atmospheric pressure hydrogen plasma," *Thin Solid films*, **506-507**: pp. 350-354.


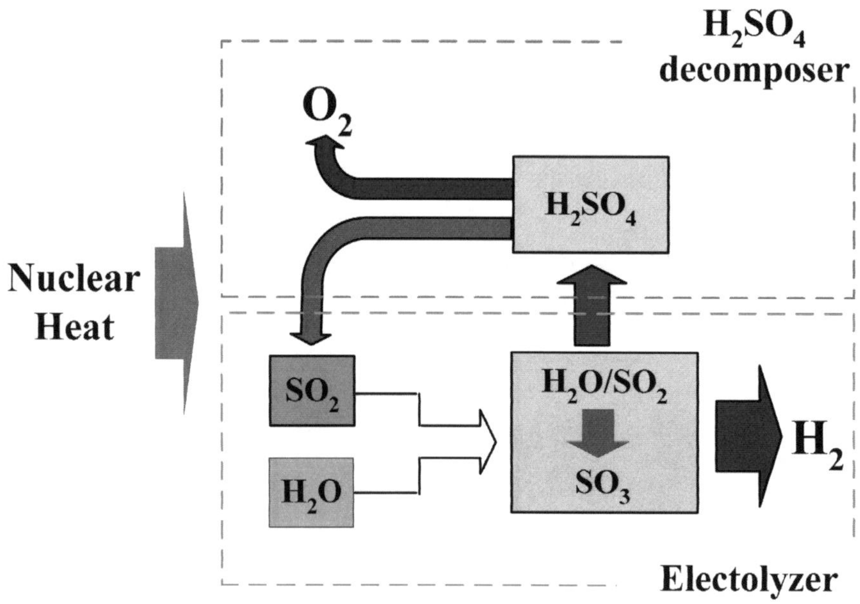


Fig. 1. A scenario of hydrogen production using $H_2O/SO_2$ mixture as the hydrogen source and utilizing nuclear heat to decompose $H_2SO_4$ for $SO_2$ recirculation.

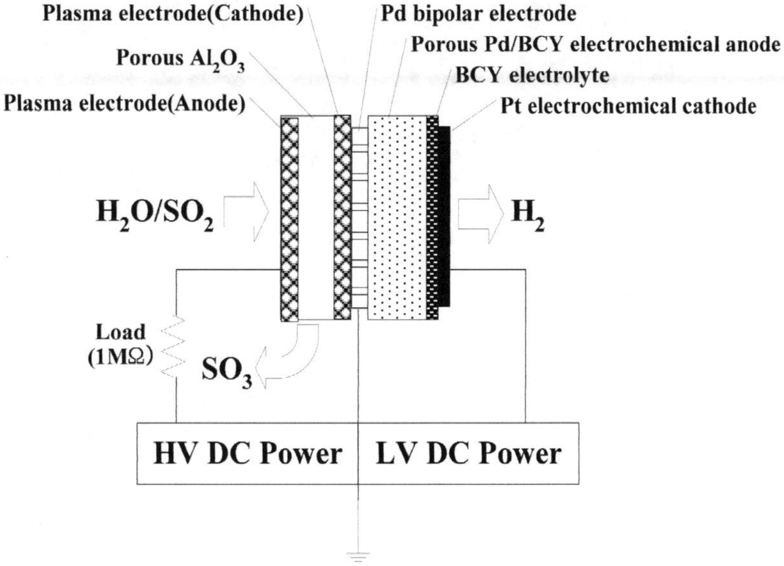


Fig. 2. Configuration of the PAHPA (plasma-assisted hydrogen pump assembly) system.

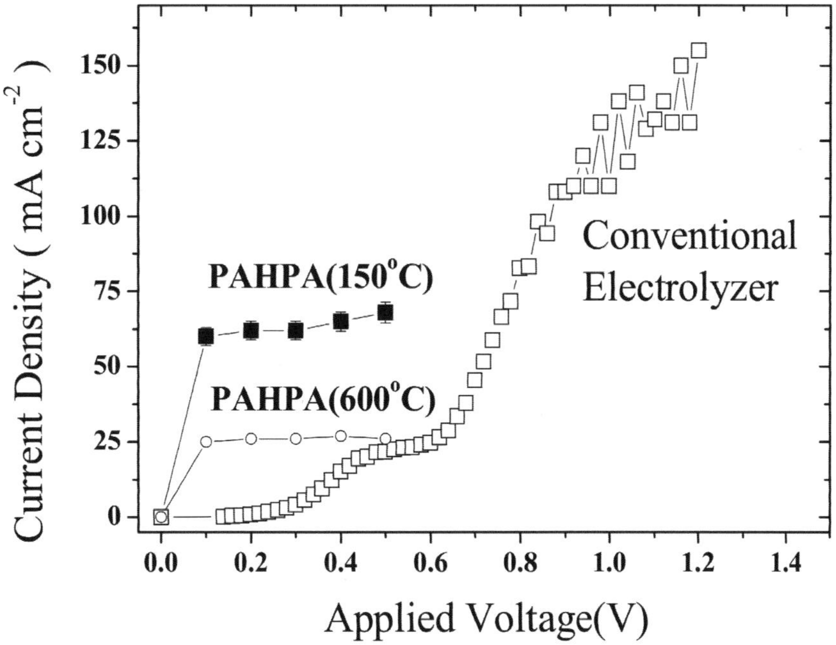


Fig. 3. The current density due to proton transport of the hydrogen pump as a function of the applied voltage.

# DESIGN OF COMPOSITE SULFURIC ACID DECOMPOSITION REACTOR, CONCENTRATOR AND PREHEATER FOR HYDROGEN GENERATION PROCESSES

*Sarah M. Connolly*[1], *Ernest Zabolotny*[2], *David F. McLaughlin*[1], *and Edward J. Lahoda*[1]
[1]*Westinghouse Electric Company, LLC. Pittsburgh, PA*
[2]*Consultant, Burlington, MA*

## 1. Introduction

### 1.1 Background

The sulfur cycles, including the Hybrid Sulfur Process and the Sulfur-Iodine Process, are a group of processes designed to generate hydrogen via electrolysis and thermochemistry. The Hybrid Sulfur Process (HyS), or the Westinghouse Sulfur Process (WSP), uses both thermal and electrical energy from a Pebble Bed Modular Reactor (PBMR) to produce hydrogen. The thermal energy is provided by helium that is used as a coolant in the nuclear reactor and is transferred via an intermediate heat exchange loop to a second helium stream. The thermal energy from the intermediate loop is used to decompose $H_2SO_4$ into $SO_2$, $O_2$, and $H_2O$. Oxygen is then separated from the stream and the water-$SO_x$ mixture is electrolyzed to generate $H_2$ gas and $H_2SO_4$ via the reaction:

$$2\,H_2O + SO_2 \rightarrow H_2 + H_2SO_4 \qquad \textbf{Equation 1}$$

The $H_2SO_4$ can then be recycled back into decomposition reactor.

### 1.2 Design Challenges

All of the sulfur cycle processes require the decomposition of concentrated sulfuric acid into sulfur dioxide, oxygen, and water at temperatures exceeding 800°C. Additionally, the lower condensation temperature of the decomposition products relative to the feed stream requires operating pressures near 9 MPa. The high temperatures and pressures in combination with the extremely corrosive nature of the materials present considerable challenges in designing a reactor system.

It is possible to overcome the challenges presented by both the high temperatures and the corrosive nature of the $H_2SO_4$ by using a ceramic material such as silicon carbide (SiC) or alumina ($Al_2O_3$). However, the use of ceramics then presents the problem of creating seals between the low temperature metal components and the high temperature ceramic components of the reactor. Plastic materials, such as Teflon, are only allowable at temperatures up to 300°C. The use of metal gaskets would create issues of corrosion and possible loosening under cyclic operating conditions. Additional issues include maintaining pressure differentials between the helium side of the reactor and the sulfuric acid side and creating a design that is amenable to maintenance.

## 2. Westinghouse Composite Reactor

## 2.1. Design Concept

The Westinghouse solution to this design challenge is a multi-step reactor that preheats, concentrates, and decomposes sulfuric acid. The reactor will operate under moderate pressure differentials (<1 MPa) between the heated gas and liquid sides. Ceramic material will be utilized for the high temperature and heat transfer portions of the reactor and lined metal in the low temperature distribution headers of the reactor. All seals are located in low temperature portions of the reactor, allowing for the use of Teflon or a similar material. A general schematic of the composite reactor is shown in Figure 1. The design allows for a relatively low cost of construction by using carbon or stainless steel for the pressure boundary of the vessel.

The decomposition tubes consist of a ceramic bayonet tube with an inside ceramic support tube, as shown inFigure 2. The bayonet tubes are contained in a vessel with internal insulation and structured packing to facilitate heat transfer from the helium to the bayonet tube. Inside the bayonet tube, a double-walled insulated ceramic support tube sits on the bottom of the bayonet tube and supports a ceramic liquid catch ring with sloped sides and a hole in the middle. A ceramic gas distribution plate then sits on top of the liquid catch ring, which in turn supports ceramic packing material. On top of the packing material is a set of ceramic gas riser tubes filled with packing. At the bottom of the bayonet tube, ceramic packing material sits in the annulus between the bayonet and support tubes and supports a catalyst to be used in the decomposition reaction. The entire assembly is made of ceramic, such as SiC, and is handled as a single unit.

## 2.2. Reactor Operation

The incoming liquid $H_2SO_4$ feed is approximately 50%wt at 94°C and flows downward between the ceramic gas riser tubes where it is preheated 202°C by the upward flowing gases from the decomposition products. The gas riser tubes are long enough to extend above the highest operating liquid level. The preheated liquid then flows downward through the packing where it absorbs the $SO_3$ from the decomposition products and water is stripped out into the gas stream. At the bottom of the concentrator, the liquid $H_2SO_4$ is saturated at 492°C and 85-90%wt. The saturated liquid is directed downward into the liquid catch ring via the gas distribution plate where it flows down through the support tube. Vaporization is minimized in the support tube via the use of a double-walled tube filled with an insulating material, such as silica. At the bottom of the support tube, the liquid can enter into the bayonet via holes or perforations in the support tube. The $H_2SO_4$ vaporizes and is demisted by the ceramic packing as it flows upward. As the $H_2SO_4$ vaporizes, it is decomposed into $SO_3$ and $H_2O$. Upon reaching the catalyst, the $SO_3$ is further decomposed into $SO_2$ and $O_2$. At the top of the bayonet tube, the gas will flow preferentially up through the scallops cut into the support catch and up through the packing that serves as the concentrator. At the top of the packing, the gas flows up through the ceramic gas riser tubes where it preheats the incoming liquid.

**Figure 1 – Overall Schematic of the Composite Design Sulfuric Acid Decomposition Reactor, Concentrator and Preheater for Hydrogen Generation Processes**

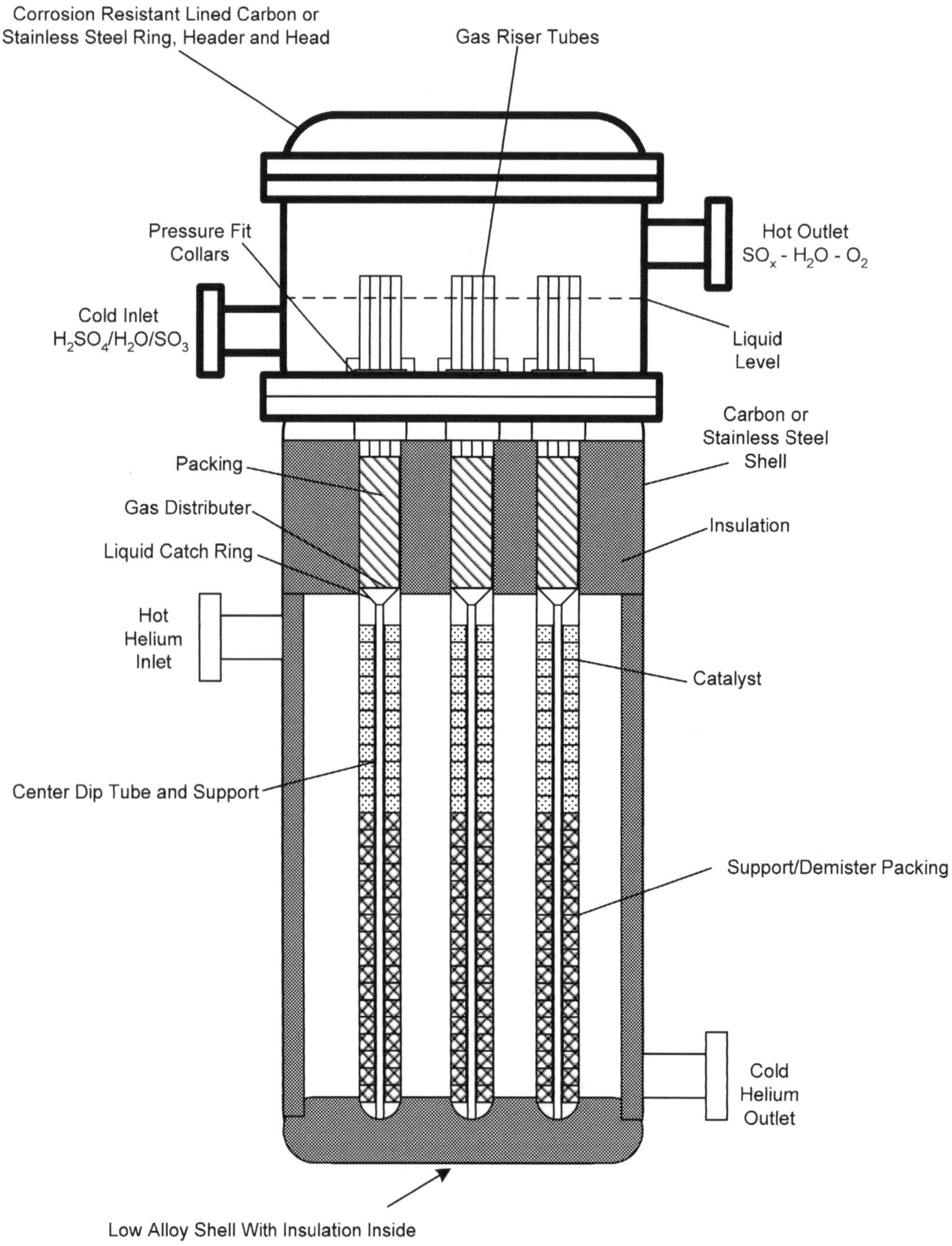

## Figure 3 - Detail of Composite Operation Tube

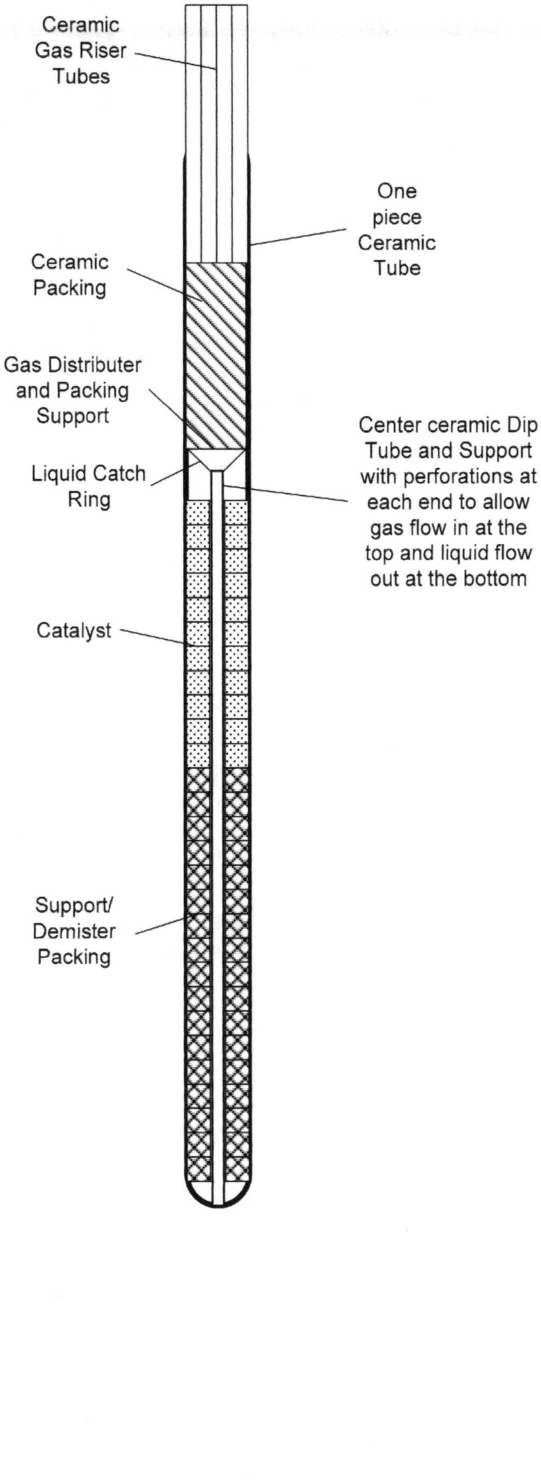


## Figure 2 - Support Ring and Support Tube Interface

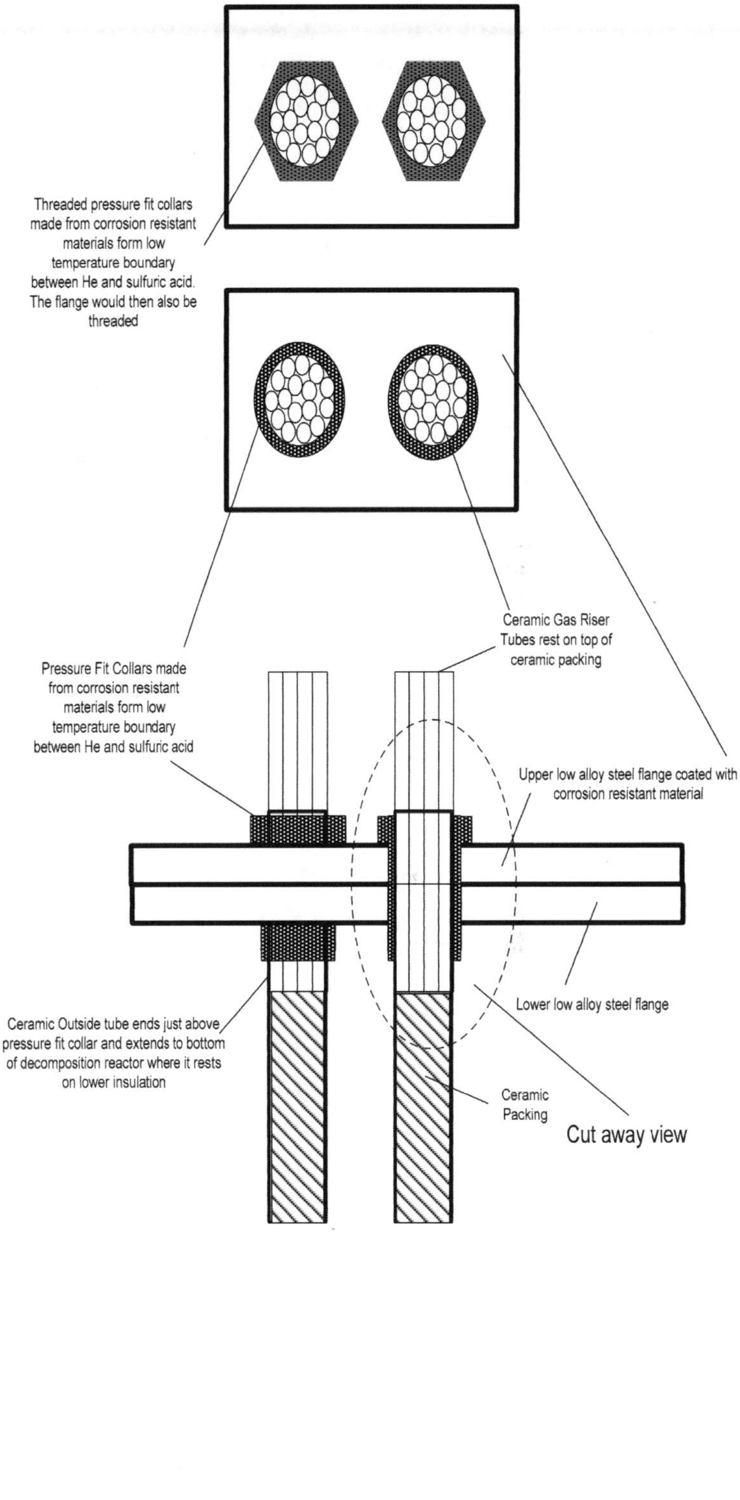

## 2.3 Alternative Methods of Seals and Support System

Figure 3 shows three views of the details of the connecting joint between the bayonet tube assembly and the metal header components of the decomposition reactor. The temperature in this area is less than 300°C, therefore various commercially available seals can be utilized. The current design uses a collar of Teflon-like material that is pressure fitted between the bayonet tube and the metal header. An alternative method is also shown in which the sealing material is in the form of an annular bolt with a smooth inside and a threaded exterior. A corrosion resistant coating or liner could be used to prevent the metal header from corroding. Additionally, an optional cooling source could be inserted in the flange area to maintain lower temperatures if desired. A third alternative method for forming the seal allows for a non-liquid seal as long as the pressure is greater in the intermediate helium loop. Cooling water tubing can be welded to the underside of the flange to allow for positive cooling on the plastic coated flange.

Two possible support systems for the bayonet tubes could be utilized. The first approach utilizes individual metal springs for each tube. This method would be dependent on the strength of the spring decreasing as the temperature increases, off-setting the differential expansion of the ceramic components versus the steel vessel. An alternative, and likely more cost effective approach is to place a support plate under the tubes which is in turn supported by metal springs. The support platform could also have indentations for each individual spring. This method would also provide for ease of installation.

# 3. Development of Temperature-Enthalpy Diagram

The T-H diagram shown in Figure 4 was developed using ChemCAD data up to the ASPEN predicted bubble point of 492°C, ASPEN data from 492°C to the dew point at 584°C, extrapolated CEA data from 492°C to 607.6°C and CEA data above 607.6°C. Data from three separate sources was utilized due to thermodynamic inconsistencies between ChemCAD and limitedly available raw thermodynamic data from Wüster(1) and Vermulean(2) above the bubble point. The merging of the data at the bubble and dew points is extremely smooth indicating good correlation between the three data sources. Calculation of the T-H line for the helium is straightforward. Figure 4 confirms that there is no pinching between the helium and the sulfuric acid systems.

The inlet temperature of the helium and the outlet temperature of the decomposition products were set in the HyS flowsheet to be 25°C apart at 897°C and 872°C, respectively. Similarly, the inlet temperature of the $H_2SO_4$ must simply be the saturation temperature at the operating pressure of 8.7 MPa. Thus, the T-H diagram allows for prediction of the outlet temperature of the helium and provides a detailed path between the endpoints.

**Figure 4 - T-H Diagram for Decomposition Reactor at 8.7 MPa**

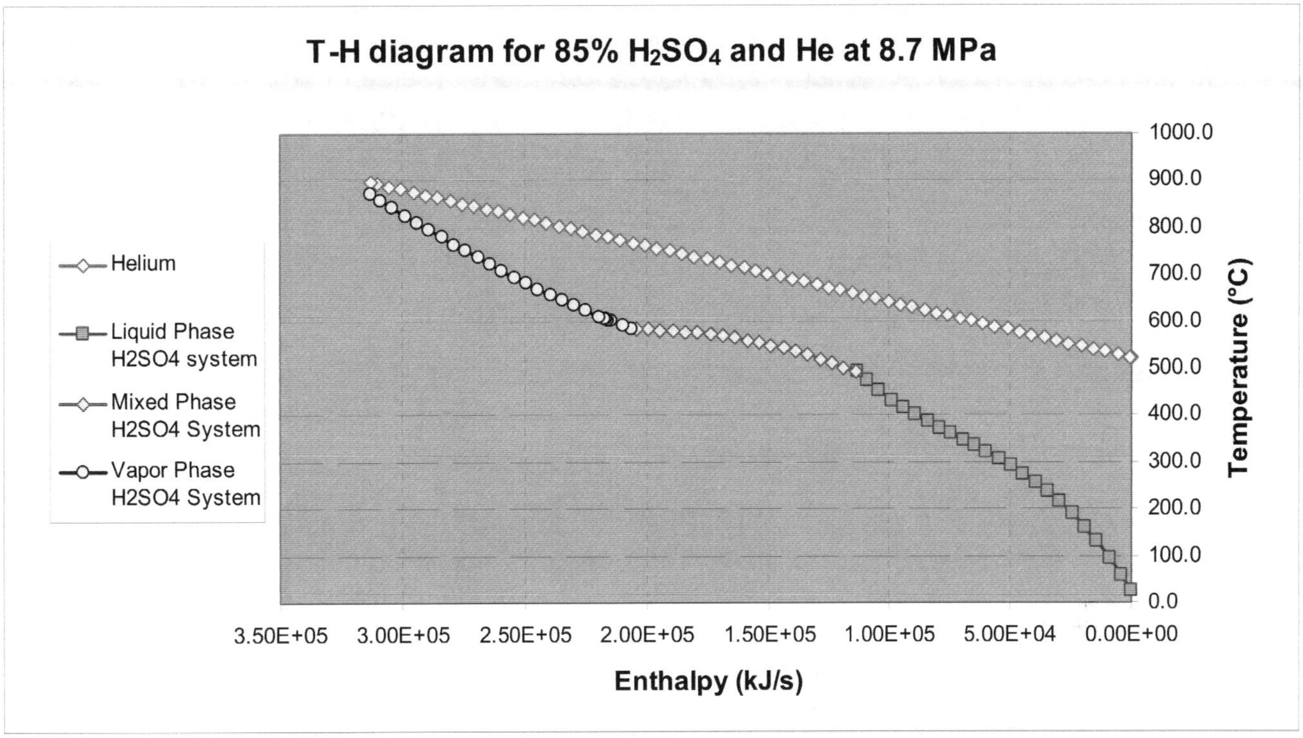


## 4. Decomposition Reactor Model

### 4.1. Design Equations

The model for the decomposition reactor is set up to balance the enthalpy and the heat transfer between the helium side, the bayonet annulus in which the decomposition reactions take place, and the liquid flowing downward through the support tube. The system relates six total temperatures – the inlet and outlet temperatures for the helium side, the bayonet, and the support tube. The inlet temperature for the support tube is known, as it is the saturation temperature of the liquid $H_2SO_4$. Additionally, the inlet temperature of the helium stream and the outlet temperature of the decomposition products are set by the WSP flowsheet. Thus, the three unknown temperatures – outlet temperature of support tube, outlet temperature of the helium, and the inlet temperature to the bayonet – serve as the variables for the system. Hence, three equations relating these temperatures via enthalpy and heat balances are required to model the system and are shown in Appendix C. Separate boiling-phase Nusselt correlations were needed in the bayonet and support tubes, as well as a need to account for the packing/catalyst in the bayonet tubes. Equations for the described correlations can be found in Appendix A.

### 4.2. Model Development

The current model for the decomposition reactor partitions the total required reactor tube length into 20 equal sections and balances the three previously stated design equations for each section. The requirement for convergence is that the temperature leaving the bottom of the support tube ($T_{outS}$) and the temperature going into the bayonet ($T_{inD}$) must be the same at the bottom of the tube (ie in the 20$^{th}$ section). The model allows the user to vary tube diameters, clearances, and lengths as well as mass flow rates and insulation thicknesses until the heat transfer requirements are met and pressure drops on both the helium side and inside the bayonet are kept below approximately 15 psi. The optimization limits for various tube geometries were obtained from either literature values or previous calculations.

The primary optimization parameter was the minimization of the total number of required tubes while still maintaining acceptable pressure drop losses and necessary heat transfer. The number of tubes was chosen as the primary optimization variable because initial cost estimates show that the internal components of the reactor will comprise the majority of the total reactor cost due to the necessity for corrosion resistant materials at high temperatures and pressures.

### 4.3. Packed Decomposition Reactor

To increase the heat transfer on the helium side of the reactor, it was proposed that using structured ceramic packing rather than the standard baffled approach would increase the heat transfer coefficient and eliminate the need for using a correction factor in the log-mean temperature difference due to crossflow. The assumed structured packing used for calculation purposes was Flexeramic® Type 38. This type of packing is typically used in mass transfer applications, however the effects on heat transfer and pressure drop are relevant.

The effect of the packing was to increase the heat transfer coefficient for the helium by an order of magnitude and to more than double the overall heat transfer coefficient between the helium and the bayonet. This effect, combined with the removal of the correction factor in the log-mean temperature difference, led to a significant improvement in the heat transfer between the helium and the bayonet when compared to the baffled approach. The final optimized model utilizes only 1056 bayonet tubes distributed within two vessels with inner diameters of 3.6 meters, as shown in Table 1. Additionally, the height of the tubes was reduced from nine meters (29.5 ft) to 4.26 meters (14 ft). The length was set at exactly 4.26 meters because SiC tubes are commercially available at this particular length and would thus further promote cost effectiveness and ease of construction. The pressure drops in the helium and bayonet were both within acceptable limits at 12.6 psi and 10.7 psi, respectively.

**Table 1 - Optimization Table for Packed Decomposition Reactor**

**OPTIMIZATION TABLE**

| | | |
|---|---|---|
| **inside shell diameter ($D_s$)** | 3.6 | m |
| **number of bayonet tubes required** | 1056 | |
| **number of vessels** | 2 | |
| **total length of tubes** | 4.26 | m |
| **clearance between bayonets** | 0.019 | m |
| ID of support tube ($d_s$) | 0.020 | m |
| thickness of support tube | 0.004 | m |
| OD of support tube | 0.028 | m |
| **thickness of insulation** | 0.0100 | m |
| **ID of annulus** | 0.048 | m |

| | | |
|---|---|---|
| **thickness of annulus** | 0.0048 | m |
| **ID of bayonet ($d_b$)** | 0.11 | m |
| **thickness of bayonet** | 0.0110 | m |
| **Pitch:OD bayonet ratio** | 1.144 | |
| **catalyst void volume (ε)** | 50% | |
| **catalyst/packing particle diameter ($d_p$)** | 0.010 | m |
| **mass flow rate $H_2SO_4$ ($m_S$)** | 0.0900 | kg/s |
| **mass flow rate Helium per finger ($m_H$)** | 0.152 | kg/s |
| **pressure drop on shell side** | 12.6 | psi |
| **pressure drop in bayonet** | 10.7 | psi |
| **pressure drop in support tube** | 0.04 | psi |

The temperature profile for the packed decomposition reactor is shown in Figure 5. Recall the zero point on the x-axis corresponds to the top of the tube. Again, the temperature change in the support tube is kept to a minimum to prevent any possible vapor lock in the tube. Additionally, the total pressure drop in the helium, bayonet and support tubes as a function of tube length is shown in Figure 6. There is significant improvement of the pressure drops in the helium and bayonet when compared to the baffled reactor model. In each case the pressure drop in the support tube is negligible.

The packed design allows for significantly improved heat transfer and thus reduces the total number and length of required bayonet tubes, which significantly reduces the overall cost of the composite reactor. Therefore, the packed design is preferable to the original baffled design.

**Figure 5 - Temperature Profile for Packed Reactor**

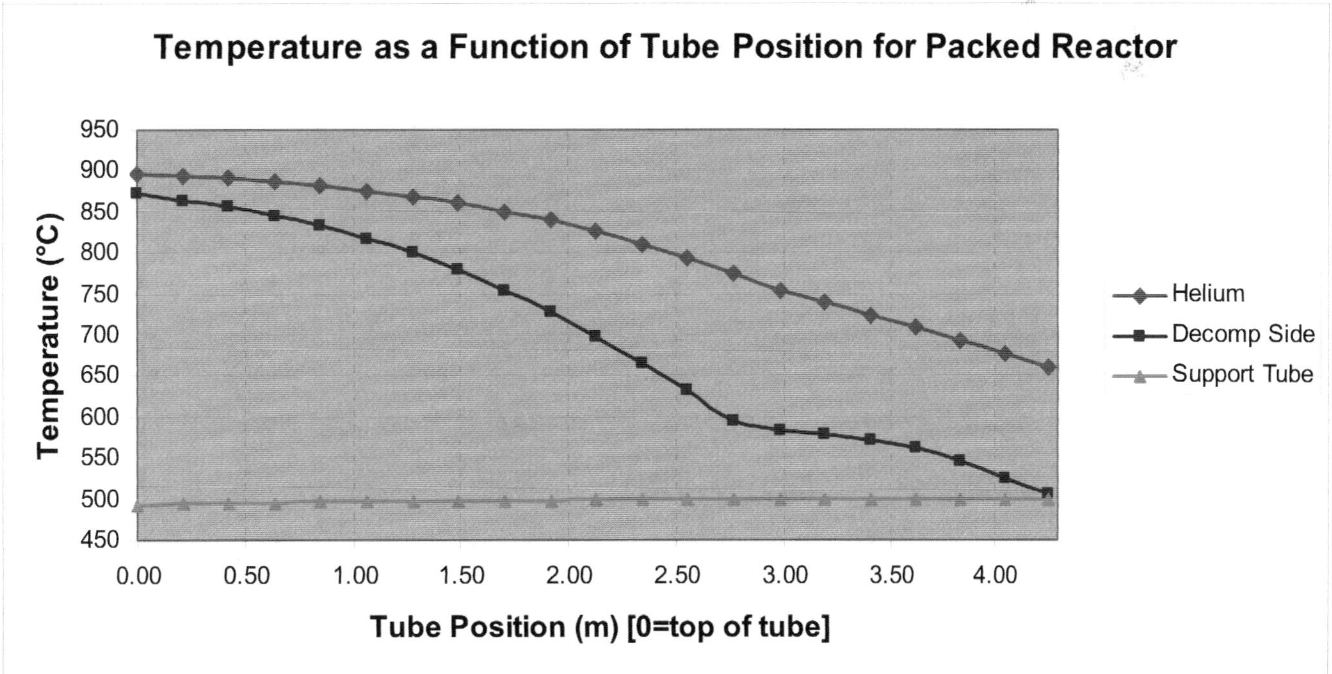

**Figure 6 - Pressure Drop in Packed Reactor**

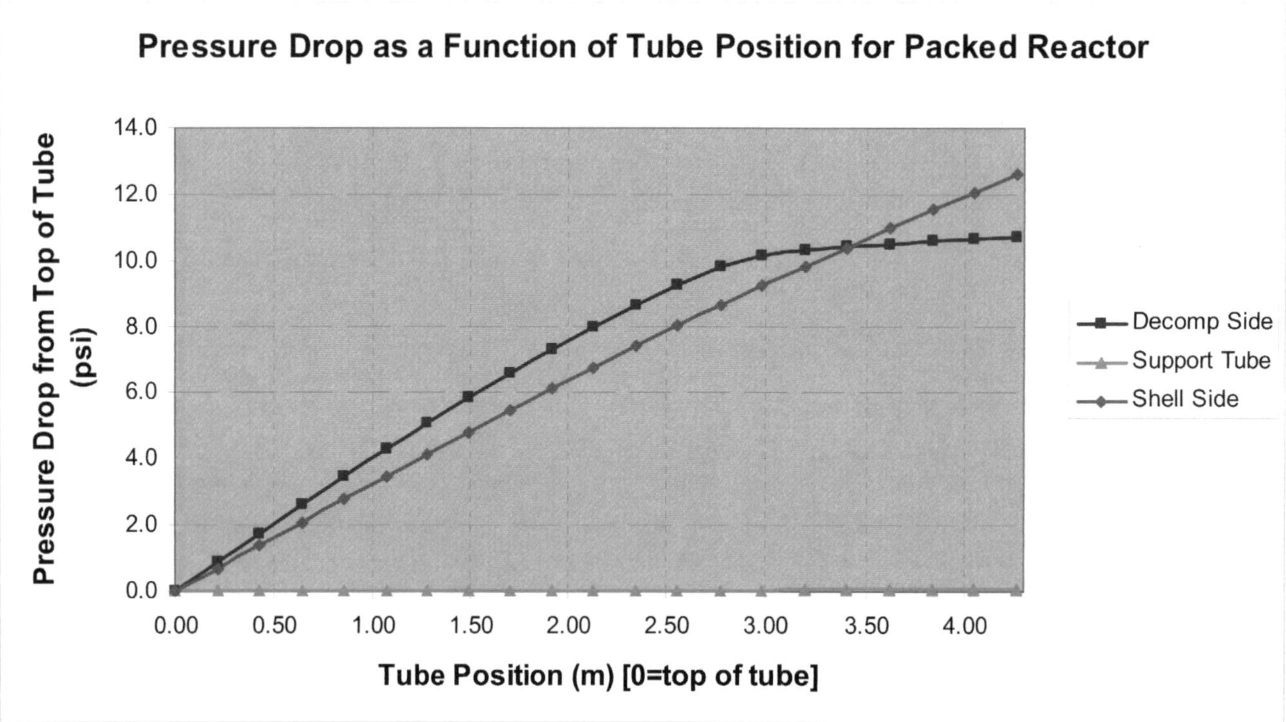


### 4.4. Pressure Drop Calculations

Similarly to the heat transfer correlations, various methods of predicting the pressure drop through the various sections of the decomposition reactor were required. Specific equations for the pressure drops are available in Appendix B. The pressure drop through the bayonet portion was calculated using the Ergun equation with characteristic length and velocity relations as reported in Mills(**Error! Bookmark not defined.**). The pressure drop in the support tube was calculated using a friction factor correlation for the transitional regime and a pressure drop equation for flow through circular pipes as reported in Welty(**Error! Bookmark not defined.**).

The method of calculating pressure drop on the shell side depends on the design approach. For the baffled design, the pressure drop was calculated using a shell-side friction factor correlation and equation described by Kern[(3)]. For the packed design, the method for dry pressure drop in structured ceramic packing detailed by Uresti-Melendez and Rocha(**Error! Bookmark not defined.**) was utilized.

## 5. Optimization of Gas Riser Tubes, Pre-Heat, Vessel Size, and Concentrator

### 5.1. Liquid Pre-Heat and Gas Riser Tubes

Further optimizations were all calculated relative to the tube geometries and mass flow rates determined by the packed design decomposition model. Figure 7 shows the heat transfer profile for the required length in order to pre-heat the incoming liquid 50%wt $H_2SO_4$ from 94°C to 202°C in the gas riser tube area. The calculated required length (0.52 m) is used for a pressure drop calculation in Table 2. The gas riser tubes are assumed to be packed with 0.5 cm ceramic spheres to facilitate heat transfer.

The calculation of the gas velocity in the riser tubes as shown in Table 3 is based on a non-ideal equation of state using estimated pseudocritical properties. The use of Kay's Rule[4] allows for the estimation of a compressibility factor for gas mixtures. The calculated gas velocity is 0.30 m/s (1.0 ft/s).

Table 2 details the geometrical and flow calculations for the gas riser tube area. The available pressure drop based on the liquid pre-heat length is balanced with the actual pressure drop by varying the blanked liquid flow space between the gas riser tubes. Note that the height of the gas riser tubes is above height of the required liquid pre-heat length to prevent liquid from entering the gas riser tubes.

**Figure 7 - Temperature Profile of Pre-heater**

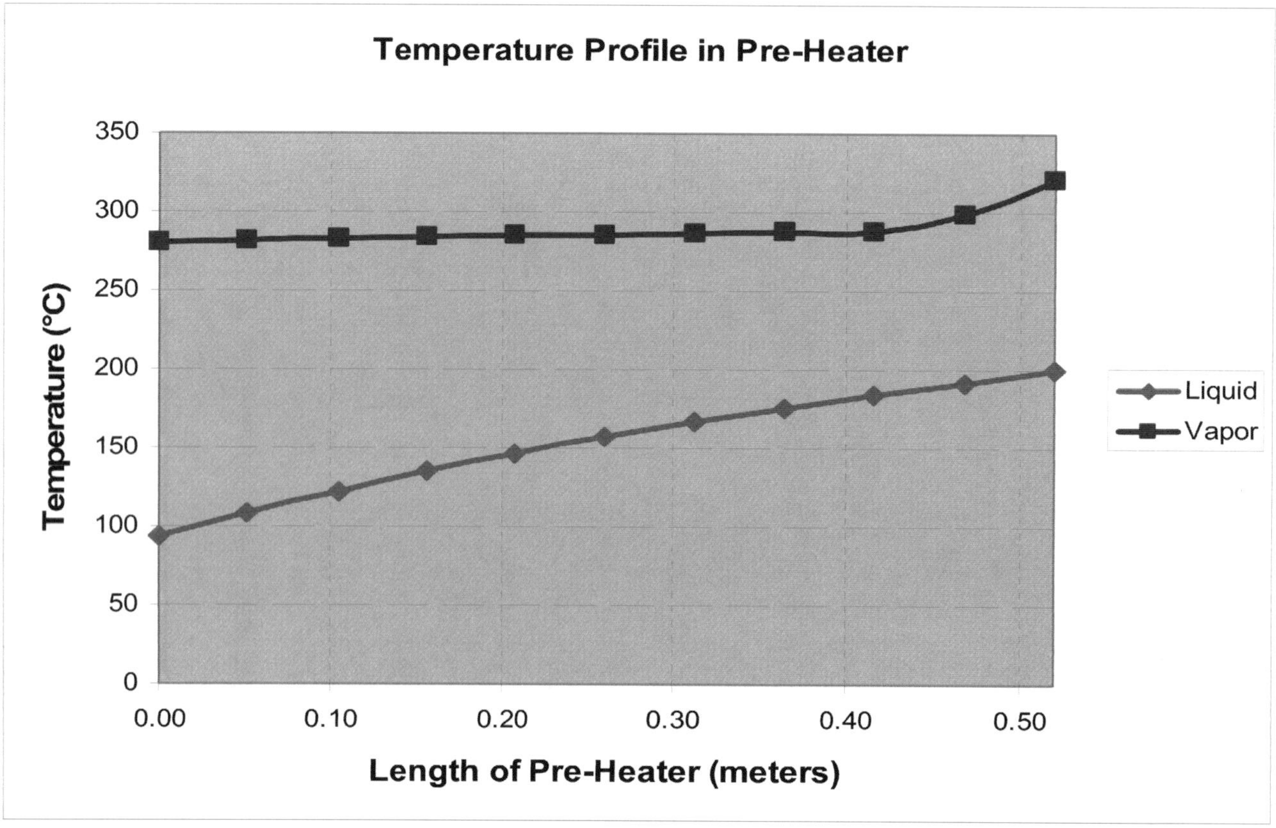

## Table 2 - Flow and Geometric Calculations for the Gas Riser Tube Area

### Calculation of Pressure Drop Through Liquid in Pre-Heat

| | | | | |
|---|---|---|---|---|
| ID of bayonet (D) | 0.11 | m | | |
| ID of gas riser (d) | 0.0254 | m | | |
| wall thickness (t) | 0.00127 | m | | |
| length of side of triangle | 0.028 | m | | |
| height of triangle | 0.024 | m | | |
| area of triangle | 0.000 | $m^2$ | | |
| area of bayonet tube | 0.010 | $m^2$ | | |
| number of gas riser tubes | 14.00 | | | |
| Blanked Liquid Flow Spaces | 15% | | | |
| length of tube | 0.3048 | m | | |
| total area of gas riser tubes | 0.341 | $m^2$ | | |
| | | | | |
| INLET LIQUID: | | | | |
| flow rate (m) | 0.073 | kg/s | | |
| concentration $H_2SO_4$ (Str 3) | 50 | wt% | 15.5 | mol% |
| avg molecular weight of solution | 30.42 | g/mol | | |
| temperature (Str 3) | 94 | °C | | |
| density (ρ) | 1361.9 | $kg/m^3$ | | |
| viscosity (μ) | 0.0012 | kg/(m*s) | | |
| | | | | |
| cross-sectional area of flow (A) | 8.75E-04 | $m^2$ | | |
| wetted perimeter | 1.574 | m | | |
| equivalent diameter ($D_{eq}$) | 2.22E-03 | m | | |
| | | | | |
| velocity of liquid flow = m/ρA | 0.0610 | m/s | | |
| | | | | |
| Reynolds number = $\rho D_{eq} v/\mu$ | 154 | laminar | | |
| | | | | |
| assumed roughness (e) | 0.00015 | in | | |
| $D_{eq}/e$ | 5.84E+02 | | | |
| fanning friction factor ($f_f$) | 1.04E-01 | | | |
| | | | | |
| assumed liquid length | 20.47 | in | | |
| | | | | |
| Liquid $\Delta P = 2 * f_f * (L/D_{eq}) * \rho * v^2$ | 0.00035 | MPa | 0.05 | psi |
| Available pressure drop $\Delta P/\Delta Z = 150\mu * v_c/L_t^2 + 1.75\rho v_c^2/L_t$ | | | 0.05 | psi |

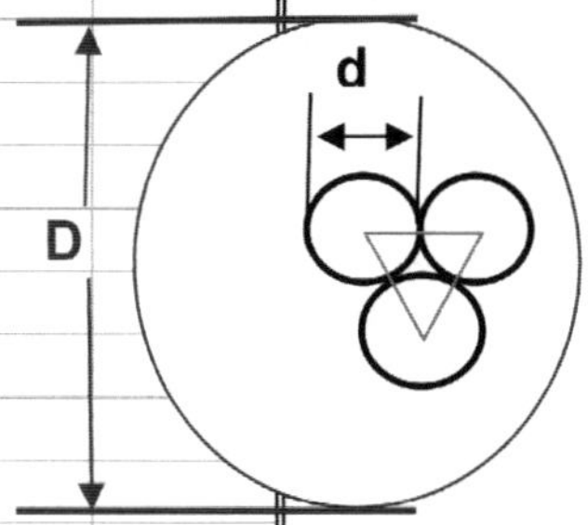

### Table 3 - Gas Velocity Calculation Using Non-Ideal Equation of State

****Calculates velocity of gas in riser tubes using a non-ideal equation of state

| | | | | |
|---|---|---|---|---|
| **Overall Pressure** | 8.7 | MPa | | |
| **Temperature of gas** | 550 | °C | | |
| **mol frac $SO_2$ in gas** | 0.13 | | | |
| **mol frac $O_2$ in gas** | 0.06 | | | |
| **mol frac $H_2O$ in gas** | 0.81 | | | |
| **T'c (K)** | 589 | | | |
| **P'c (Pa)** | 1.93E+07 | | | |
| **T'r (K)** | 1.397 | | | |
| **P'r (Pa)** | 0.452 | | | |
| **compressibility factor (z)** | 0.930 | | | |
| **calculated gas velocity** | 1.0 | ft/sec | 0.300 | m/s |
| **gas volume flow** | 0.050 | $ft^3$/sec | 1.42E-03 | $m^3$/s |
| **gas molar flow** | 1.938 | mol/s | | |
| **mol of liquid** | 2.389 | mol/s | | |
| **flow rate (m)** | 0.073 | kg/s | | |

## 5.2. Vessel Sizing and Concentrator Height

The calculation of the inner diameter of the vessel is based on the total number of bayonet tubes as determined by the decomposition model. The tubes will have a triangular pitch arrangement with a minimum pitch-to-diameter ratio of 1.125. The user can input a desired number of vessels to obtain the required vessel diameter as shown in Table 4.

### Table 4 - Calculation of Vessel Diameter

**Calculation of ID of Vessels**

| | | |
|---|---|---|
| **Area of Unit Cell** | 0.0099 | $m^2$ |
| **Area of Tube** | 0.0068 | $m^2$ |
| **Total Cross-Sectional Area** | 20.84 | $m^2$ |
| **Inside Diameter of Vessel** | 3.64 | m |
| **Inside Diameter of Vessel** | 11.95 | ft |
| **Number of Vessels** | 2 | |

## 5.3. Calculation of Required Catalyst Volume

The required catalyst volume to decompose the $SO_3$ into $SO_2$ was calculated to ensure that the annular space in the bayonet tubes were large enough to accommodate the catalyst. The calculation was based on kinetic data from Lin and Flaherty[5]. The catalyst used was iron oxide. Assuming an $SO_3$ conversion of 50%, the required catalyst volume space is approximately 0.005 $m^3$ and the available volume in the annular region of the bayonet tubes is 0.029 $m^3$. Thus, the volume in the annular region exceeds the required volume by an order of magnitude and achieving the necessary residence time will not pose an issue.

## 6. Concentrator Model

## 6.1. ASPEN Composition Profiles and Enthalpy Data

Calculations to determine the required height of the concentrator requires performing simultaneous heat and mass balances. At the bottom of the concentrator portion of the reactor, any $SO_3$ that has not been decomposed will be absorbed by the liquid $H_2SO_4$ leaving the bottom of the concentrator. This process is highly exothermic and thus to accurately model the system both the absorption of the $SO_3$ as well as the heat transfer between the hot gas and the cooler liquid phases must be accounted for.

To simplify the mass transfer portion of the model, the composition profile was determined using the stage profiles for the concentrator as calculated in the ASPEN flow sheet. Using this composition data, the model is simply reduced to a heat transfer problem. The required enthalpy data to perform heat transfer analysis was also based on the ASPEN flow sheet.

## 6.2. Concentrator Model Development and Design Equations

A variety of types and sizes of ceramic packing are available for use in the concentrator. This particular model assumes that the packing is 13 mm Berl Saddles. Packing properties were obtained from Perry's(6). The pressure drop was calculated by the method described in Eckert(7) for countercurrent gas and liquid flow through packing. The Nusselt correlation used is the same as that shown in Appendix A for flow through packed beds. The concentrator was split into ten sections based on the compositional data from the ASPEN flow sheet. The design equations for the concentrator model are given in Appendix C.

## 6.3. Concentrator Optimization

The total required length for the concentrator portion of the reactor is 2.49 meters with a pressure drop of 0.01 psi. The pressure drop is based on a correlation by Eckert. The small pressure drop can be attributed to the relatively small gas mass flow rate. The temperature profile for the concentrator is shown in

Figure **8**. Similarly, composition profiles for the liquid and the vapor are shown in Figure 9. Note that zero on the x-axis corresponds to the bottom of the concentrator.

**Figure 8 - Temperature Profile in Concentrator**

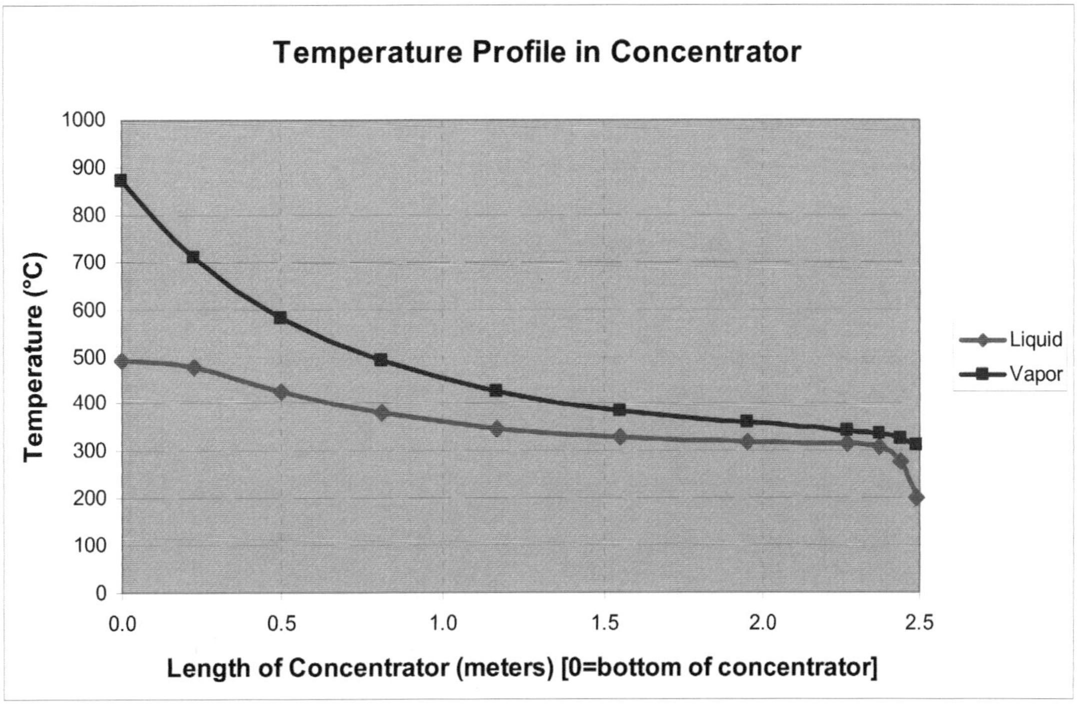

## Figure 9 - Composition Profile in Concentrator

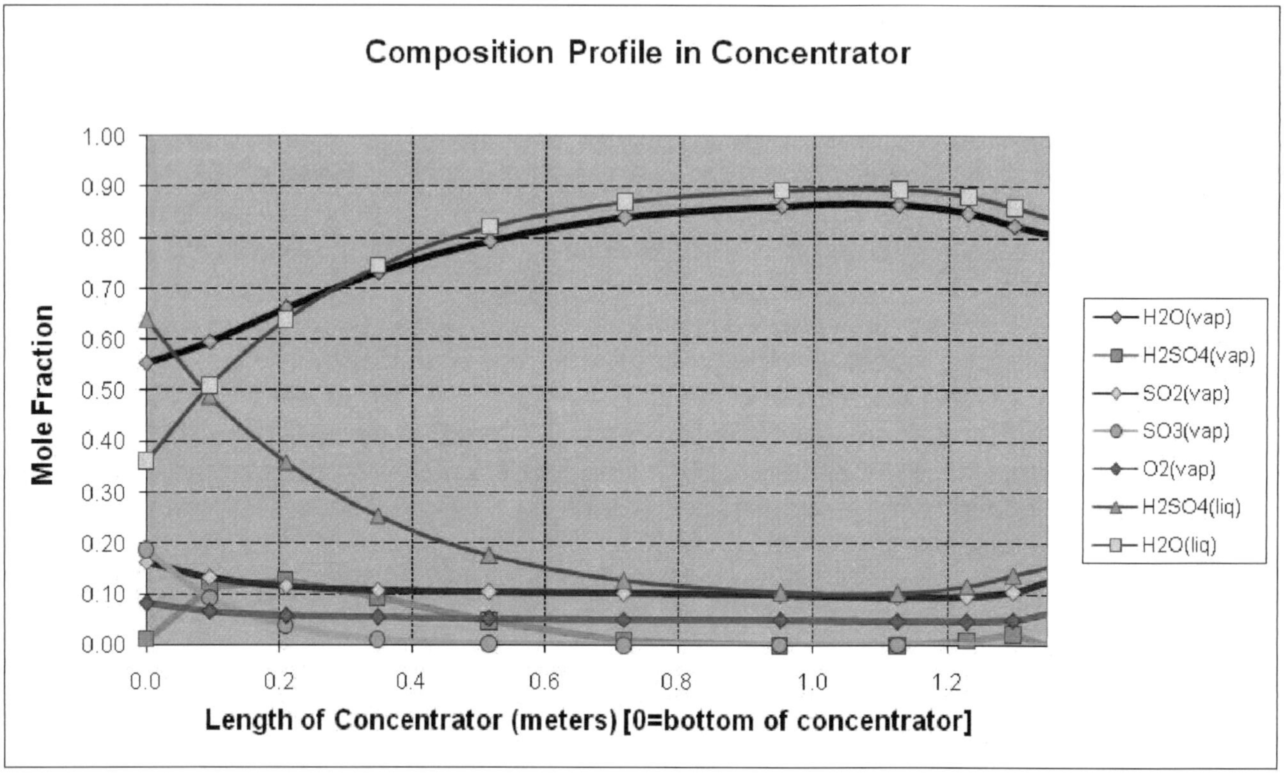

## Appendix A: Heat Transfer Correlations

i. Boiling Correlation(**Error! Bookmark not defined.**)

$Nu_{boiling} = .024*Re^{0.8}*Pr^{0.43}\,[1+VF(\rho_L/\rho_g-1)]^{0.5}$

ii. Liquid Flow Through Support Tube(**Error! Bookmark not defined.**)

$Nu_{liq} = .023*Re^{0.8}*Pr^{1/3}\,\Phi^{0.14}$

iii. Flow Through Packed Bed(**Error! Bookmark not defined.**)

$Nu_{packed} = (0.5*Re^{1/2}+0.2*Re^{2/3})Pr^{1/3}$

iv. Donohue equation(**Error! Bookmark not defined.**)

$$\frac{h_s OD_b}{k} = 0.2\left(\frac{OD_b G_e}{\mu}\right)^{2/3}\left(\frac{c_p\mu}{k}\right)^{1/3}\left(\frac{\mu}{\mu_w}\right)^{0.14}$$

v. Reynolds number for various sections of reactor

a. Packed helium flow(**Error! Bookmark not defined.**)

$Re_{shell} = \rho*S*v_c/\mu$

$v_c = v_s/\varepsilon/\sin\theta$

$h = k*Nu/S*(1-\varepsilon)/\varepsilon$

b. Flow Through Packed Bed(**Error! Bookmark not defined.**)

$Re_{packed} = \rho*L_c*v_c/\mu$

$v_c = v_s/\varepsilon$

$h = k*Nu/L_c$

c. Support tube(**Error! Bookmark not defined.**)

Reynolds number= $\rho*d_s*v/\mu$

## Appendix B: Pressure Drop Calculations

i. Packed Bed Flow(**Error! Bookmark not defined.**)

$$\frac{\Delta P}{\Delta Z} = \frac{150\mu v_c}{L_c^2} + \frac{75\rho v_c^2}{L_c}$$

ii. Flow through Pipe(**Error! Bookmark not defined.**)

$$\frac{1}{\sqrt{f_f}} = 4\log_{10}\frac{d_s}{e} + 2.28 - 4\log_{10}\left(4.67\frac{d_s/e}{Re\sqrt{f_f}} + 1\right)$$

$$\Delta P = 2f_f\frac{Z}{d_s}\rho v^2$$

iii. Packed Helium side (structured Packing)(**Error! Bookmark not defined.**)

$$\frac{\Delta P}{\Delta Z} = \frac{0.0453\rho_g}{S\varepsilon^2(1-\varepsilon)(\sin\theta)^2} + \frac{6.6441\mu_g}{S^2\varepsilon(1-\varepsilon)\sin\theta}$$

v. Countercurrent Gas and Liquid Flow[7]

X=Liquid Rate/Gas Rate*$(\rho_{gas}/\rho_{liq})^{1/2}$

Y=Gas Rate$^2$*F*$\psi$*$\mu^{0.2}/\rho_{liq}/\rho_{gas}/g_c$

$\Delta P$ per unit length determined using table

## Appendix C: Design Equations

i. Reactor Model

$\dot{m}_{He}\Delta H_{He} + \dot{m}_{H2SO4}(\Delta H_D + \Delta H_S) = 0$

$\dot{m}_{H2SO4}\Delta H_D = U_{D/S}A_2\Delta T_{LMD/S} - U_{He/D}A_1\Delta T_{LMHe/D}$

$\dot{m}_{H2SO4}\Delta H_S = U_{D/S}A_2\Delta T_{LMD/S}$

ii. Concentrator Model

$\dot{m}_{gas}*\Delta H_{gas} + \dot{m}_{liq}*\Delta H_{liq} = 0$

$U_{conc}*A_{conc}*\Delta T_{LMconc} = \dot{n}_{vap}*\Delta H_{vap} + \dot{m}*C_p*\Delta T$

## Nomenclature

$A_1$ [$m^2$] – area available for heat transfer between the helium side and the bayonet
$A_2$ [$m^2$]- area available for heat transfer between bayonet and support tubes
$A_{conc}$ [$m^2$]– area available for heat transfer in the concentrator
B – number of baffles
$c_p$ [J/kg/°C]– heat capacity
$D_{eq}$ [m]- equivalent diameter of the baffled shell
$d_s$ [m]– diameter of support tube
F [$ft^{-1}$]– packing factor
$f_f$ – Fanning friction factor
$g_c$ [$kg*m/N/s^2$] = 1
$G_c$ [$kg/m^2/s$] – mass velocity for cross flow
$G_e$ [$kg/m^2/s$] – weighted average mass velocity in cross flow
h [$W/m^2/°C$] - heat transfer coefficient
$H_{in}$ [kJ/kg] – enthalpy at the inlet temperature
$H_{out}$ [kJ/kg] – enthalpy at the outlet temperature
$h_s$ [$W/m^2/°C$] - heat transfer coefficient on shell side
k [W/m/°C] – thermal conductivity
$L_c$ [m] – characteristic length of packing
$\dot{m}_{gas}$ [kg/sec] – mass flow rate of vapor phase in concentrator
$\dot{m}_{He}$ [kg/sec] - mass flow rate of the helium
$\dot{m}_{H2SO4}$ [kg/sec] - mass flow rate of the $H_2SO_4$
$\dot{m}_{liq}$ [kg/sec] – mass flow rate of liquid phase in concentrator
$\dot{n}_{vap}$ [mole/sec] – molar flow rate of evaporated water
$Nu_{boiling}$ – Nusselt correlation for boiling phase
$Nu_{gas}$ – Nusselt correlation for gas phase
$Nu_{liq}$– Nusselt correlation for liquid phase
$OD_b$ [m] – outer diameter of bayonet
Pr – Prandtl number = $\mu c_p/k$
S [m] – side dimension of structured packing corrugation
$T_{inD}$ [°C] - inlet temperature from the bayonet tube
$T_{inHe}$ [°C] – inlet temperature to the helium side
$T_{inS}$ [°C] – inlet temperature to the support tube
$T_{outD}$ [°C] - outlet temperature from the bayonet tube
$T_{outHe}$ [°C] – outlet temperature on the helium side

$T_{outS}$ [°C] – outlet temperature from the support tube
$U_{conc}$ [W/m$^2$/°C] – overall heat transfer coefficient in the concentrator
$U_{D/S}$ [W/m$^2$/°C] – overall heat transfer coefficient between the bayonet and support tubes
$U_{He/D}$ [W/m$^2$/°C] – overall heat transfer coefficient between the helium and the bayonet
$v_c$ [m/s] - characteristic velocity
VF – vapor fraction
$v_s$ [m/s] - superificial velocity
Z [m] – height
$\varepsilon$ – void fraction of packing
$\mu_g$ [kg/m/s] - viscosity of gas
$\Delta H_D$ [kJ/kg] – change in enthalpy in the bayonet tube
$\Delta H_{gas}$ [kJ/kg] – change in enthalpy of vapor phase in concentrator
$\Delta H_{He}$ [kJ/kg] – change in enthalpy of the helium
$\Delta H_{liq}$ [kJ/kg] – change in enthalpy of liquid phase in concentrator
$\Delta H_S$ [kJ/kg] – change in enthalpy in the support tube
$\Delta H_{vap}$ [J/mol] – heat of vaporization of water
$\Delta T_{LM}$ [°C or K] – log mean temperature difference
$\Delta T_{LMconc}$ - log mean temperature difference in the concentrator
$\Delta T_{LMD/S}$ – log mean temperature difference between the bayonet and support tubes
$\Delta T_{LMHe/D}$ - log mean temperature difference between the helium side and the bayonet
$\Delta P/\Delta Z$ [Pa/m] – pressure drop per unit height
$\Phi$ - ratio of viscosity in the bulk fluid to viscosity at the wall
$\theta$ [radians] – angle with the horizontal for corrugation channel
$\rho_L$ [kg/m$^3$] - density of liquid
$\rho_g$ [kg/m$^3$] - density of vapor
$\mu$ [kg/m/s]– viscosity
$\psi$ – density of water/density of liquid

**References:**


1. Giazitzoglou, Z. and Wüster, G., *"Das binäre System $H_2SO_4$; p,v,T-Daten und Dampfdruckkurven in weiten Konzentrationsbereichen bei erhöhtem Druck"*, Ber Bunsenges. Phys. Chem., 85: 127-132, 1981.
2. Perry's Chemical Engineers' Handbook, 6th Ed., McGraw-Hill, 1984, Table 3-13, p. 3-65.
3. Kern, Donald Q. Process Heat Transfer. McGraw-Hill, 1950, Ch 7.
4. Felder, Richard M. and Ronald W. Rousseau. Elementary Principles of Chemical Processes, 3rd Ed. John Wiley & Sons, Inc., 2000.
5. Lin, S.S. and R. Flaherty. *"Design Studies of the Sulfur Trioxide Decomposition Reactor for the Sulfur Cycle Hydrogen Production Process"*, Int. J. Hydrogen Energy (8), 32: 589-596, 1983.
6. Perry's Chemical Engineers' Handbook, 7th Ed., McGraw-Hill, 1997, Table 14-7b, p. 14-43.
7. Eckert, J.S. *"Selecting the Proper Distillation Column Packing"*. Chemical Engineering Progress, 1970, 66 (3).

# MEASUREMENTS OF THE SIMULTANEOUS SOLUBILITY OF OXYGEN AND SULPHUR DIOXIDE IN WATER FOR THE HYBRID SULPHUR THERMOCHEMICAL CYCLE

Andrew Shaw[1], Bruce Ewan[1] and Ray W. K. Allen[1],
[1] Chemical and Process Engineering, University of Sheffield, Mappin St, Sheffield, S1 3JD, United Kingdom

Corresponding author: a.shaw@shef.ac.uk

## ABSTRACT

In recent years, the HyS process (formerly known as The Westinghouse process) has emerged as one of the top two routes for the thermochemical manufacture of Hydrogen from water. The work considered here focuses on the oxygen separation stage of the process where, in order for a science based design approach to be taken, thermodynamic data are required for multicomponent phase equilibrium relationships between water, sulphur dioxide and oxygen; $H_2O$-$SO_2$-$O_2$. A method of making flash calculations, useful in the separation equipment design, has been proposed and coded into Mathematica®. Further, a vapour liquid equilibrium still has been designed and built to make measurements of the multicomponent solubility of sulphur dioxide and oxygen in water. Preliminary results of total sulphur dioxide dissolved in water show very good agreement with literature data. At 24.6 °C and a total pressure of 0.249 atm the solubility was found to be 0.215 mol $SO_2$ / kg $H_2O$ and at a total pressure of 1.93 atm and 39.1 °C the solubility was 1.54 mol $SO_2$ / kg $H_2O$.

## INTRODUCTION

Increasingly, hydrogen looks to be a realistic replacement for carbon based fuels in vehicles. Of the principal production routes, 96% are accounted for by fossil fuel based methods (1) primarily based on natural gas. When hydrogen is produced using energy derived from fossil fuels there is little or no environmental advantage. A sustainable method of hydrogen production on a scale large enough to satisfy the requirements of the transport industry is required.

There are a growing number of schemes that produce hydrogen by sustainable methods. The Westinghouse or HyS process is one such process employing a hybrid thermochemical cycle based on the decomposition of sulphuric acid. The products from the decomposition stage are sulphur dioxide, oxygen and water. Oxygen is a process by-product and must be removed leaving sulphur dioxide and water which are sent to an electrolyser where the following reactions occur

| | | |
|---|---|---|
| Cathode: | $2H^+ + 2e^- \rightarrow H_2$ | $E^0 = 0.00$ volts |
| Anode: | $H_2SO_3 + H_2O \rightarrow 2H^+ + H_2SO_4 + 2e^-$ | $E^0 = -0.17$ volts |

The objective of this research is to investigate the equilibrium of water, sulphur dioxide and oxygen to aid the design of the oxygen separation equipment. To achieve this equilibrium calculations using the ensemble form of Henry's law and experiments to measure the simultaneous solubility of oxygen and sulphur dioxide in water are being performed. Initial results from the apparatus designed to measure multicomponent equilibrium are presented here.

## SULPHUR DIOXIDE AND OXYGEN SOLUBILITY IN WATER

The solubility of sulphur dioxide in water has been the subject of many studies; a review has been reported elsewhere (2). It is well known that sulphur dioxide in water dissociates into ions forming a weak electrolyte. The solution consists of hydrated $SO_2$ as well as sulphite ($SO_3^{2-}$) and bisulphite ($HSO_3^-$, $HOSO_2^-$) ions. It is generally agreed that undissociated sulphurous acid $H_2SO_3$ is not thought to exist in aqueous solution (3, 4). Table 1 gives a summary of the literature containing data on the solubility of sulphur dioxide in water.

The presence of oxygen in the liquid phase causes complications due to the potential for oxidation of the sulphite species to sulphate. Data applicable to the HyS process is scarce although a study was published by Hartmann (5). More general studies on sulphite and bisulphite oxidation applicable to acid rain formation are available with many concluding that pH has a role in the oxidation process. At lower pH the rate of oxidation falls although a quantitative value is unknown. The ultimate aim of the proposed experiment is to develop an understanding of the simultaneous solubility of oxygen and sulphur dioxide and to determine the extent of the oxidation of sulphur species in the liquid phase.

**Table 1:** Table of publications containing data on the solubility of sulphur dioxide in water

| Author | Hartmann | Beuschlein | Johnstone | Hudson | Maass | Sherwood | Rumpf |
|---|---|---|---|---|---|---|---|
| Year | 1984 | 1940 | 1934 | 1925 | 1928 | 1925 | 1992 |
| Reference | (5) | (6) | (7) | (8) | (9) | (10) | (11) |
| Temperature Range (°C) | 25 - 50 | 23 - 113 | 25 - 50 | 10 - 90 | 10 - 27 | 0 - 50 | 20 - 120 |
| Pressure Range (atm) | 3.5 - 50 | 0.064 -1.89 | $2.7\times10^{-4}$ - 0.0135 | 0.996 - 1.27 | 0.32 - 3.36 | $3.29\times10^{-4}$ - 0.955 | 0.35 - 25 |

## CALCULATIONS

The calculation of vapour liquid equilibria for water-sulphur dioxide binary solutions was performed using the method outlined by Zemaitis (12) written in Mathematic® software. The calculation uses the ensemble form of Henry's law to express the vapour liquid equilibria of each component. The Henry's law expressions are combined with K equations describing the liquid phase reactions and balance equations for the vapour mole fractions, sulphur species, hydrogen species and electroneutrality. Vapour phase fugacity coefficients are calculated using Nakamura's method (13) and liquid phase activity coefficients are calculated using equations given by Beutier and Renon (14).

## EXPERIMENTAL

Figure 1 shows the equilibrium still arrangement for measuring the solubility of sulphur dioxide and oxygen in water. The apparatus consists of two pots connected by pipe work and two valves; the overall volume is 1005.8 $cm^3$. The inside of the apparatus has been coated with a DuPont® thin film fluoropolymer, B5513 Green. All valves have been cleaned to be compatible with oxygen rich environments and the apparatus hydraulically tested to a pressure of 60 × $10^5$ Pa. Heating is supplied by mat and tape heaters held on the outside of the pots individually controlled using LabVIEW software. The internal temperature is measured using Teflon® coated thermocouples as indicated in Figure 1. The temperature difference between the pots is less than 1°C at 40°C. Pressure is monitored using LabVIEW data acquisition and a Sensortechnics CTE 9000 series pressure transducer calibrated with a Bundenberg dead weight tester. The pH is monitored using pH electrode having a chemically resistant body inserted through a compression fitting in the side of the apparatus.

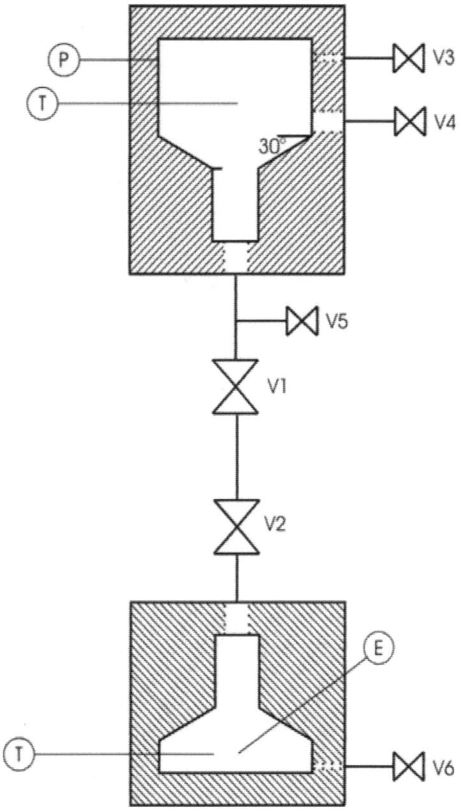


**Figure 1:** Schematic of equilibrium still: T, Thermocouples; P, Pressure Transducer; E, pH probe

During operation the apparatus is filled with deionised and distilled water so that the phase interface falls between valves V1 and V2. The conductivity of water after deionising is 2 × $10^{-7}$ S cm. The amount of dissolved oxygen and nitrogen in water as a result of equilibrium with the atmosphere is sufficiently small that any effect is less than 0.5 % of experimental results therefore no attempt is made to degas the water further than distilling. Having filled the apparatus with water the top and bottom pots are isolated from each other using valves V1 and V2. The top pot is placed under vacuum before a known weight of liquid sulphur dioxide is supplied in through valve V4. Finally a known weight of oxygen can be added from a high pressure sample cylinder through valve V3. When all the components have been added to the apparatus the heaters are turned on and the equipment temperature monitored using the two internal thermocouples. Once the internal temperature reaches the set point valves V1 and V2 are opened. The whole apparatus is mounted on a rotating stand so that mixing can be achieved by oscillation through 180°. The apparatus is balanced so that rotation can be performed quickly causing chaotic motion with large interfacial areas for mass transfer. Once equilibrium is reached the pots are rotated to their starting orientation as shown in Figure 1. The gas and liquid phases are separated from each other by closing valve V1 and V2. This traps representative liquid and gas samples without affecting the equilibrium position. Analysis can then be performed on both phases and an overall mass balance used to check the consistency of results.

The liquid phase is analysed by taking a sample volume, equivalent to one percent of the total liquid sample, from valve V6. The effect of removing this volume has been estimated from computations and is found to be less than 0.1 % of the equilibrium solubility. The sulphurous acid sample is added to excess iodine which is then back titrated with sodium thiosulphate to determine the original concentration of total dissolved sulphur species.

**RESULTS**

Initial experiments have been conducted to determine the ability of the apparatus to measure equilibrium. The two component system sulphur dioxide-water was used as a test case and the results compared with those found in the literature. Table 2 shows the results obtained to date. Pressures are given as absolute values and concentrations are given as molalities, moles $S_{IV}$ / kg $H_2O$, where $S_{IV}$ represents the total amount of sulphur in solution.

$$S_{IV} = SO_2.aq + HSO_3^- + SO_3^{2-}$$

**Table 2:** Solubility of sulphur dioxide in water from experiments at temperatures and total pressure measurements indicated

| Temperature | Total Pressure (absolute) | Solubility |
|---|---|---|
| °C | atm | moles $S_{IV}$ / kg $H_2O$ |
| 24.6 | 0.249 | 0.215 |
| 23.9 | 1.28 | 1.52 |
| 39.1 | 1.93 | 1.54 |

Figures 2 and 3 give comparisons of the results obtained with experimental results found in the literature and the Henry's law calculations from Mathematica®. The 23.9 and 24.6 °C data points are compared with the experimental results of Maass (9) at 25 °C. Maass gave solubility of sulphur dioxide as a function of total vapour pressure (the sum of partial pressures of water and sulphur dioxide) which allows direct comparison with the results obtained in this work. The result at 39.1 °C is compared to the results of Rumpf and Maurer (11) who also gave values as a function of total vapour pressure.

The effect of temperature deviations between this work and the comparison results have been estimated using the vapour liquid equilibrium calculations. The program was used to calculate the solubility at both temperatures. The difference was then used to calculate the percentage change in solubility in moving from the higher temperature to the lower. In reference to 25 °C the solubility increases by 1.5 % at 24.6 °C and 3.6% at 23.9 °C. In reference to 40 °C the solubility increases by 2.7% at 39.1 °C.

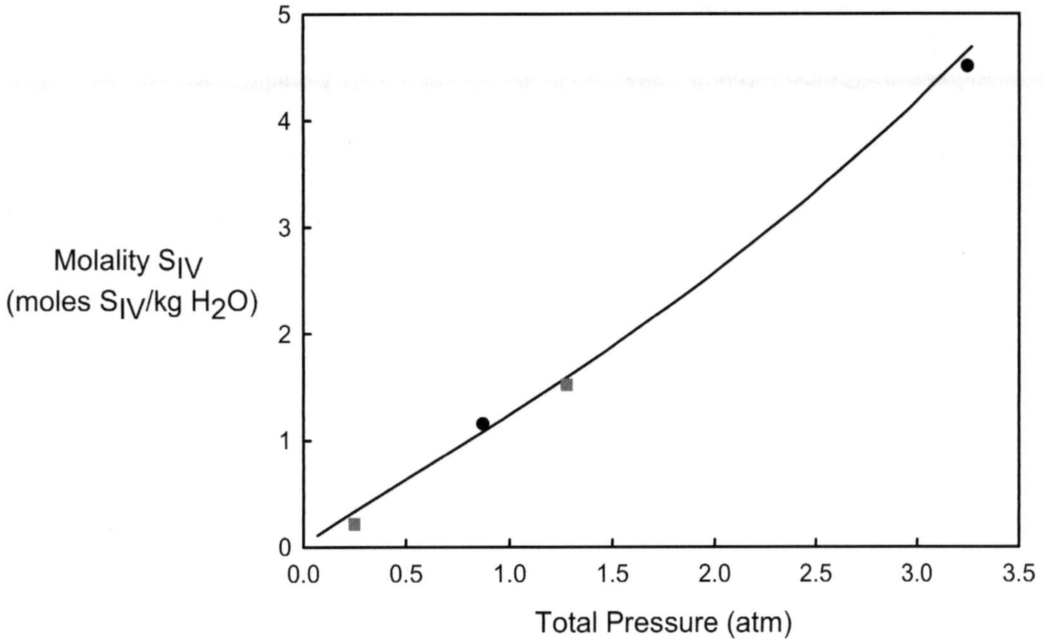


**Figure 2:** Comparison of values for sulphur dioxide solubility in water at 25 °C: ●, Experimental results, Maass (1928); ■, Experimental results, this work; —, Calculation results, this work.

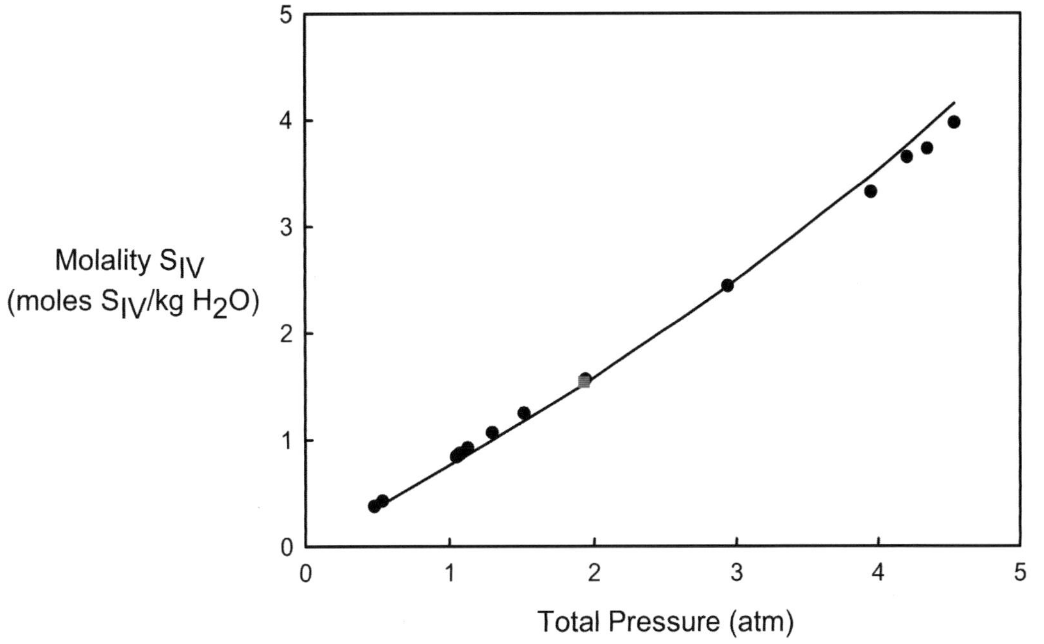


**Figure 3:** Comparison of values for sulphur dioxide solubility in water at 40 °C: ●, Experimental results, Rumpf and Maurer (1992) & Hudson (1925); ■, Experimental results, this work; —, Calculation results, this work.

## CONCLUSIONS

An experimental set up has been designed and built which is capable of measuring vapour liquid equilibrium of sulphur dioxide, oxygen and water mixtures up to 30 atm and 120 °C. Preliminary results of sulphur dioxide solubility in water show good agreement with literature data and calculations. It was concluded that the novel design of apparatus can measure equilibrium to acceptable precision in the range of interest. Further studies can now be conducted with the addition of oxygen to the sulphur dioxide water mixtures to determine equilibrium solubility and oxidation effects. Finally the data will be compared with an extended calculation method to include oxygen solubility and used to design separation equipment for the HyS process.

## REFERENCES


1. Ewan, B. C. R. & Allen, R. W. K. (2005) A figure of merit assessment of the routes to hydrogen, International Journal of Hydrogen Energy. 30, 809-819.
2. Shaw, A. C., Ewan, B. C. R. & Allen, R. W. K. (2006). Sulphuric acid decomposition reactions in the sulphur iodine and Westinghouse processes for hydrogen generation. Paper presented at the World Hydrogen Energy Conference, Lyon, France.
3. Bishenden, E. & Donaldson, D. J. (1998) Ab Initio Study of SO2 + H2O, Journal of Physical Chemistry A. 102, 4638-4642.
4. Zhang, Z. & Ewing, G. E. (2002) Infrared spectroscopy of SO2 aqueous solutions, Spectrochimica Acta Part A: Molecular and Biomolecular Spectroscopy. 58, 2105-2113.
5. Hartmann, H. (1984) Determination of the gas- and liquid-mass transfer coefficients as well as equilibrium solubilities for the sulfur dioxide/oxygen-water system. II Commission European Communities Report.
6. Beuschlein, W. L. & Simenson, L. O. (1940) Solubility of sulfur dioxide in water, Journal of the American Chemical Society. 62, 610-12.
7. Johnstone, H. F. & Leppla, P. W. (1934) The solubility of sulfur dioxide at low partial pressures. The ionization constant and heat of ionization of sulfurous acid, Journal of the American Chemical Society. 56, 2233-8.
8. Hudson, J. C. (1925) The solubility of sulfur dioxide in water and in aqueous solutions of potassium chloride and sodium sulfate, Journal of the Chemical Society, Abstracts. 127, 1332-47.
9. Maass, C. E. & Maass, O. (1928) Sulfur dioxide and its aqueous solutions. I. Analytical methods, vapor density and vapor pressure of sulfur dioxide. Vapor pressure and concentrations of the solutions, Journal of the American Chemical Society. 50, 1352-68.
10. Sherwood, T. K. (1925) Solubilities of sulfur dioxide and ammonia in water, Journal of Industrial and Engineering Chemistry (Washington, D. C.). 17, 745-7.
11. Rumpf, B. & Maurer, G. (1992) Solubilities of hydrogen cyanide and sulfur dioxide in water at temperatures from 293.15 to 413.15 K and pressure up to 2.5 MPa, Fluid Phase Equilibria. 81, 241-60.
12. Zemaitis, J. F., Clark, D. M., Rafal, M. & Scriviner, N. C. (1986) Handbook of Aqueous Electrolyte Thermodynamics, AIChE, New York.
13. Nakamura, R., Breedveld, G. J. F. & Prausnitz, J. M. (1976) Thermodynamic properties of gas mixtures containing common polar and nonpolar components, Industrial & Engineering Chemistry Process Design and Development. 15, 557-64.
14. Beutier, D. & Renon, H. (1978) Representation of ammonia-hydrogen sulfide-water, ammonia-carbon dioxide-water and ammonia-sulfur dioxide-water vapor-liquid equilibriums, Industrial & Engineering Chemistry Process Design and Development. 17, 220-30.

# Mechanical Design and Fabrication of a High-Temperature Sulfuric Acid Decomposer

**Roger X. Lenard and Fred Gelbard**
**Sandia National Laboratories**

## Abstract

A sulfuric acid decomposer suitable for use in with a high temperature heat source was required by the Sulfur-Iodine project at Sandia National Laboratories. The design requirements for the decomposer were that the decomposer needed to be resistant to corrosion, thermal creep, and provide a contamination-free product stream. The product stream upon output from the decomposer is $O_2$, $SO_2$ and water vapor. The input stream is sulfuric acid that can exhibit a variety of temperatures and concentrations. The idealized decomposer would be able to be fabricated from existing materials, could be easily assembled and disassembled, provide excellent heat transfer and recovery, exhibit ease in loading and unload catalyst, and finally be amenable to mâss production.

A design was generated that employed materials resistant to corrosion and could be joined together. An overall assembly was designed that allowed ease of assembly/ disassembly. The initial design did not fully address ease of loading catalyst, although a second generation model did address this issue. Two models were fabricated, a 27 inch and a 48 inch decomposer bayonet. At this time, the 27" decomposer bayonet has been tested, and produced excellent results on the first tests. Subsequent tests continued to show satisfactory performance. This paper addresses the mechanical design and integration as well as the installation of the 27" bayonet into a test fixture. A companion paper will discuss the test results.

## Background

Decomposing sulfuric acid is an important precursor step in many approaches to producing hydrogen from water. In the decomposition process, sulfuric acid must be heated to approximately 850 C and passed over a catalyst in order to decompose it into $SO_2$, $O_2$ and water vapor. The reaction products are later separated to form sulfurous acid, the excess water vapor is condensed and the oxygen released. However, hot sulfuric acid is extremely corrosive; few materials are compatible with it. Some of the preferred materials cannot be bonded together with substances that will withstand the environment. Other materials are extremely brittle, some have poor thermal conductivity, and no metal is compatible with both the hot acid and the sulfur dioxide. The author was challenged to design and fabricate a decomposer that exhibited the following features:

1. Operate at high temperatures, with a based temperature greater than 250 C
2. Be easily assembled and disassembled
3. Allow easy removal and filling of particulate catalyst of various shapes and sizes

4. Be relatively impervious to corrosion (~ years without noticeable corrosion)
5. Be leak-tight and able to withstand internal pressures of $\leq 10^6$ Pa
6. Allow line-replaceable parts

## Decomposer Description

This paper describes a sulfuric acid decomposer that is comprised of an external and internal bayonet made from Hexoloy brand Silicon Carbide, and inner alumina rod, tantalum flanges and connector disks, a tantalum inlet tube and a silicon carbide exit tube. They are compression sealed to a Kelrez gasket via tightening a large nut into a receptor boss. The Silicon Carbide outlet tube is brazed to the tantalum disk using a high temperature Ticusil braze. A tantalum flange ring is bonded to the outer silicon carbide bayonet by a similar brazing process. The inner alumina rod is centered by two set of centering pins made from 1/8 inch alumina rod. The central silicon carbide tube is manufactured in two parts. The upper part has a centering land with gas passages cut into its perimeter. The tube is a thin wall Hexaloy with catalyst placed between the inner and outer silicon carbide walls. The lower section fits into the upper tube. The lower tube has a centering land at the top of the tube with a receptor recess ground into the silicon carbide. In operation, sulfuric acid enters the decomposer at the bottom through a 1/4 inch tantalum tube, is heated with an external heat source and decomposes.

This was a challenging task since no acid decomposer had been developed that met some much less all of the above requirements. The decomposer is comprised of a base plate, a threaded hold-down and thread nut, a decomposer bayonet, an inner assembly, a inlet/ outlet disk, a gasket , a central rod, an inlet tube and an outlet tube.

### *Base Plate*

The base plate is shown in Figure 1. In this incarnation it is made to hold a single decomposer bayonet for experimental purposes, although in can be fabricated in a manifold configuration to hold multiple bayonets. It is fabricated from weldable stainless steel. With no other attachments, there is a 1.5" (plus or minus) hole drilled in the center of the plate.

**Figure 1. Base Plate**

The base plate also has an attachment that is welded on in this incarnation. This attachment is called the threaded boss, shown in Figure 2.

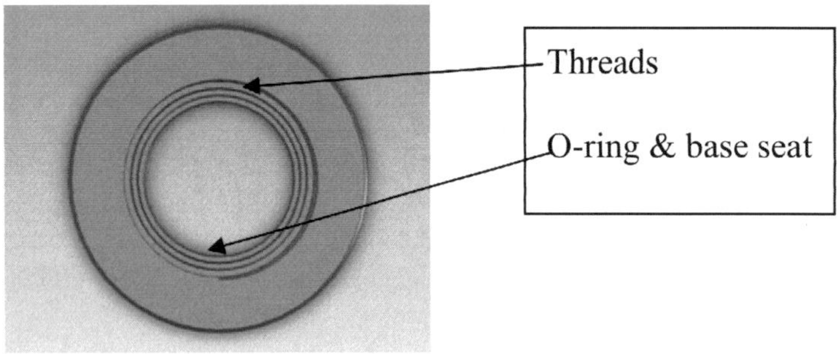


**Figure 2. Threaded Boss**

The threaded boss contains threads and a base seat that is machined that an "O"-ring seal can be placed in to seal the bottom portion of the threads from any external pressure. The base plate requires a support as shown in Figure 3.

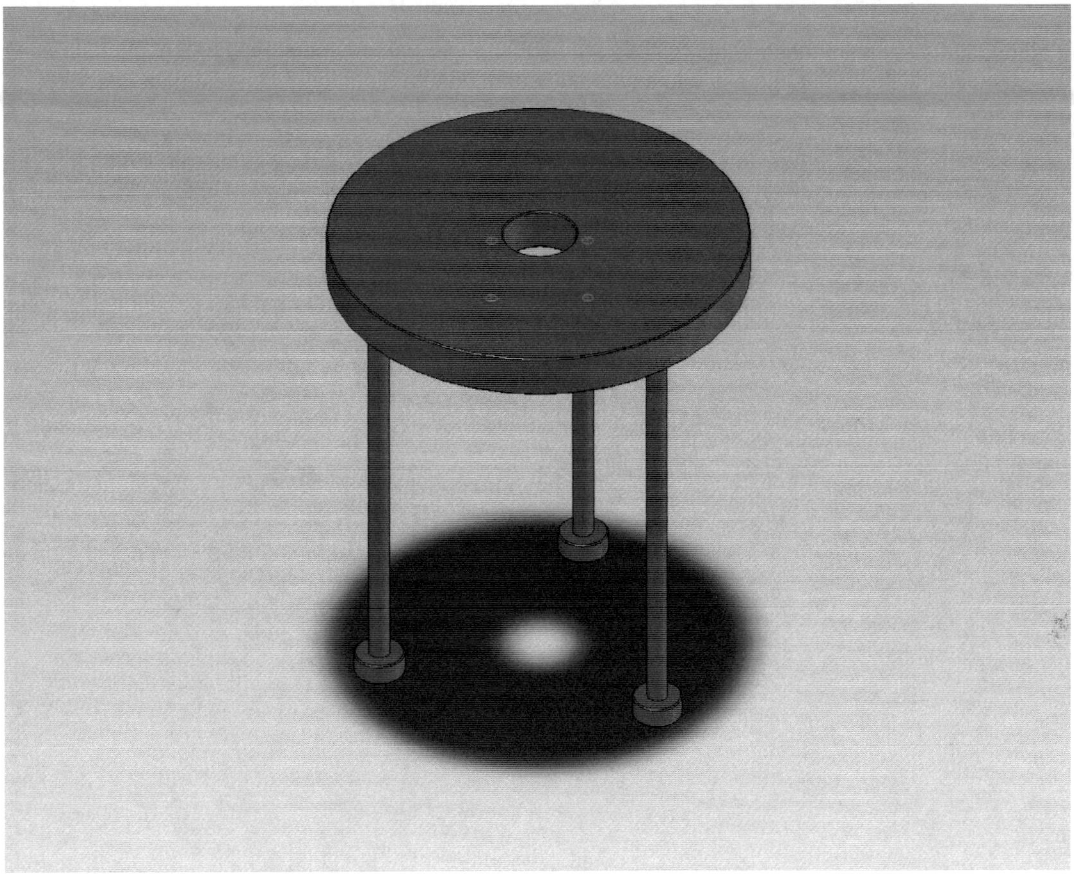

**Figure 3. Base Plate with Stand**

When the base plate is complete with threaded boss and some of the internals it appears as shown in Figure 4. There is also a mechanism for supporting the silicon carbide exit tube that is attached to the bottom of the base plate. The support mechanism is shown in a white-grey color. This attachment is affixed to the base plate via threaded #8 screws that have blind holes drilled into the base plate.

**Figure 4. Base Plate with Mounting Hardware**

## *Decomposer Bayonet*

The decomposer bayonet is a hollow cylindrical tube closed one end and is a device for entraining and decomposing the sulfuric acid. The bayonet mean must be capable of operating with corrosive liquids and gases at high temperatures, it must provide a mechanism for sealing the internal reagents from the external environment and it must seal the external environment from the chemical supplies. The bayonet in the present incarnation is from 27" to 54" long, although in future plants it could be as long as 14 feet, perhaps longer. The decomposer bayonet with threaded mounting and nut is shown in cross-section in Figure 5. In Figure 7, there is a close-up view of the mount and seal of this bayonet configuration. Note there are two O-ring seals. There is also a mounting flange made of tantalum that is connected to the bayonet tube. The figure also shown the entry and exit tubes for the acid and reaction products.

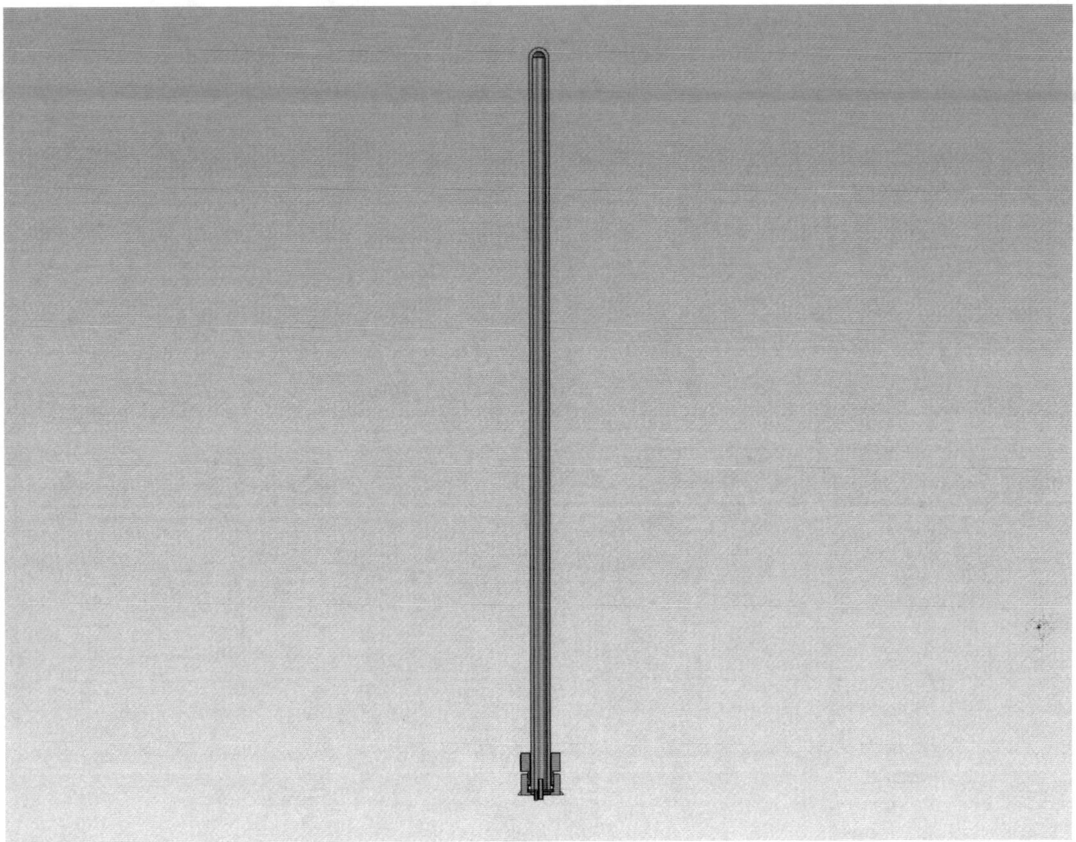

**Figure 5. Cross reference of bayonet and mounting base**

Also included on the bayonet is a mounting flange shown in Figure 6, below. It is a tantalum ring that is brazed or otherwise connected to the bayonet, made from silicon carbide. Other materials are being investigated. It is contacted by the nut and compressed onto the disk plate shown below. This provides a completely sealed unit for acid decomposition.

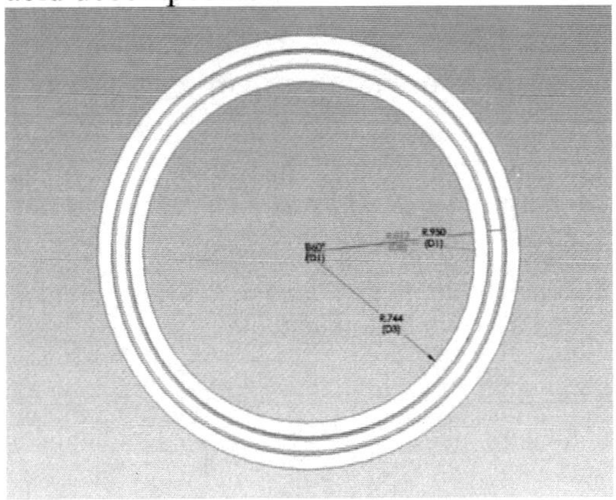


**Figure 6. Tantalum mounting flange**

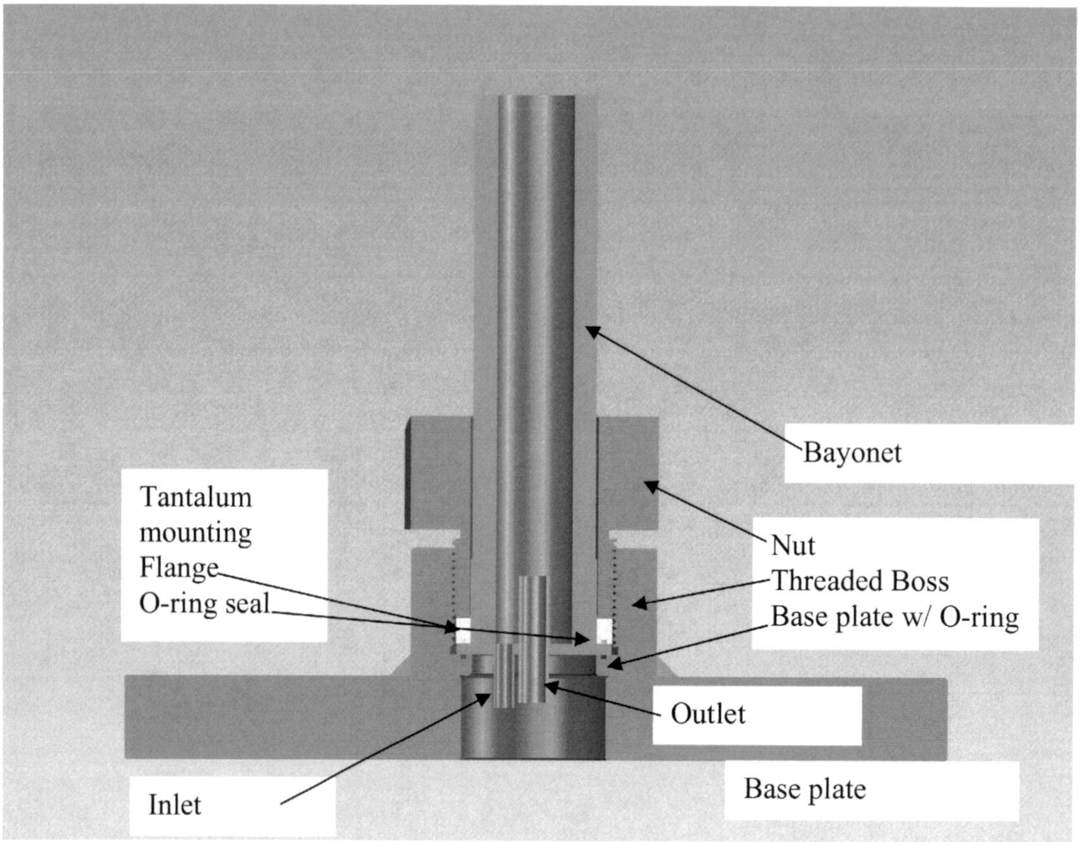


**Figure 7. Cross section view of mounting plate and bayonet**

## ***Base Disk***

The base disk in this incarnation is a piece of machined tantalum 1.5 inches OD and 1/8" thick. It is shown in Figure 8. It has two holes, a central hole 3/8$^{th}$ inch ID for receiving a 3/8" OD silicon carbide exit tube. The exit tube is brazed or otherwise connected to the base disk. The 3/8$^{th}$ inch hole is center drill on the base disk . There is a 1/4$^{th}$ inch hole drilled adjacent to the 3/8$^{th}$ inch center hole, offset but 1/16$^{th}$ of an inch from the outer edge of the central hole. A ¼" tantalum tube is e-beam welded into that hole, and is flush mounted to the top of the base disk.

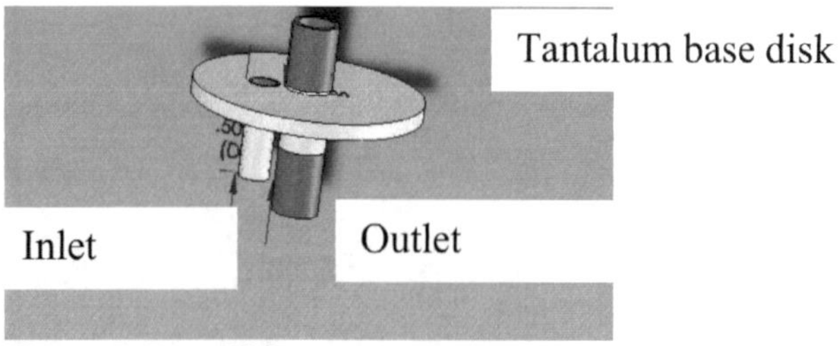


**Figure 8. Base Disk**

## *Inner Assembly of the Decomposer*

The inner tube is a method of separating the reaction products from the reactants after the reaction has occurred. It is also provides a means for extracting useable heat energy from the product stream. A cross-section view is shown in Figure 9.

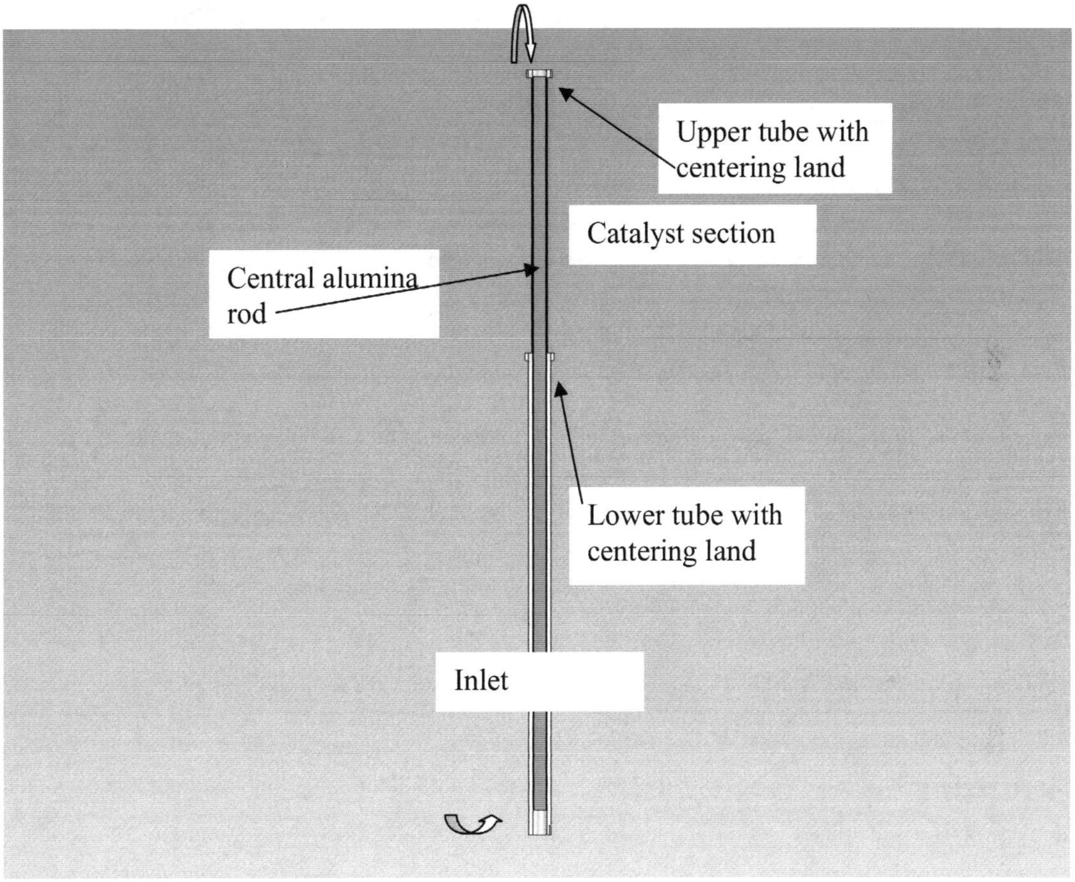


**Figure 9. Inner tube assembly**

The upper and lower portions of the inner tube assembly are constructed from silicon carbide. The upper portion slides into the lower portion. Inside these tubes is an alumina rod that is centered inside the silicon carbide tubes. It has close tolerances so that the down coming gas stream exhibits a high Reynold's number, hence good heat transfer to the outer wall of the inner tube assembly. This allows incoming $H_2SO_4$ to be preheated by energy from the product stream. As the sulfuric acid proceeds up the tube, it is vaporized, and the gas transits slots cut into the lower land assembly. From there the vaporized sulfuric acid is further heat and enters the catalyst region, where decomposition is completed. The products of the decomposition make a U-turn at the top of the assembly inside the decomposer bayonet and proceeds downward between the alumina tube and the inner wall of the inner tube assembly. The inner tube is placed with a graphoil center sleeve onto the exit tube with the lower end of the inner tube sealing against the upper portion of the exit tube. Thus, incoming acid is separated from the outgoing gaseous reaction products.

### *Testing and Operations*

The high temperature decomposer has been tested several times at the time of this paper's publication. Results have generally been good, but with some problems encountered. The first test was aborted when it was discovered that the inner bushing sealing the product exit tube to the inner tube had pushed the central tube during assembly and the acid bypassed the heating section. This error was fixed and a second test was run. The second test ran until acid was depleted, and reasonably good data were generated. No corrosion or acid attack of any of the components was noted. The initial exit products were slightly dark in color, although the product stream eventually became clear. We suspect that this was a result of the acid cleaning some of the parts rather than some systemic corrosion or acid attack. However, some of the sensors were not fully calibrated and it was decided to perform a fully calibrated, long-term test. Before performing this test, the entire assembly was removed, the inner tube and catalyst removed and the device reassembled. During the test, the bayonet struck a sidewall or was subjected to some other side impact. After running the test, it was discovered that the silicon carbide bayonet had fractured at the tantalum attachment ring. The bayonet was removed and emptied of all constituents. The end was cut off and the tantalum ring re-attached.

We re-assembled the system and performed a test run. Things operated very well, but after approximately an hour of operation, with exceptional results, an electrical short occurred that aborted the heaters. However, we had great data to that point.

## Summary

A high temperature sulfuric acid decomposer was designed, fabricated and tested. The decomposer allows relative ease of assembly and disassembly. It allows separation of the various components so that compatible elements of the decomposer are exposed to the corresponding reactants and products. Fabrication proceeded in an efficient manner as did assembly. Tests results to date have been acceptable.

# 224f Experimental Testing Of A High-Temperature

**Fred Gelbard** and Roger X. Lenard. 6771, Sandia National Laboratories, 16 Dunkin Road, Edgewood, NM 87015

A high temperature sulfuric acid decomposer suitable for use with a high temperature heat source was designed and fabricated for the Sulfur-Iodine project at Sandia National Laboratories. The design requirements for the decomposer were that the decomposer needed to be resistant to corrosion, thermal creep, and provide a contamination-free product stream. The product stream upon output from the decomposer is O2, SO2 and water vapor. The design and fabrication of the decomposer is the subject of a companion paper; this paper discusses experimental results obtained to date.

The decomposer was placed in a fume hood and surrounded by electrical heating elements. Sulfuric acid is entrained into the decomposer through a ¼" tantalum tube. The stream enters an annular passageway between the outer SiC bayonet and an inner SiC insert. The inner insert contains the product gases, and also transfers usable heat from the product stream into the incoming acid stream. The acid is transported up the bayonet, gaining heat as it rises, passing by a centering land, thence entering a high temperature region ~850C containing 21 g of platinum coated titania catalyst. The acid stream decomposes into O2, SO2 and water vapor. The product stream then enters an annular passageway between the inner SiC insert and an alumina rod that forms a narrow gap between the inner SiC tube and the outer perimeter of the alumina rod. This arrangement provides very good heat transfer characteristics and a large fraction of the process heat is recaptured.

The products exit the decomposer and enter a condensing unit where water is condensed from the process stream, the resultant gas stream enters a drier where any residual water vapor is desiccated, and the gas stream is analyzed with a variety of instruments, including an oxygen monitor, as well as flow rate sensors. Process gases are then exhausted to the atmosphere. Data to date indicate satisfactory operation of the decomposer with no observable corrosion of any of the parts. Heat recovery is excellent, and the process stream to date indicates near theoretical equilibrium sulfur dioxide production rates. Work to fabricate a 48" decomposer bayonet is in progress. Results will be updated as progress continues.

# Meeting U.S. Liquid Transport Fuel Needs with a Nuclear Hydrogen Biomass System

Charles W. Forsberg
Oak Ridge National Laboratory; P.O. Box 2008; Oak Ridge, TN 37831-6165, USA
Tel: (865) 574-6783; Fax: (865) 574-0382; E-mail: forsbergcw@ornl.gov

## Abstract

The two major energy challenges for the United States are replacing crude oil in our transportation system and eliminating greenhouse gas emissions. A domestic-source greenhouse-gas-neutral nuclear hydrogen biomass system to replace oil in the transportation sector is described. Some parts of the transportation system can be electrified with electricity supplied by nuclear energy sources that do not emit significant quantities of greenhouse gases. Other components of the transportation system require liquid fuels. Biomass can be converted to greenhouse-gas-neutral liquid fuels; however, the conversion of biomass to liquid fuels is energy intensive. There is insufficient biomass to meet U.S. liquid fuel demands and provide the energy required to process the biomass to liquid fuels. With the use of nuclear energy to provide heat, electricity, and hydrogen for the processing of biomass to liquid fuels, the liquid fuel production per unit of biomass is dramatically increased, and the available biomass could meet U.S. liquid fuel requirements.

## 1. Introduction

A path forward is defined to completely replace the use of fossil fuels in the transport system of the United States. The alternative system is based on the use of biomass and nuclear energy. The nuclear energy is used to produce electricity, heat, and hydrogen with some of the electricity used directly in the transport system. Nuclear energy is also used to provide heat and hydrogen for the processing of biomass into liquid transport fuels. The combined system is robust, can be implemented with near-term technologies with a smooth transition to more advanced technologies, and can expand to meet growing transport requirements. Components of the system are economic today.

Liquid fuels (gasoline, diesel, and jet fuel) today are produced from fossil fuels. Oil is the primary feedstock in most of the world. Liquid fuel production is a major economic, national security, and environmental challenge. Two-thirds of the U.S. demand for crude oil is met by imports. Most of the world's oil reserves are in the Mideast, an area that is politically unstable. Last, the burning of liquid fuels in transport systems is one of the primary sources of greenhouse gas emissions worldwide. Thus, there is a need for alternative transport fuels.

The fuel cycle for liquid fuels includes obtaining the feedstock; conversion of that feedstock to liquid fuels; transport of the liquid fuels to the user; and burning the liquid fuel in a car, truck, or airplane. Each step consumes energy. Figure 1 shows the greenhouse gas releases per vehicle mile[1] from a diesel-powered SUV for each step in today's fossil fuel cycle for liquid fuels. The greenhouse gas releases are roughly proportional to the energy consumed in each step. With the production of liquid fuels such as diesel from fossil fuels, the total fuel-cycle energy consumption and carbon dioxide releases to the atmosphere are 130 to 200% of the energy consumption and carbon dioxide released from the vehicle.

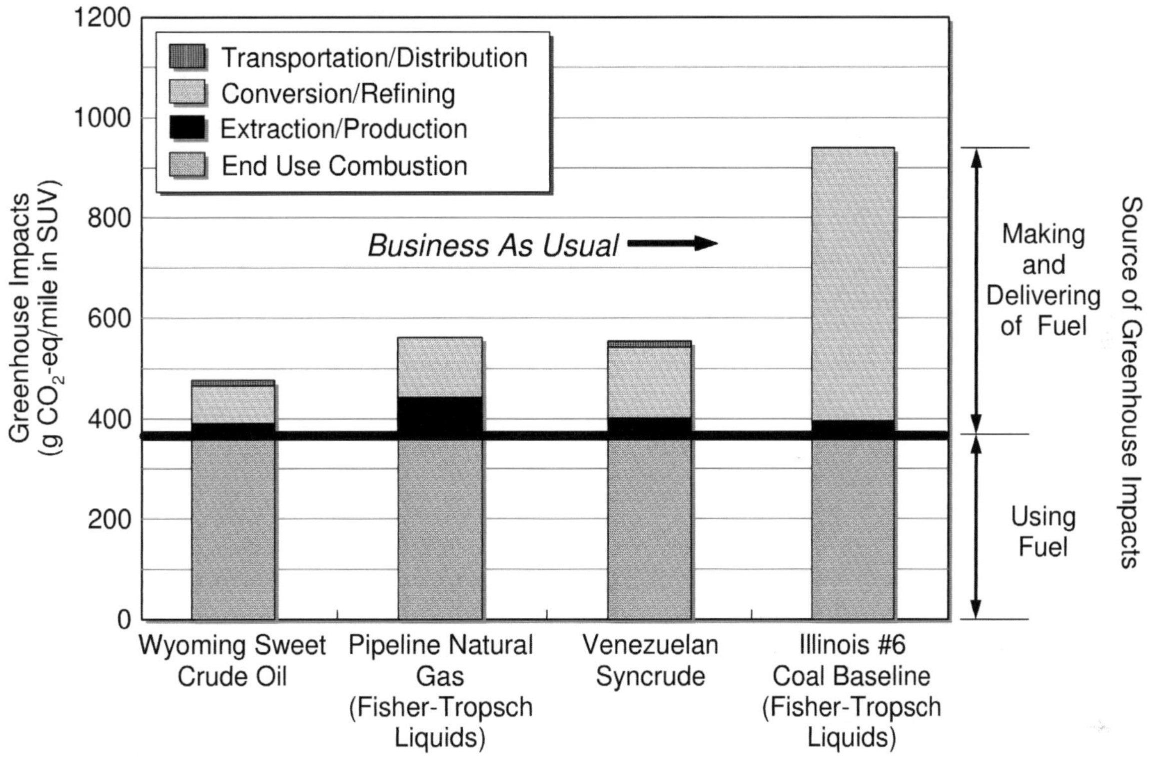


Fig. 1. Greenhouse gas releases per vehicle mile for diesel fuel produced form different sources

The process of extracting high-quality sweet (low-sulfur) crude oil, converting it to diesel fuel, and transporting the diesel fuel to the fuel pump consumes relatively little energy and releases relatively small quantities of carbon dioxide. In contrast, if liquid fuels are made from coal, more energy is used in the production process than is available in the final fuel. As the stocks of high-grade crude oil are exhausted and liquid fuel is made from lower-grade resources, much more energy is needed to make a gallon of liquid fuel.

The production of liquid fuels from biomass has a somewhat different fuel cycle. Biomass is produced by sunlight, carbon dioxide from the atmosphere, and water. The biomass is converted to liquid fuels. The carbon dioxide from burning biomass fuels recycles the carbon dioxide back to the atmosphere. Like the fossil fuel cycle, significant energy required to convert biomass feedstocks to usable liquid fuels. Studies[2] indicate that the United States could produce about 1.3 billion tons of dry biomass per year for conversion to liquid fuels without major cost or availability impacts on the production of food or fiber. The energy value of that biomass can be viewed from several perspectives.

- *Energy value*. The energy content ($16 \cdot 10^6$ Btu/ton) of the biomass, if burned, would be equal to burning 9.8 million barrels of diesel fuel per day; however, additional energy would be required to grow, collect, and transport that biomass to furnaces.

- *Diesel fuel*. If all of the carbon (carbon fraction: 0.45) in the biomass were converted to diesel fuel, 12.4 million barrels of diesel fuel could be produced per day. This scenario assumes that nonbiomass energy sources provide the needed energy for operation of the biomass-to-fuel plants and to produce the hydrogen needed for the conversion process. The energy value of this diesel fuel exceeds that of the biomass if burned because of the non-biomass energy and hydrogen inputs in the biomass-to-fuel plant.

- *Fuel ethanol*. If the biomass were converted to fuel ethanol (89.8 gal ethanol/ton), the energy value of the ethanol would be equal to about 4.7 million barrels of diesel fuel per day. This calculation assumes that some of the biomass is converted into ethanol and that the remainder of the biomass provides the energy for the biomass-to-ethanol facilities. The energy value of the product ethanol is only half that of the original biomass.

The quantities of liquid fuels that can be produced from the nation's biomass depend upon whether an external energy source provides the energy for the conversion of biomass to liquid fuels. There is insufficient biomass to meet the nation's fuel demands if biomass is used as a feedstock for production of liquid fuels and as an energy source to operate the biomass-to-liquid-fuel plants. Nuclear energy—in the form of electricity, heat, and hydrogen—can provide this energy and enable biomass to become the primary source of liquid transport fuels.

This paper addresses four related issues: liquid fuel needs, the potential of using electricity for transportation, the viability of converting biomass into liquid fuels, and nuclear-energy inputs. The analysis assumes that hydrogen is used for production of biomass liquid fuels but not used directly as a transport fuel. From the perspective of the average person, no radical changes in the transport sector are assumed. The technologies are all relatively near-term technologies.

## 2. Liquid fuel needs

### 2.1. Fuel type

Liquid fuels are chosen for use on the basis of two factors: (1) they can be easily made from crude oil, and (2) they have major advantages in terms of high energy density and safety. Energy density determines the distance between refuelings. Table 1 shows the properties of various possible future fuels[3] with different types of engines. These data show significant safety and vehicle range advantages in using liquid fuels such as diesel. The major advantage of direct use of hydrogen as a fuel is that it can be used in fuel cells with significantly higher efficiency. However, there are major technical, economic, and institutional challenges that will take significant time to overcome—including the difficulty in transitioning to a new fuel infrastructure. For this analysis, it is assumed that liquid fuels are the primary transport fuel.

Table 1.
Comparisons of different fuel systems for automobiles

| $H_2$ Storage Mechanism[a] | Engine Type and Eff (%)[b] | Est. Miles for a Tank of Fuel | LFL[c] (vol %) | UFL[d] (vol %) | Toxicity[e] | Storage Pressure (bar) | Storage Temp. |
|---|---|---|---|---|---|---|---|
| Compressed $H_2$ | Fuel Cell: 70 | 219 | 4 | 74.2 | MAH | 700 | Room |
| Liquefied $H_2$ | Fuel Cell: 70 | 264 | 4 | 74.2 | MAH | 1 | 4 K |
| Metal hydride | Fuel Cell: 70 | 132 | 4 | 74.2 | MAH | 1 | Room |
| Liquefied $NH_3$ | Hybrid: 40 | 234 | 15.5 | 27 | IDLH: 500 ppm | 10 | Room |
| Compressed $CH_4$ | Hybrid: 40 | 416 | 5 | 15 | AH | 700 | Room |
| Liquefied $CH_4$ | Hybrid: 40 | 418 | 5 | 15 | AH | 1 | 109 K |
| Methanol | Hybrid: 40 | 285 | 6 | 36 | TWA: 200 ppm; IDLH: 6000 ppm | 1 | Room |
| Ethanol | Hybrid: 40 | 285 | 3.3 | 19 | TWA: 1000 ppm IDLA: 3300 ppm | 1 | Room |
| $LiBH_4$ | Fuel Cell: 70 | 245 | 4 | 74.2 | MAH | 1 | Room |
| Diesel hybrid | Hybrid: 40 | 800 | 0.77 | 5.35 | Low (>1369 ppm for 8 h) | 1 | Room |

[a]Metal hydrides at 5% $H_2$/lb metal at 8 lb/ft$^3$; Liquefied $NH_3$ = liquefied ammonia; $LiBH_4$ (lithium borohydride) as a 50% slurry in water.
[b]Eff = engine efficiency.
[c]Lower flammability limit.
[d]Upper flammability limit.
[e]MAH = minor asphyxiation hazard; AH = asphyxiation hazard; IDLH = Immediately Dangerous to Life or Health; TWA = Time Weighted Average, typically over 8 hours.

In parts of the world with high fuel costs, advanced diesels have become the preferred engine for cars and light trucks because of their excellent fuel economy. Engine technology is driving the fuel system toward diesels and ultimately toward hybrid diesels. In hybrid vehicles, an electric motor-generator-battery system provides the energy for rapid acceleration, recovers energy from braking, and provides energy for air-conditioning and other auxiliary systems when the vehicle is at stop lights. The internal combustion engine operates at the power level and engine speed that results in the highest fuel economy. This dramatically improves fuel efficiency because the efficiency of an internal combustion engine is strongly dependent upon engine speed and load.

For various reasons, ethanol is the primary biomass fuel being produced today: the technology already exists; the fuel is economic; and with an octane number of 113, the ethanol has special characteristics as an octane enhancer. In the United States, gasoline is the primary fuel for cars and light trucks. To obtain the required octane number for proper engine performance, an octane enhancer is added. Historically, the enhancer was tetra-ethanol lead, a compound that was banned because it caused heavy metal poisoning of children. The industry then turned to MTBE; however, this compound has caused serious groundwater contamination. Today, ethanol is becoming the preferred octane enhancer because of its lower environmental risks. Although, in effect, ethanol is both a fuel and a chemical additive, its value as an octane enhancer exceeds its fuel value. Fuel ethanol does have some drawbacks as well. As noted in Table 1, the range of diesel fuel vehicles is over twice that of a vehicle using fuel ethanol. This reflects the high oxygen content of ethanol, which takes up space in the fuel tank.

2.2. Fuel demand

The United States consumes[4] slightly over 20 million barrels of oil per day, about two-thirds of which is used for transportation (Fig. 2). In the figure the ethanol that is blended into the gasoline (the primary use of ethanol today) is included as part of the gasoline pool. Gasoline ($9.0 \cdot 10^6$ barrels/day) is the primary transport fuel, followed by diesel ($2.9 \cdot 10^6$ barrels/day), and then jet fuel ($1.6 \cdot 10^6$ barrels/day).

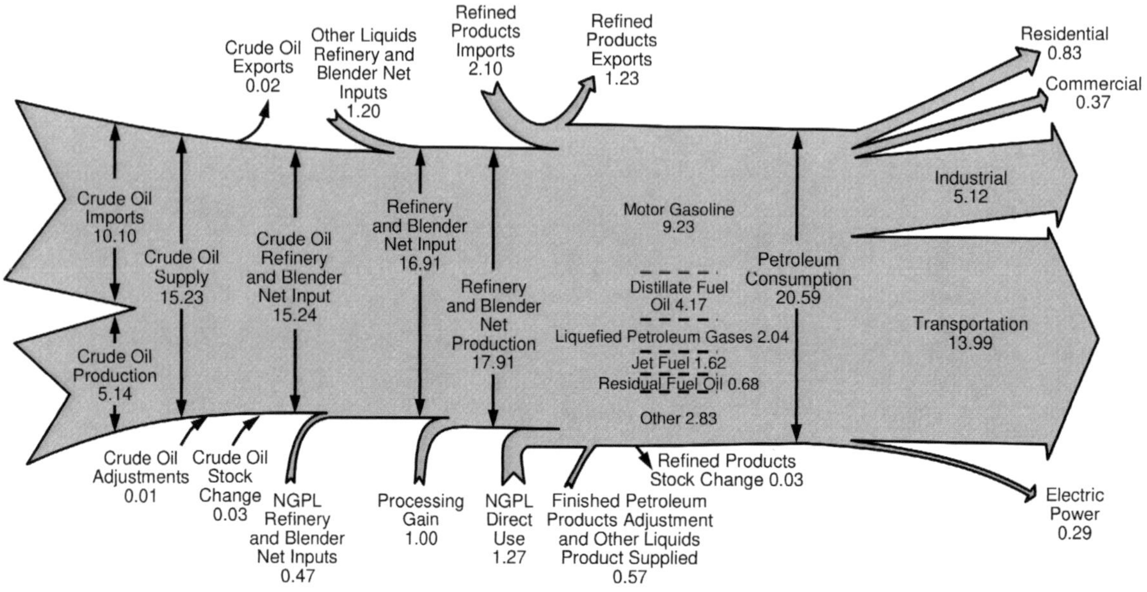


Fig. 2. U.S. oil production and consumption in 2006 (NGPL = natural gas plant liquids)

Future fuel demand is strongly dependent upon mileage standards, fuel prices, and technology. The U.S. vehicle fleet has relatively low fuel efficiency relative to those of other major industrial countries. As a consequence, more potential for rapid improvements in fuel efficiency exists. For example, because of the higher fuel prices in Europe, most new vehicles sold there are powered by more efficient diesel engines. This technology could rapidly penetrate the U.S. car and light truck markets because of (1) the large manufacturing experience base in Europe and Japan and (2) the similar infrastructures for gasoline and diesel engines. Honda's recent announcement that it will offer high-efficiency diesels in all car lines by 2010 suggests that such a change is already under way in response to higher fuel prices. Such a transformation could boost fuel economy measured in miles per gallon by ~33%. However, diesel fuel contains about 11% more energy, so the improvement in energy efficiency is ~22%.

Simultaneously, the introduction of hybrid vehicles implies major improvements in fuel economy. Full hybrid vehicles obtain about 40% better mileage than the equivalent gasoline or diesel vehicle. Hybrid vehicle technology is at an earlier stage of development than diesel technology; thus, there is the potential for more improvements. The two technologies are complementary. The use of diesel improves engine efficiency whereas hybrid vehicles enable the engine to operate at its most efficient speed and load. The U.S. demand for liquid fuels could be reduced significantly with relatively small changes in the market. Hybrid diesels could reduce oil consumption by over 3 million barrels per day.

### 3. Electric transportation options

Three electrification technologies that could reduce total demand for liquid transport fuels by 50% or more are in direct economic competition with liquid fuels.

*Plug-in hybrid electric vehicle [PHEV].* The PHEV is a hybrid car or light truck that has a larger battery system and functions as an electric vehicle for short trips and a hybrid vehicle (battery and liquid fuel) on longer trips. Because most vehicle trips are short trips, high-performance PHEVs can dramatically reduce liquid fuel consumption. It has been estimated that in principle, a 35-mile "all electric" range for cars and light trucks would reduce gasoline consumption by 74%, or about 6.7 million barrels of oil per day.[5] In practice, given that vehicles will not always be recharged, potential fuel reductions for cars and light trucks would be unlikely to exceed 50%.[6] Prototype PHEVs are operational, and Toyota Corporation has announced that the next generation of hybrid vehicles will include PHEVs.

PHEVs have major potential advantages relative to traditional vehicle systems. Electricity is less expensive per vehicle mile than liquid fuels. Because PHEVs are multifuel (liquid fuel or electricity), they offer protection against local or national supply disruptions (hurricanes, war, boycotts). The vehicles also enable leveling of the electrical load with nighttime recharging of the batteries. With the use of non-carbon-emitting electricity production, no greenhouse gas is released when the vehicle is used in its electric mode. Last, the plug-in hybrids develop the electric vehicle infrastructure for hydrogen fuel cells that produce electricity for all-electric drive systems. The major challenge is to develop reliable affordable batteries or other energy storage devices.

PHEVs would enable nuclear energy to meet a major fraction of U.S. transportation needs. If the average gasoline fuel economy for cars and trucks is 20 miles per gallon, a million barrels of oil per day supports 8.2 million miles of travel per day. The same number of travel miles per day could be obtained with the electricity from twenty-three 1000-MW(e) reactors with a 90% capacity factor, assuming an electric fuel economy of 0.603 kWh/mile.[5] The fuel cost is low. At \$0.10 per kilowatt, the electricity costs are \$0.0603 per vehicle mile. For a car with a fuel economy of 20 miles per gallon, gasoline would have to sell for \$1.21 per gallon to match this price.

*Ground freight transport.* Recent analysis[3] indicates that liquid fuel consumption in the freight transportation system could reduced by over 80% by the combination of electrification of the railroads (as in Europe) with large-scale intermodal rail-truck systems. Most of the long-distance truck transport would be replaced by containerized freight that travels long distances by rail, with local delivery by truck. This type of transport would change about 5% of America's total energy demand from the use of liquid fuels to the use of electricity. If nuclear energy were used to produce the electricity, about fifty 1000-MW(e) reactors would also be required.

*Air transport.* In the 1970s, the French government decided to build an electrified high-speed super train system to connect major metropolitan areas and reduce consumption of liquid fuels. That system has demonstrated that high-speed trains can replace air travel over distances of up to 500 miles because of lower costs, higher point-to-point speeds, and greater comfort. Simultaneously, rail stations have been built at major airports to provide point-to-point transport.

## 4. Biomass options

### 4.1 Biological resources

It is projected that by 2030 up to 30% of the liquid fuels consumed in the United States could be made from biomass[2,7] with an ultimate production capability twice as large. Long-term studies[8] indicate that biofuels could provide about 30% of the global demand in an environmentally acceptable way without impacting food production. Table 2 shows the estimated annual sustainable biomass production for the United States[2] to be about 1.3 billion dry tons.

Table 2.
Sustainable annual biomass availability for liquid fuel production in the United States

| Agriculture | | Forest Residues | |
|---|---|---|---|
| Source | Millions of dry tons | Source | Millions of dry tons |
| Crop residues | 428 | Manufacturing residue | 145 |
| Perennial crops | 377 | Logging debris | 64 |
| Grains to biofuels | 87 | Fuel reduction treatments | 60 |
| Process residues | 106 | Fuel wood | 52 |
| | | Urban wood waste | 47 |
| Total agriculture | 998 | Total forest | 368 |

Almost all of the biomass except "grains to biofuels" and some of the "process residues" are cellulosic feedstocks. The large biomass resources are crop residues (primarily corn stover) and perennials such as switch grass and poplar trees grown specifically for energy use on marginal lands. Today the primary biomass used to produce liquid fuels in the United States is corn (starch) that is converted to ethanol; however, as shown in Table 2, the resources of corn and other grains are limited. To make a major contribution to liquid fuel demands, biomass-to-fuels futures must be built on the much more abundant cellulosic feedstocks. Not included in this table are more advanced options such as growing algae.

Plants are complex, possessing six biomass components that are potentially important in production of liquid fuels.

*Cellulose.* Cellulose is the most common form of biomass and the principal constituent of wood. It is 40 to 60% of cellulose-rich feedstocks and is a biopolymer of glucose with a repeating unit of $C_6H_{10}O_5$. Hydrolysis can reduce cellulose to a cellobiose repeating unit ($C_{12}H_{22}O_{11}$) and ultimately to glucose ($C_6H_{12}O_6$). It is structured to be difficult to break down. This feature serves as a defense mechanism for plants, because only certain species of animals can digest cellulose.

*Hemicellulose.* Hemicellulose is the second most common form of biomass and typically compromises 20 to 40% of most biomass. It is also a sugar biopolymer. However, unlike the other sugars, it is a highly branched chain of five- and six-carbon sugars. Like cellulose, only certain animals have the capability to digest it.

*Lignin.* Lignin (Fig. 3), the component in cells that cements the cell structures together, is typically 10 to 25% of all lignocellulosic biomass. It is a highly polymeric, cross-linked aromatic structure with molecular weights near 10,000 and is derived primarily from coniferyl alcohol ($C_{10}H_{12}O_3$) via condensation polymerization. Lignin is believed to be the primary biological precursor to crude oil and because its molecular structure in many respects could be considered polymerized gasoline.

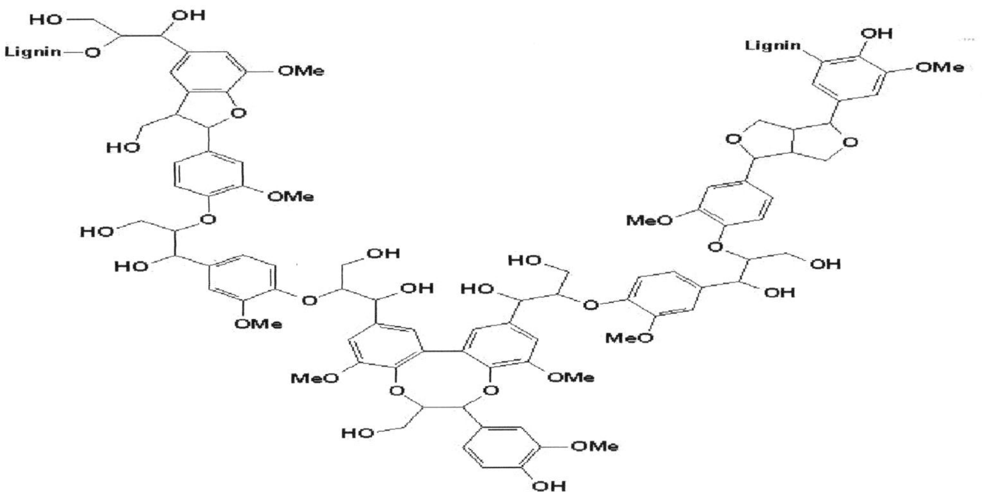

Fig. 3. Section of lignin polymer

*Monomeric sugars.* Sugarcane, sugar beets, and several other plants have high concentrations of monomeric sugars. Traditional fermentation can directly convert these simple sugars into alcohol. This is the primary method that has been used to produce alcohol for human consumption for thousands of years. It is also the basis of fuel ethanol production from sugarcane in Brazil and in a few other locations where the combination of land, labor, and climate provides favorable economic conditions. Glucose ($C_6H_{12}O_6$) is the most common monomeric sugar. However, the availability of these feedstocks is limited because they are also used for food.

*Starch.* Starch is a biopolymer of glucose, a monomeric sugar, with a repeating unit of $C_{12}H_{16}O_5$. It is the primary component of corn and other grains. Starch cannot be directly fermented to alcohol. An enzyme is required to break it down into simple sugars, which can then be fermented to alcohol. The resource availability of starch is an order of magnitude larger than that of monomeric sugars but is also constrained because starch is a food for humans and many farm animals.

*Oils.* A few plants, such as soybeans, produce oils—making them a biological fuel resource that requires little processing to be converted into a transport fuel.

All plants contain significant amounts of cellulose, hemicellulose, and lignin. Cellulose-rich feedstocks typically contain 40–60% cellulose, 20–40% hemicellulose, and 10–25% lignin. A few plants contain significant quantities of monomeric sugars, starches, or oils. Databases provide the composition of many types of biomass, including secondary biomass such as municipal wastes.[9] In areas with appropriate climate and soils, plants can be chosen to produce sugars and oils. However, those plants will also have significant quantities of cellulosic materials because of the central role of these materials in plant structures. Consequently, a major component in any biomass-to-fuel future will be composed of cellulosic feedstocks.

The use of biotechnology to increase the biomass yields of plants for fuels is at an early stage of development, as is the use of biotechnology to alter the plant contents. Major work is under way to develop plants with (1) higher biomass yields and (2) selected biomass constituents. Over a period of a few decades, dramatic increases in biomass production may be possible with plants optimized for fuel production. For example, the corn yield per acre in the United States has increased by a factor of 6 over 80 years.[10] However, the biotechnology for optimizing plants for liquid fuel production is still at a very early stage of development.

### 4.2 Biomass conversion options

Biomass can be converted to three classes of liquid fuels: ethanol, biodiesel, and hydrocarbon fuels. The conversion options and the energy requirements of the process facilities are described.

#### 4.2.1 Ethanol

The emphasis today in biomass-to-liquid-fuel production is on the production of fuel ethanol. The benefits of fuel ethanol have long been understood; however, the development of new biotechnologies is beginning to make this option technically and economically viable in large parts of the world. There are three primary feedstocks.

- *Monomeric sugars.* In Brazil sugarcane is the primary feedstock to produce ethanol from simple sugars. The sugarcane is squeezed to separate the sugar water from the cellulose-rich cane called bagasse. The sugar is then fermented to produce alcohol, and the bagasse is burned to provide the energy for the ethanol plant.

- *Starch.* In the United States corn is the primary carbohydrate that is converted to ethanol. Corn kernels contain carbohydrates (starch) and proteins. In the corn-to-ethanol process, enzymes convert the carbohydrates to sugar, which is then fermented to produce ethanol. About two-thirds of the corn kernel is starch that becomes ethanol. The nonfermentable components, which consist primarily of proteins and the by-products of fermentation, become animal food or are converted to other useful products. Because of the value of the protein as an animal food, these by-products are not burnt to produce energy to operate the plant. In most cases, natural gas is used to provide the energy to operate the plant. The rapid growth in fuel ethanol production is a consequence of three factors: (1) the development of low-cost enzymes to convert carbohydrates to sugars, (2) the need for an octane enhancer to replace MTBE, and (3) various incentives to encourage biomass to liquid fuel plants.

- *Cellulose and hemicellulose.* Because of feedstock availability, the primary source of fuel ethanol for the future is expected to be ethanol from cellulose and hemicellulose. Enzymes convert these feedstocks to sugars, which are then fermented to produce ethanol. A number of pilot plants are operating, the technology is advancing rapidly, and costs are decreasing. It is currently proposed to burn the lignin associated with the cellulose and hemicellulose to provide the energy to operate these plants.

The conversion of these feedstocks into ethanol is an energy-intensive process. Two forms of energy are consumed.

- *Fermentation.* The fermentation process converts sugars to ethanol using yeast. The yeast consumes sugar for energy and releases carbon dioxide in the process of converting the remaining sugars to ethanol.
- *Processing.* For the biomass to be converted to ethanol, the plant requires significant energy beyond that consumed internally in the fermentation process.

Consider the conversion of corn into ethanol—the primary production process for fuel ethanol in the United States. The non-solar-energy input to grow the corn and convert it to ethanol is typically about 70 to 80% of the energy value of the ethanol.[11] Most of that energy is supplied by burning fossil fuels. About one-half the energy input is in the form of low-temperature, low-pressure (150-psi) steam[12] used within the ethanol plant. The fermentation of sugars yields a mixture of water and alcohol. With corn, the mixture is typically >13% alcohol by volume. Above ~15%, the alcohol is toxic to the yeast. Distillation, an energy-intensive process using low-pressure steam, is required to separate the ethanol from the water. Smaller quantities of steam are required to sterilize the feed before fermentation and drying of various secondary products. After steam, the second-largest energy input for corn ethanol is fertilizer, particularly ammonia fertilizers.

Most U.S. ethanol plants use natural gas to produce steam and meet other energy needs. However, this process releases greenhouse gases and reduces the benefits of substituting ethanol for fossil fuels for liquid fuel production. The economics can be favorable because liquid fuels for transportation are more valuable than natural gas or other fuels. Ethanol production from all sources also consumes natural gas in the form of fertilizer. Natural gas is the feedstock for ammonia fertilizer production.

For ethanol plants based on corn, the fossil fuel consumption per gallon of ethanol can be cut in half by providing low temperature steam from existing or future nuclear reactors.[13-14] Because nuclear power plants are a non-greenhouse source of energy, this reduces the carbon dioxide releases from the entire corn-to-ethanol production process in half.

Nuclear power plants produce steam, which is then converted into electricity. Because ethanol plants require low-temperature, low-pressure steam, steam from the nuclear reactor would first go through high-pressure turbines to produce electricity and then be sent to the ethanol plant. In the ethanol plant, the steam would be condensed and warm water would be returned to the nuclear power plant. Almost all of the heat would come from condensing the steam. Modern steam systems would allow more than a mile of separation between the reactor and the ethanol plant; thus, the ethanol plants would be located beyond any security perimeter. The cost of such steam[13] is estimated to be \$3–\$4/million Btu, about half the cost of natural gas. The low cost is possible because the ethanol plant needs low-temperature steam; not the more valuable high-temperature steam.

Outside of the United States,[15] steam from nuclear plants has been used for district heating (45 reactors), desalting (10 reactors), and industrial purposes (25 reactors). However, this application of nuclear energy has not been used in the United States because of historically low natural gas prices and because nuclear plants were located in rural areas where there was no large demand for steam. With the rapidly increasing size of ethanol plants, the steam demands are now sufficiently large for the United States to consider using steam from nuclear plants for ethanol production. A large ethanol plant producing 100 million gallons of ethanol per year requires about 80 MW(t) of steam. The combination of increased natural gas prices, the growth in the size of ethanol plants, and the siting of ethanol plants in rural areas is now creating the option to use nuclear heat for ethanol production.

Current plans for conversion of cellulosic biomass to ethanol involve separation of the cellulosic components and the lignin components. Enzymes convert the cellulosic feedstocks to sugars, which are fermented to produce alcohol, while the lignin that can not be converted to ethanol is burned for energy. To maximize liquid fuel production per unit of biomass, the lignin must also be converted into liquid fuels. This can be done by using nuclear energy to provide the energy for ethanol production that is to be provided by burning lignin and than commercializing methods to convert the available lignin to liquid fuels.

Several processes to convert this lignin to hydrocarbon fuels are under development.[16-18] Because of the chemical structure of lignin (Fig. 3), the products are either high-value gasoline or high-value, high-octane (>100) oxygenates. Most of these processes require hydrogen for hydrotreating, a product that can be produced by nuclear reactors. If lignin is converted to liquid fuel and steam from nuclear plants replaces the lignin that was to be burnt as a fuel, the energy content of the liquid fuels per unit of biomass feedstock can potentially be increased by 50%, depending upon technological parameters and upon the specific biomass and its lignin content.

Use of nuclear energy for biomass-to-ethanol plants can reduce greenhouse gas releases in half and boost fuel production per unit of biomass by up to 50%. In this scenario, biomass would not be used as an energy source for biomass processing. The nuclear energy inputs into biomass-to-ethanol production would be 30 to 60% of the energy value of the liquid fuels that are produced. Most of the energy input is in the form of low-temperature heat; however, significant quantities would be hydrogen for conversion of lignin into gasoline and for use in the production of fertilizer. For ethanol from corn, the economics are favorable today. For ethanol from cellulose, the critical technology is development of methods to convert lignin to liquid fuels.

There are limitations on the use of nuclear energy to supply heat for liquid fuels from biomass. Biomass is expensive to transport. As a consequence, ethanol plants are located where biomass is available or where the by-products can be sold. This limits the potential sale of steam from nuclear plants to those plants near large sources of biomass or to river locations where low-cost barge transport may allow long-distance transport of biomass.

4.2.2. Biomass to hydrocarbon fuel

To maximize the energy value of the liquid fuel per unit of biomass, the final liquid fuel should be a hydrocarbon (Table 1). The energy value[16,19-20] of these liquid fuels per unit of biomass input is 3 to 4 times greater than that achieved by using biological processes to produce liquid fuels. This is a consequence of two factors.

- *Full carbon utilization.* All of the carbon is converted to fuel. None of the carbon is oxidized to carbon dioxide by the fermentation process used to produce ethanol.
- *High-energy fuel.* The biomass is fully converted into a hydrocarbon [$(CH_2)_n$] rather than into ethanol.

With these options, the energy and hydrogen content of the hydrocarbon liquid fuel are significantly greater than those of the initial biomass per carbon atom. This option requires significant external energy input in the forms of hydrogen, heat, and electricity. There are two technological approaches.

The first option is the Fisher-Tropsch process, which can convert all the carbon in the biomass to liquid fuels[20] when an outside source of hydrogen is supplied. This is the classical process for the conversion of fossil fuels to liquid fuels. There are multiple commercial Fisher-Tropsch plants that convert coal or natural gas to liquid fuels, primarily diesel fuel. It is a brute-force process that can process any carbon stream. The Fisher-Tropsch process has three major chemical reactions.

Oxidation of carbon: $$2C + O_2 \rightarrow 2CO$$

Water-gas-shift reaction: $$CO + H_2O \leftrightarrow CO_2 + H_2$$

Fischer-Tropsch: $(n/2 + m)H_2 + mCO \rightarrow C_mH_n + mH_2O$

Recent analysis[20] has provided estimated quantities of hydrogen required for conversion of biomass to diesel fuel. About 0.95 kg of hydrogen is required per gallon of diesel fuel produced. This assumes that heat is also supplied to the liquid fuel plant for drying of biomass (that is, hydrogen is not used to dry biomass). Large commercial electrolyzers[21] consume ~50 kWh per kilogram of hydrogen produced. Based on this analysis, ninety-two 1000-MW(e) nuclear reactors with a 90% capacity factor would be required to provide the hydrogen to produce 1 million barrels of diesel fuel per day from biomass. This assumes full utilization of the carbon in that biomass.

Alternatively, there is the potential to add hydrogen directly to biomass[22-23] to produce liquid fuels (hydrogenation). Sugar molecules are chains of carbon atoms similar to liquid fuels, except that many of the hydrogen atoms have been replaced by hydroxides (–O–H). Several processes, in the early stages of development, are being developed to replace the hydroxides with hydrogen atoms. If successfully developed, these processes have the potential for substantially lower costs than Fisher-Tropsch because (1) theoretically less hydrogen is required per unit of liquid fuel produced and (2) these alternative processes can potentially be implemented on a smaller scale.

For production of liquid fuels from biomass, plant size is a major issue. Biomass is bulky and heavy; thus, high costs are associated with its transport for any distance. For this reason, ethanol plants are distributed across the Midwest Corn Belt. A trade-off exists between the economics of scale for the biomass-to-fuel plants and the costs of biomass transport. The central requirement for both options is the need for hydrogen.

All of these estimates assume an average biomass. With a large-scale biomass-to-hydrocarbon fuel industry, there would be strong incentives to engineer the biomass to maximize hydrocarbon fuel yields by whatever process is used.

### 4.2.3 Biodiesel

The United States produces about 90,000 barrels per day of biodiesel from waste oils (cooking fat, etc.) and plant oils such as soybeans. With little processing, these oils can be converted to acceptable fuels.[11] The process energy input[24] is much less than that required for ethanol. However, the availability of such oils is limited. Oils are extracted from seeds; however, the biomass in the seeds after oil extraction is several times the quantity of oil. To maximize liquid fuel production, the biomass remaining after oil extraction will need to be converted to fuel by one of the processes outlined above. The development of plants to produce biodiesel provides a method to reduce process energy inputs per unit of liquid fuel, but does not alter the need for external energy for biomass processing to maximize liquid fuel production.

## 4.3. Biomass futures

The biomass is available to meet the current liquid fuel needs of the United States if outside sources of energy can provide the heat and hydrogen for the conversion processes. Ethanol production with lignin conversion to liquid fuels with limited quantities of hydrogen is the first step. The second step is the production of hydrocarbon fuels that requires massive quantities of hydrogen. At the same time, there are strong reasons to believe that the ultimate capacity to produce liquid fuels may be much larger as the biotechnology and processing technologies are developed.

### 5. Hydrogen-fueled vehicles

The long-term transport option is the direct use of hydrogen for transportation. The nuclear-hydrogen biomass futures described herein create much of the infrastructure (hydrogen production systems) and technology (PHEVs) required for direct use of hydrogen. As such, a nuclear hydrogen biomass transport fuel system is an enabling system for other hydrogen futures.

A nuclear hydrogen biomass system is likely to encourage the earlier deployment of hydrogen-fueled vehicles in niche markets such as municipal buses, ferries, and other such applications. Onboard hydrogen storage is the single largest technical challenge for hydrogen-fueled vehicles; however, these classes of vehicles have "excess" space and thus do not require development of high-performance storage systems. However, the economic viability of such applications is strongly dependent upon having the hydrogen supply infrastructure and associated technologies.

In some markets where it will be extremely difficult for any hydrogen fuel technology to penetrate. The classic example is air travel. Studies[25] show large penalties for hydrogen-fueled aircraft. Weight restrictions imply that hydrogen would be stored as liquid hydrogen. However, the low density of liquid hydrogen relative to that of other liquid fuels implies larger storage tanks. The larger storage tanks result in a larger aircraft, with large increases in air friction and the need for more fuel to overcome this air friction. For such markets, nuclear hydrogen biomass liquid fuels may be the preferred long-term fuel option.

### 6. Nuclear energy inputs

If biomass is to meet the nation's need for liquid fuels, large quantities of external heat and hydrogen are required for the process facilities. There are many choices of nuclear reactors and hydrogen production options (electrolysis, high-temperature electrolysis, thermochemical, etc.) to supply heat and hydrogen. For part of the transportation market, liquid fuels will be in competition with electricity used in rail transport and for PHEVs.

If there were a national consensus, the experiences of France indicate that a transition away from oil could be conducted within 20 to 30 years. France suffered a series of oil-related shocks during WWII, the Suez crisis, the Algerian war of independence, and the 1973 oil embargo. The entire French economy was based on oil, including the generation of electric power. This led to a national consensus in April of 1975 that the country had to cut oil consumption. That consensus led to three programs: (1) use of nuclear energy to generate most of the electricity, (2) an industrial energy conservation program, and (3) the electric-powered French super trains to replace aircraft for travel distances up to a few hundred miles. These programs were implemented within about 25 years with drastic reductions in French dependency on oil. A transition away from crude oil as a transport fuel in the United States would require a similar level of effort relative to the total economy. The difference is more technological options now exist to eliminate crude oil in our transport system. Furthermore, these options are better understood than was the strategy that the French undertook when France made the decision to launch a program to reduce oil dependence.

## 7. Conclusions

A domestic greenhouse-neutral nuclear hydrogen biomass system to fully replace oil in the transportation sector is technically feasible today and could be implemented over a period of several decades. Nuclear energy would be used to produce electricity, heat, and hydrogen. Some of the electricity would be directly used for transportation. The liquid fuels would be made from biomass, with the energy input for biomass processing provided by nuclear reactors. By using nuclear energy to support the energy-intensive processing of biomass, liquid fuel production per unit of biomass is maximized while greenhouse impacts are minimized. Components of this strategy are economic today.

The conclusions are robust. Liquid fuel demands for transport could be reduced in half by combinations of several options such as diesel engines and plug-in hybrids. Independently, the biomass liquid fuel options could meet existing liquid fuel demands without reductions in oil demand. Rapid technological changes are occurring with the development of biological plants for fuel production, methods to process biomass, and plug-in hybrid vehicles, as well as in other areas. Consequently, the specific combination of biomass, nuclear energy, and liquid fuels for transportation will be determined by the results of this development work.

## References


1. Perlack RD, Wright LL, Turhollow AF, Grahan RL, Stocks BJ, Erbach DC. *Biomass as Feedstock for a Bioenergy and Bioproducts Industry: The Technical Feasibility of a Billion-Ton Annual Supply*, DOE/GO-102995-2135, ORNL/TM-2005/66, U.S. Department of Energy, Washington, D.C., April 2005.
2. Marano JJ, Ciferno JP. *Life-Cycle Greenhouse-Gas Emissions Inventory for Fischer-Tropsch Fuels,* Energy and Environmental Solutions, LLC, for the U.S. Department of Energy National Energy Technology Laboratory, Pittsburgh, Pa., 2001.
3. Lahoda EJ, Forsberg CW, McLaughlin DF, "A Low-Greenhouse-Impact Hydrogen-Based Liquid-Fuels Future," *Proc. American Institute of Chemical Engineers Annual Meeting, San Francisco, California,* November 12–17, 2006.
4. United States Energy Information Administration, *Annual Energy Review*, 2006.
5. Uhrig RE. "Using Plug-in Hybrid Vehicles," *The Bent of Tau Beta Pi*, **XCVI** (2) pg 13-19 (Spring 2005).
6. Sanna L. "Driving the Solution: The Plug-in Hybrid Vehicle," *EPRI Journal* (Fall 2005).
7. Somerville C. "The Billion-Ton Biofuels Vision," *Science*, **312**, 1277 (June 2, 2006).
8. Koonin SE. "Getting Serious About Biofuels," *Science*, **311**, 435 (January 26, 2006).
9. U.S. Department of Energy, Energy Efficiency and Renewable Energy, "Biomass Feedstock Composition and Property Data Base," http://www1.eere.energy.gov/biomass/feedstock_databases.html.

10. U.S. Department of Agriculture, National Agricultural Statistics Service, "Trends in U.S. Agriculture," 2003, www.usda.gov/nass/pubs/trends/.
11. Hill J, Nelson E, Tilman D, Polasky S, Tiffany D. "Environmental, Economic, and Energetic Costs and Benefits of Biodiesel and Ethanol Biofuels," *Proceedings of the National Academy of Sciences of the United States of America*, **103** (30), 11206–11210 (July 25, 2006).
12. McAloon A, Taylor F, Yee W, Ibsen K, Wooley R, *Determining the Cost of Producing Ethanol from Corn Starch and Lignocellulosic Feedstocks*, NREL/TP-580-28893, National Renewable Energy Laboratory, Golden, Colo., October 2000.
13. Forsberg, CW, Rosenbloom S, and Black R, "Fuel Ethanol Production Using Nuclear-Plant Steam," *Proc. International Conference on Non-Electrical Applications of Nuclear Power: Seawater Desalting, Hydrogen Production, and Other Industrial Applications, Oarai, Japan, April 16–19, 2007*, IAEA-CN-152-47, International Atomic Energy Agency, Vienna, Austria, 2007.
14. Forsberg CW, Rosenbloom S, Black R, "Nuclear Steam for Fuel Ethanol Production from Corn," *Nuclear News*, **50** (3), 67–69 (March 2007).
15. International Atomic Energy Agency, *Market Potential for Non-Electric Applications of Nuclear Energy*, Technical Report Series No. 410, Vienna, Austria, 2002.
16. Ragausk AJ et al., "The Path Forward for Biofuels and Biomaterials," *Science*, **311**, 484–489 (January 27, 2006).
17. U.S. Department of Energy, Energy Efficiency and Renewable Energy, "Biomass Program: Lignin-Derived Co-Products," 2007, http://www1.eere.energy.gov/biomass/lignin_derived.html.
18. Montague L. *Lignin Process Design Confirmation and Capital Cost Evaluation*, NREL/SR-510-31579, National Renewable Energy Laboratory, Golden, Colo., October 2003.
19. Shinnar R, Citro F., "A Roadmap to U.S. Decarbonization," *Science*, **313**, 1243 (September 1, 2006).
20. Agrawal R, Singh NR, Ribeiro FH, and Delgass WN. "Sustainable Fuel for the Transport Sector," *Proceedings of the National Academy of Sciences of the United States of America*, **104** (12), 4828–4833 (March 20, 2007).
21. Norsk Hydro Electrolysers, Notodden, Norway, AS, www.hydroelectrolysers.com.
22. Robinson JM et al., "The Use of Catalytic Hydrogenation to Intercept Carbohydrates in a Dilute Acid Hydrolysis of Biomass to Effect a Clean Separation from Lignin," *Biomass and Bioenergy*, **26**, 473–483, 2004.
23. Anagho SG, Ngomo HM, Edwor JO. "Bituminous Oils from the Hydrogenation of Cellulosic Wastes: A Kinetic Study," *Energy Sources*, **26** (4), 415–425 (March 2004).
24. U.S. Department of Energy, Energy Efficiency and Renewable Energy, "Alternative Fuels Data Center," http://www.eere.energy.gov/afdc/altfuel/biodiesel.html.
25. Airbus Deutschland GmbH, *Liquid Hydrogen Fueled Aircraft—Systems Analysis*, European Community Competitive and Sustainable Growth Subprogram, September 23, 2003.

# 299b A Hybrid Sulfur Cycle Flowsheet Using A Pem Electrolyzer And A Bayonet-Type Decomposition Reactor

**Maximilian B. Gorensek**, Computational Science and Statistics Department, Savannah River National Laboratory, Aiken, SC 29808

Recent developments for the HyS cycle include new information on the performance characteristics of PEM electrolyzers and the availability of a bayonet-type decomposition reactor made out of silicon carbide. Flowsheets incorporating these changes have been prepared and their performance evaluated. The results will be presented and discussed.

# 299c Improved Process Flowsheet For The Hybrid Sulfur Process For Hydrogen Production

**Maximilian B. Gorensek**, Computational Science and Statistics Department, Savannah River National Laboratory, Aiken, SC 29808, David F. McLaughlin, Science and Technology, Westinghouse Electric, LLC, 1344 Beulah Road, Pittsburgh, PA 15235-5083, William A. Summers, Hydrogen Technology Center, Savannah River National Laboratory, Savannah River Site, 773-42A/137, Aiken, SC 29808, and Edward J. Lahoda, Science and Technology Department, Westinghouse Electric, LLC, 1344 Beulah Road, Pittsburgh, PA 15235-5083.

Westinghouse Electric Company (WEC) and the Savannah River National Laboratory (SRNL) have independently developed different process flowsheets for the hybrid sulfur process for hydrogen generation using an advanced nuclear reactor heat source (such as the pebble-bed modular reactor, or PBMR). The two efforts have differed in design philosophy, the process simulators used to develop the heat and material balances (CHEMCAD™ and Aspen Plus™, respectively), and to a lesser extent the basic thermodynamic data underlying the simulations. A collaboration is now underway between the two groups to simplify and optimize the two designs, and to consolidate them into a single reference flowsheet. The revised flowsheet attempts to optimize unit operation cost, fabricability, and maintainability (e.g. minimizing pressure drops in very high temperature components); provide the electrolysis units with a uniform sulfuric acid feed concentration (to optimize electrolyzer performance and minimize development requirements); match assumed electrolysis conversion efficiency with the most recent experimental data from SRNL; rationalize differences in thermochemical data; minimize the total number of unit operations (for cost optimization); and maximize SO2 recovery following sulfuric acid decomposition by modifying system operating pressures.

# 299d Process Design And Economics For The Hybrid Sulfur Process

**William A. Summers**, Hydrogen Technology Center, Savannah River National Laboratory, Savannah River Site, 773-42A/137, Aiken, SC 29808

The Hybrid Sulfur Process is a leading candidate for producing hydrogen by thermochemical water-splitting using heat from advanced nuclear reactors. This paper will discuss the process flowsheet, major equipment and hydrogen production economics for this cycle.

# 299e S-I Process Economics Using H2A Analysis

**Robert T. Buckingham**, General Atomics, 3550 General Atomics Court, San Diego, CA 92121

An economic analysis of the S-I process using extractive distillation has been conducted. The H2A framework as outlined by the US Department of Energy has been used as a basis for all economic assumptions and calculations. Influences on the plant cost of engineering materials used in the process will be discussed, along with a review of major factors affecting the cost of hydrogen.

# Improving the Efficiency of the Sulphur Iodine Thermochemical Cycle for Hydrogen Production Using Membrane Separations

*R. H. Elder, G. H. Priestman and R.W.K. Allen*
*Department of Chemical and Process Engineering, University of Sheffield, Mappin St, Sheffield. S1 3JD UK*

The sulphur iodine thermochemical cycle is a promising method for hydrogen production; however current calculations of process efficiency give low results leaving scope for improvement. This paper investigates the application of membrane separations to the $HI_x$ processing stage of the cycle, with a view to improving the efficiency. Simulations have been performed, indicating the potential benefits of dewatering the $HI_x$ prior to the reactive distillation column. The membrane performance requirements have also been investigated. An experimental programme has been carried out, assessing the applicability of Nafion® membranes for water separation by pervaporation. Flux and selectivity data are presented for HI and $HI_x$ solutions, showing the feasibility and practicability of applying the membrane. Experimental flux measurements as a function of temperature, feed concentration and permeate side pressure are reported. The fluxes obtained demonstrate that the separation is feasible; though a higher flux is desirable to reduce the membrane area required. Separation factors are found to be extremely high, with almost pure water permeating through the membrane.

## Introduction

The sulphur iodine cycle has attracted much international interest over the last decade. Research into ways of improving the overall cycle efficiency has been centred in the USA, Japan and Europe. A diagram of the sulphur iodine cycle is given in Figure 1. The third section, where the HI solution is decomposed to form hydrogen and iodine, dominates efficiency calculations as it has a low rate and a high energy demand. The solution entering the third section is not pure HI, but a mixture of HI, $H_2O$ and $I_2$ in the molar ratio 1:5.1:3.9, termed $HI_x$. This solution needs to be separated in order for the HI to be broken down. The presence of an azeotrope, the slow decomposition rate of HI and the lack of accurate thermodynamic data makes this separation very difficult. Current flowsheets employ either extractive distillation using phosphoric acid, or reactive distillation in a column; however the overall process efficiency remains low at around 38%. Work at JAEA in Japan has focussed on using electroelectrodialysis and a hydrogen permselective membrane reactor (Nomura et al., 2005) to improve the efficiency. Results are promising and a test loop is under construction (Kubo et al., 2004). This paper focuses on using pervaporation to remove water from the $HI_x$ solution. Simulations are presented which show the potential impact of using such a membrane separation process, followed by experimental results which indicate the feasibility of membrane application. Over the last couple of years researchers at INL (Orme and Stewart, 2007, Stewart and Orme, 2005) and JAEA/KIER (Choi et al., 2006)} have also investigated the pervaporation of HI solutions. Results presented here are compared to those obtained elsewhere.

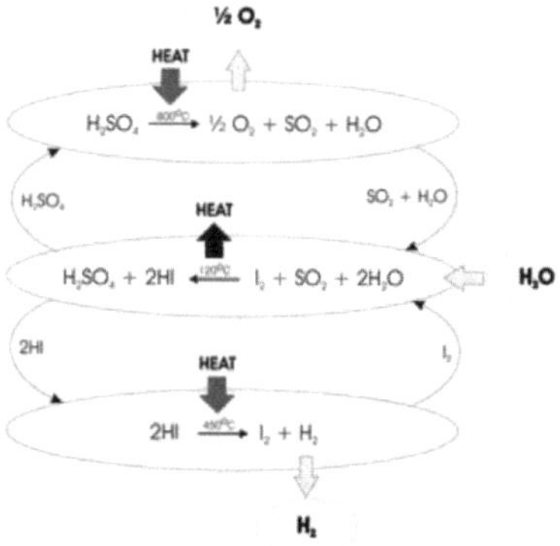


Figure 1: The sulphur iodine cycle

## Simulations

Simulations have been carried out using ProSimPlus to investigate the effect of applying a membrane separation unit to the $HI_x$ processing section. It was previously reported that applying a membrane separation unit to the column reboiler or a column sidestream has negligible effect on efficiency (Elder et al., 2005). For this reason only feed dewatering is considered here. The basic flowsheet used includes the pumping power required by the vacuum pump and energy integration and is described elsewhere (Elder, 2007). The simulation results are presented in Figure 2.

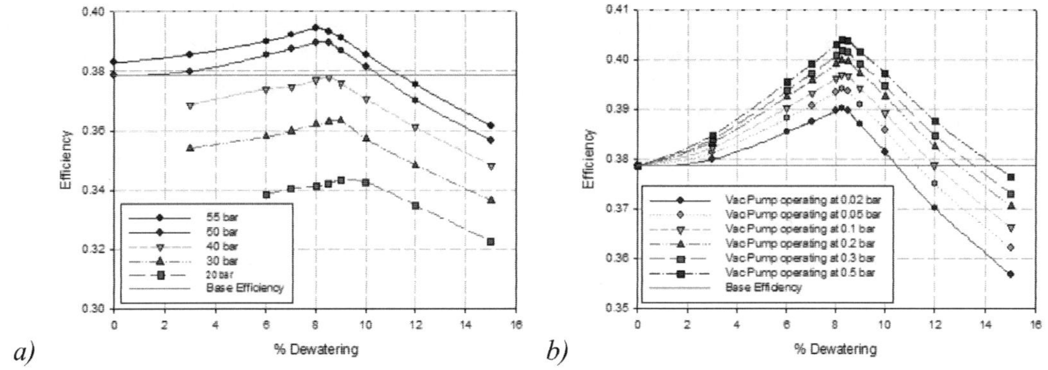


Figure 2: ProSimPlus simulation results *a)* the effect of changing column pressure and *b)* the effect of permeate side pressure

Figure 2a shows that applying a membrane separation unit to the feed of a 50 bar distillation column can increase the process efficiency from a base level of 0.379 to a maximum of 0.390 for 8.25% dewatering. The maximum occurs due to a trade off of the different energy requirements of the system, the most significant being an increase in the vacuum pump power as more water is removed, and a decrease in the energy required after heat integration up to 8.25% dewatering after which it increases slightly. Increasing the column pressure is seen to increase efficiency. This is because the decomposition of HI is modelled as an equilibrium reaction in ProSimPlus. As the pressure changes, the equilibrium of the reaction shifts. At lower column pressures less decomposition of HI occurs and so less hydrogen is produced. However, at lower column pressure the pumping requirements are less and the mechanical and structural considerations are easier. Note that for low dewatering values at low column pressures the column does not solve in ProSimPlus. The reason for this is not obvious and further stability testing is required. It could be due to the presence of two liquid phases, something which ProSimPlus cannot cope with, or with deficiencies within the thermodynamic package which lead to the solution being inaccurately modelled. The thermodynamic model used needs to be improved in order to get results across the whole spectrum. The maximum pressure investigated is 55 bar. At pressures greater than this the system becomes unstable and, once again, the column does not solve.

Figure 2b shows the effect of changing the vacuum pump pressure and therefore the permeate side pressure. As the pressure increases, that is less suction is applied, the vacuum pump uses less energy and the efficiency increases. However, the driving force for pervaporation is also reduced. The largest effect is seen at low pressures as the energy used by the pump is proportional to $\ln(1/P_p)$. With a permeate side pressure of 0.02 bar the maximum efficiency observed is 0.390, whereas for a pressure of 0.2 bar the efficiency increases to 0.400, and for 0.5 bar the efficiency is 0.404. The permeate side pressure used therefore has a significant effect on efficiency when dewatering is employed. To determine the optimum conditions requires data on the way in which the membrane flux varies with the applied suction.

## Experimental Programme

The experimental programme was designed to first investigate the stability of candidate membranes in HI solutions and then carry out flux and selectivity testing with those deemed suitable. A discussion of the stability testing can be found elsewhere (Elder, 2007). Nafion® membranes, which are commonly used in pervaporation systems, where found to be stable in azeotropic HI solutions. Both Nafion® 117 and Nafion® 212, obtained from Ion Power, are tested in the flux and selectivity testing described here. The two membranes vary only in thickness. Before use, the membranes are prepared by immersion in 10 wt% nitric acid at 40°C for 16 hours, followed by immersion in deionised water for 8 hours. This is then repeated. The membranes are stored in deionised water to ensure they are swollen when put into the apparatus.

The aim of the flux and selectivity testing is to establish if the achievable fluxes are high enough to enable a practical membrane area, and to establish how much, if any, impurity there is in the permeate. Selectivity factors of greater than 500 can be regarded as giving almost pure water and satisfy process requirements. Anything higher than this is a bonus, but is unnecessary. A specially designed rig is used for the experiments, which, due to the nature of the $HI_x$ solution, has all wetted surfaces made from PTFE or glass. A schematic of the system is given in Figure 3.

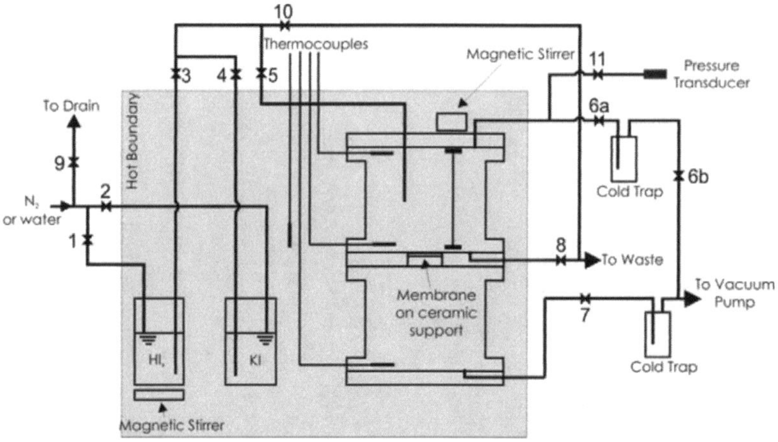


Figure 3: Schematic of the flux and selectivity testing set up

The rig is designed so that both online diagnostics, in this case Coherent Anti stokes Raman Spectroscopy (CARS), and offline diagnostics can be used to monitor the selectivity. The results presented in this paper use offline titrations with sodium thiosulphate and sodium hydroxide to analyse the permeate samples. CARS is planned for future experiments. During experiments the rig is placed in an oven at 120°C. This allows the Bunsen section exit conditions to be emulated as closely as possible. Experiments are run for 6 to 8 hours, with permeate samples being taken every 2 hours. The duration of experiments is limited by the time available in a day as safety constraints prevent experiments being left unattended overnight. The rig is designed with two chambers, separated by the membrane, a vacuum being applied to the bottom chamber to drive pervaporation. The plates are made from solid PTFE, whereas the two chamber sections are machined from Stainless steel and are PTFE coated (Fluorocarbon Bakeware Systems, Sheffield). The coating is renewed approximately every 5 weeks as small amounts of iodine absorb into the thin layer of PTFE and cause it to bubble off. Both the top and bottom compartments have windows to enable laser access for CARS. The exposed membrane area is $1.257 \times 10^{-3}$ $m^2$. For each experiment the flux and separation factor are calculated. The separation factor is defined in equation 1, where y and x corresponds to permeate and feed side mole fractions respectively.

$$\text{Separation factor} = \frac{y_{H_2O} / y_{HI}}{x_{H_2O} / x_{HI}} \qquad (1)$$

The effect of temperature, feed concentration, permeate side pressure and membrane thickness are all investigated.

## Experimental Results

Figure 4 shows the fluxes obtained when a Nafion® 117 membrane is used at different temperatures with an azeotropic feed. As the experiment proceeds the fluxes decrease slightly as the feed becomes more concentrated. As expected, the higher the temperature, the higher the flux. Literature data from INL (Stewart and Orme 2005, Stewart *et al.* 2007) and JAEA/KIER (Choi *et al.* 2006) are also plotted on the graph to allow a comparison. The experimental values obtained are significantly higher than those found by other researchers, although they used less suction pressure. The experiments carried out here used a permeate side pressure of 20 mbar. At INL (Orme et al., 2005, Stewart et al., 2007) fluxes of 0.07 $kg/m^2h$ at 60°C and 0.15 $kg/m^2h$ at 109°C were found using a permeate side pressure of 133mbar. At 25°C the flux observed of 0.024 $kg/m^2h$ is slightly lower than that observed at KIER of 0.04 $kg/m^2h$ (Choi et al., 2006). This difference, however, is much smaller than in the cases at higher temperature. It is also more difficult to collect samples at room temperature when the flux is so low, which makes any errors more significant than at higher temperatures.

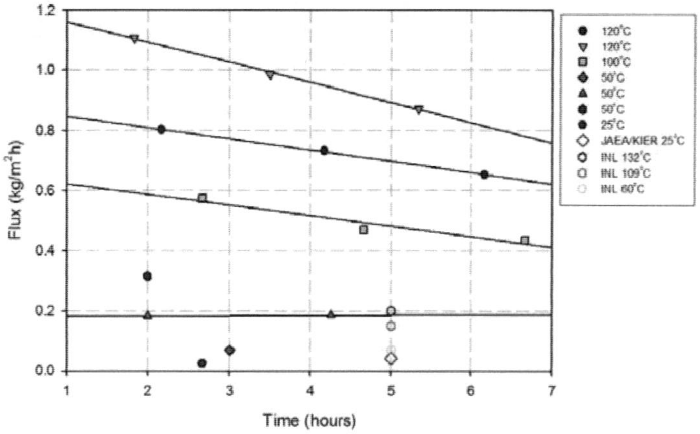


Figure 4: Variation of flux with temperature using an azeotropic feed solution

Figure 5 shows the average fluxes and separation factors obtained when using a Nafion® 117 membrane with different feed compositions. Each point plotted is the average of three samples taken during the experiment. The variation in flux during an experiment is shown by the error bars. The red squares on the flux graph denote experiments conducted at 100°C for 2.8 mol%, 9 mol% and 15.7 mol% HI (the azeotrope). To make the weaker solutions, water is added to the azeotrope. As the water concentration increases above the azeotrope, the flux increases dramatically. The green circles denote experiments carried out at 120°C. These experiments use the azeotropic solution with increasing amounts of iodine added until the $HI_x$ concentration of 39 mol% iodine is reached. These were the first experiments to be carried out using such a concentrated $HI_x$ solution in a membrane application.

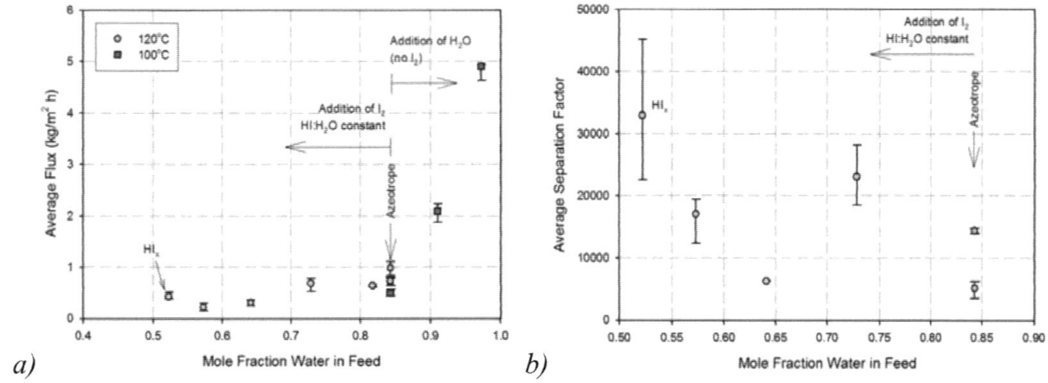


Figure 5: *a)* Fluxes and *b)* separation factors obtained for different feed compositions. The points plotted are average values with the error bars denoting the maximum and minimum values observed.

Basic models suggest that the flux is driven by the difference in species chemical potential across a given membrane. For a given temperature and membrane, ignoring any variation in molecular species interactions, this can be approximated by the concentration difference. In these experiments there is a vacuum on the downstream side of the membrane, so the concentration difference can be approximated to the feed concentration. The relevant flux and water concentration values are given in Table 1 and it is seen that both the flux and the water content of the feed are approximately 1.6 times lower for the $HI_x$ solution than for the azeotrope. This also suggests that concentration polarisation next to the membrane has not been a limiting factor, indicating that the stirring in the rig is effective.

Table 1: Fluxes obtained for different solution concentrations

| Feed | Composition mol% water | Flux $kg/m^2h$ |
|---|---|---|
| Azeotrope | 84 | 0.72 |
| $HI_x$ | 52 | 0.43 |

To the right of the azeotrope, the flux increases significantly more than the concentration probably corresponding to the non-linear behaviour in chemical potential for the mixture. The separation factors observed for $HI_x$ solutions containing iodine show extreme variation, both within a single experiment and between experiments. However, they are all extremely high, greater than 5000. This shows that the permeate is almost pure water. A separation factor of 5000 equates to less than $10^{-3}$% of the HI being removed in the permeate. The large variation in separation factors is due to differences in very small numbers obtained during the titrations. There are no equivalent results with such significant iodine available for comparison in the literature. Orme and Stewart (2007) used a solution with a $HI:H_2O:I_2$ ratio of 1:11:1 at 134°C. Under these conditions the flux seen was 0.7 $kg/m^2h$ and the separation factor was around 5000. The flux is similar to the ones observed above, despite the fact that significantly more water is present in the feed, suggesting that the fluxes should be higher. The results observed in these experiments are very promising and show that the azeotrope in the $HI_x$ solution can be broken using pervaporation.

As the fluxes obtained using azeotropic HI solution and $HI_x$ solutions were very similar, the experiments to investigate the effect of permeate pressure were carried out using the azeotropic HI solution and not $HI_x$. This makes the experimental conditions much easier. Figure 6a shows the flux results obtained when using a Nafion® 117 membrane with different permeate side pressures. The points plotted are average values from the samples taken, with the error bars denoting the maximum and minimum values observed. The top chamber of the rig is at approximately one bar for all experiments, so the pressure differential across the membrane is 1 bar minus the permeate side pressure. Each point plotted is the average of three samples taken during the experiment. Initially it was expected that the flux would vary significantly at low pressures. Experiments were carried out between 0.02 bar and 0.1 bar. The results showed that the flux is relatively constant within this region. Experiments at higher permeate side pressures showed that the flux does eventually decrease with increasing pressure, meaning that to get the same percent of dewatering, the membrane area would need to be larger. Whilst some of the flux values during a single experiment show very little variation, as shown by the short error bars, others show much larger variation. This could be due to small amounts of permeate remaining in the tubes before the sample bottle. Heating of the tubes was introduced to prevent condensation, but some still occurs. The average flux over a whole experiment is therefore a much more representative value than any single sample flux observed.

The two points with squares round them correspond to the points with separation factors less than 100 shown in Figure 6b, also shown with squares round them. The slightly higher flux and much lower separation factor suggest that a trace leak may have been present around the membrane during these experiments. A separation factor of 100 corresponds to around $10^{-2}$g HI in the permeate

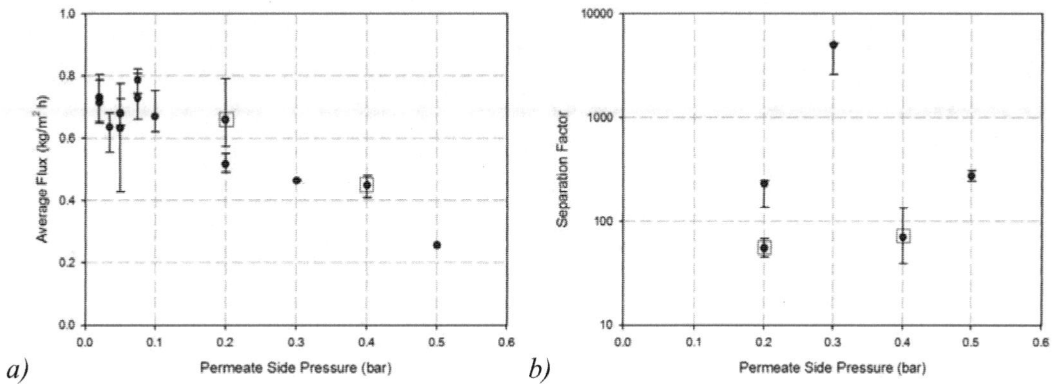


Figure 6: *a)* Fluxes and *b)* separation factors obtained using a Nafion® 117 membrane for varying permeate side pressures with an azeotropic HI feed. The points plotted are average values with the error bars denoting the maximum and minimum values observed. The points with squares round them are explained above.

over the course of a 2 hour sample, as opposed to $10^{-4}$g for a separation factor of 6000. The amount of feed solution that would need to leak round the membrane to affect the results in this way would therefore be very small, in the order of $10^{-2}$g HI azeotrope. This is roughly equivalent to a flux of 0.02 kg/m$^2$h. This is very small in comparison to the fluxes measured. The separation factors vary greatly across the range of permeate pressure tested, although are generally very high. Again, this is due to differences in very small numbers. Referring back to Figure 2, simulations showed that the process efficiency increases with increasing permeate side pressure. A trade-off clearly exists between efficiency and membrane area, as membrane area also increases with increasing permeate side pressure. A decision as to the optimal conditions will rely on cost and space considerations and is beyond the scope of this work. It appears that operating around 0.2 bar will be the most beneficial as the efficiency is significantly higher than for 0.02 bar whereas the flux is only slightly lower.

The effect of permeate side pressure on performance of the thinner Nafion® 212 membranes was also investigated. The results are shown in Figure 7, along with the fluxes obtained for Nafion® 117 to allow easy comparison. The results for Nafion® 212 show only a small deviation between the maximum and minimum fluxes observed, as seen by the small error bars. As expected, the flux decreases as the permeate side pressure increases. The fluxes obtained using Nafion® 212 membranes are significantly higher than those found with Nafion® 117 membranes. The relevant values when the permeate side pressure is 0.02 bar are given in Table 2. The flux using the thinner Nafion® 212 is approximately twice that found using Nafion® 117. Nafion® 212 is around three times thinner than Nafion® 117. Simple models predict that the flux would be three times larger. The difference could be due to the simplicity of the model, or the absorption and desorption limiting the transport as opposed to the path length through the membrane. Note also that with the thinner membrane the permeate pressure is seen to have a more continuous effect on flux, with no evidence of a constant flux at the lowest pressures.

Table 2: Fluxes obtained for different thickness membranes using a permeate pressure of 0.02 bar

| Membrane | Thickness $\mu$m | Flux kg/m$^2$h |
|---|---|---|
| Nafion® 212 | 51 | 1.34 |
| Nafion® 117 | 178 | 0.73 |

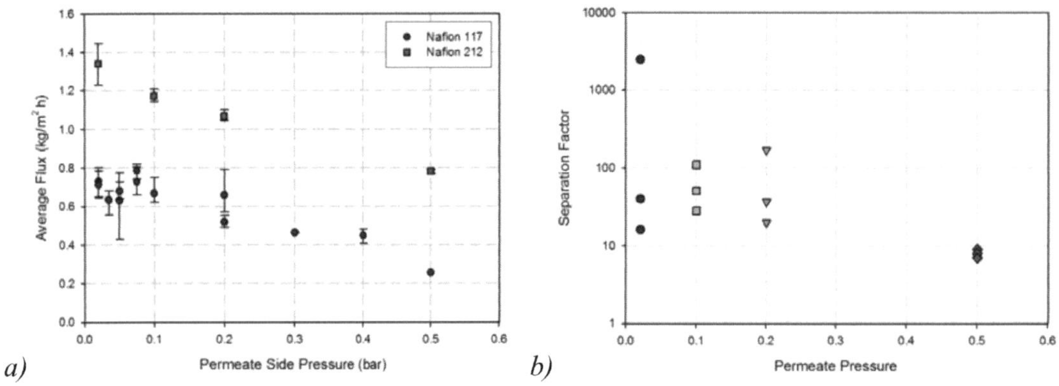


Figure 7: *a)* Fluxes and *b)* separation factors obtained using a Nafion® 212 membrane. The points plotted are average values with the error bars denoting the maximum and minimum observed.

Orme and Stewart (2007) found that using a Nafion® 112 membrane and a feed solution with a $HI:H_2O:I_2$ ratio of 1:11:1 at 134°C gave a flux of 2.5 kg/m$^2$h. Nafion® 112 has the same composition as Nafion® 212 but was made using a different process and is no longer available commercially. In that work the flux is just over three times greater than that observed with Nafion® 117. It is possible that the difference in feed solution composition causes different sorption and so accounts for the difference in the results. Further experiments need to be carried out to investigate the relationship between thickness and flux. As the flux is higher, and the separation factors are still fairly high, using Nafion® 212 is likely to be more beneficial for the SI process as the membrane area required will be lower for the same process efficiency. To date no data has been obtained for Nafion® 212 membranes with a $HI_x$ feed. Experiments using $HI_x$ solutions are significantly more difficult to carry out than those with no iodine due to major problems with iodine aggressiveness and solidification in the tubes.

The separation factor results shown in Figure 7 are very interesting. Each individual separation factor measured is plotted. Extreme variation is found between the separation factors measured during a single experiment despite the care taken in sealing the membrane. The experiment carried out at a permeate pressure of 0.02 bar shows separation factors from 16 to 2500. The reason for this is unknown, but further investigation is clearly needed so that repeatable results are obtained. As discussed earlier, separation factors around 100 are equivalent to $10^{-2}$g HI being present in the permeate. A sample weighs around 2g, so the contamination is negligible and the flux results are valid with only a small error.

Table 3: Summary of important results

| | Nafion® 117 | | Nafion® 212 | |
|---|---|---|---|---|
| | Flux (kg/m$^2$h) | Sep. factor | Flux (kg/m$^2$h) | Sep. factor |
| HI azeotrope | 0.7 | ~1000 | 1.35 | ~100 |
| $HI_x$ | 0.45 | ~30000 | ? | ? |

Table 3 shows a summary of the important results which are relevant to the operation of the SI cycle. To achieve 8.25% dewatering and a product flowrate of 1 mol/s hydrogen, approximately 2.5 mol/s water must pass through the membrane. Using the values given for $HI_x$ and a Nafion® 117 membrane, 360 m$^2$ of membrane would be needed. This is a very large area for a small amount of hydrogen. Assuming that the flux ratio of the HI azeotrope to $HI_x$ is the same for Nafion® 212 as Nafion® 117 gives a predicted flux of 0.87 kg/m$^2$h for a $HI_x$ feed and a Nafion® 212 membrane. This equates to a membrane area of 186 m$^2$ . This is still a large membrane area, but is a significant decrease in size compared to using Nafion® 117. Further experiments should be carried out to investigate determine the $HI_x$ flux through Nafion® 212 membranes. Ideally, a membrane would be found which retains the high selectivity found in the experiments carried out here, but which has a higher flux.

## Conclusions

Simulations show that dewatering of the feed stream to the reactive distillation column in the $HI_x$ processing section of the SI cycle can have a significant impact on the process efficiency. Using 8.25% dewatering could increase the process efficiency from 0.379 to 0.390 using a permeate side pressure of 0.02 bar, or 0.400 using 0.2 bar.

Experiments show that with HI and $HI_x$ feed solutions, water is transported almost exclusively across Nafion® membranes. Fluxes obtained are moderate, the highest seen being 1.35 $kg/m^2h$ when using a Nafion® 212 membrane with an azeotropic feed and a permeate pressure of 0.02 bar. Predicted membrane areas to achieve maximum efficiency are high and the challenge now is to investigate ways to increase the flux through the membrane whilst retaining the high separation factors so that the required membrane area decreases.

## Acknowledgements

The authors would like to acknowledge EPSRC for funding under the DTA scheme, the KNOO programme and the HYTHEC project, funded by the EC Framework 6 programme.

## References

Choi, H. S., Hong, S.-D., Hwang, G. J., Park, C.-S., Bae, K.-K. & Onuki, K. (2006) Dehydration through pervaporation from $HI_x$ solution ($HI$-$H_2O$-$I_2$ mixture) using a cation exchange membrane for thermochemical water-splitting iodine-sulfur process. *Korean Journal of Chemical Engineering,* 23**,** 288-291.

Elder, R. H. (2007) Improving the Efficiency of the Sulphur Iodine Cycle using Membrane Separations. *Dept of Chemical and Process Engineering, University of Sheffield.* PhD Thesis

Elder, R. H., Borgard, J. M., Priestman, G. H., Ewan, B. C. & Allen, R. W. K. (2005) Use of Membranes and Reactive Distillation for the Separation of $HI_x$ in the Sulphur-Iodine Cycle. *AIChE Annual Meeting.* Cincinnati.

Kubo, S., Nakajima, H., Kasahara, S., Higashi, S., Masaki, T., Abe, H. & Onuki, K. (2004) A demonstration study on a closed-cycle hydrogen production by the thermochemical water-splitting iodine - Sulfur process. *Nuclear Engineering and Design,* 233, 347-354.

Nomura, M., Kasahara, S., Okuda, H. & Nakao, S. I. (2005) Evaluation of the IS Process Featuring Membrane Techniques by Total Thermal Efficiency. *International Journal of Hydrogen Energy,* 30**,** 1465-1473.

Orme, C. J., Jones, M. G. & Stewart, F. F. (2005) Pervaporation of water from Aqueous HI using Nafion-117 Membranes for the Sulfur-Iodine Thermochemical Water Splitting Process. *Journal of Membrane Science,* 252**,** 245-252.

Orme, C. J. & Stewart, F. F. (2007) Pervaporation of Water from Aqueous Hydriodic Acid and Iodine Mixtures Using Nafion® Membranes. *Journal of Membrane Science,* 304**,** 156-162.

Stewart, F. F. & Orme, C. J. (2005) Stability and Performance of Nafion-117 Membranes for the concentration of HI/Water and HI/Water/Iodine mIxtures. *AIChE Annual Meeting.* Cincinnati, OH, AIChE.

Stewart, F. F., Orme, C. J. & Jones, M. G. (2007) Membrane Processes for the Sulfur-Iodine Thermochemical Cycle. *International Journal of Hydrogen Energy,* 32**,** 457-462.

Evolution of the Sulfur-iodine Flowsheet, 1977-2007

Lloyd C. Brown, General Atomics

The Sulfur-Iodine watersplitting process was invented at General Atomics in the mid 1970's. The process development proceeded through a combination of laboratory experiments and chemical process flowsheeting until 1984, when funding was eliminated, in the US, in the aftermath of the oil crisis. Our investigation was resumed in 1999, first on the basis of improved flowsheet calculations. In 2004 we resumed chemical investigations aimed at the production of a significant amount of hydrogen under engineering conditions.

The sulfur-iodine process consists of three chemical reactions:

| | | |
|---|---|---|
| [1] | $2\ H_2O(\ell_1) + SO_2(g) + I_2(\ell_2) \rightarrow H_2SO_4(\ell_1) + 2\ HI(\ell_2)$ | {75-150°C} |
| [2] | $H_2SO_4 \rightarrow H_2O + SO_2 + \frac{1}{2}\ O_2$ | {750-1000°C} |
| [3] | $2\ HI \rightarrow H_2 + I_2$ | {350-450°C} |

The viability of the process is based on the fact that, in the presence of a large excess of iodine, the two acids formed in reaction 1 (the Bunsen reaction), are concentrated in two immiscible liquid phases. The sulfuric acid phase is nominally contains 4 or 5 moles of water per mole of sulfuric acid. The hydriodic acid phase contains about 4 moles of iodine and 5 moles of water per mole of hydrogen iodide. This phase, which may contain polyiodides ($HI_3$, $HI_5$, . . .), is commonly called $HI_X$.

The process flowsheet is usually divided into three sections corresponding to these three chemical reactions and the process steps associated with the chemistry of that reaction. The history of the process development effort involved trade-offs between process complexity, with accompanying high efficiency and high capital costs, against process simplicity with low efficiency and capital costs.

Section 1 includes the chemical reaction making two acids in two separate phases, the removal of the exothermic heat of the reaction, the recovery and reuse of that heat, and the removal of any sulfur species from the hydriodic acid phase. Also, the oxygen is cleaned up and vented from Section 1. The separation of iodine from the sulfuric acid phase may take place in Section 1 or in Section 2. Major questions in Section 1 are the reactor configuration (co-current or countercurrent), are the $SO_2$ and $O_2$ separated by liquefaction of the $SO_2$ water absorption or reactive scrubbing.

Section 2 is where the high temperature heat enters the process and decomposes the sulfuric acid. The major other unit operations in Section 2 deal with the recuperation of the heat and reuse of heat for low temperature processes, including concentration of the sulfuric action prior to reaction.

Section 2 appears simple, on the surface, but it is probably the most complicated of the three sections. This is because the second liquid phase incorporates a lot of iodine and water. The equilibrium of reaction 3 tends to the left so excess iodine must be removed. Hydrogen iodide that is bound to water is not free to react so the water must be separated as well. Alternate means of separating the HI include use of phosphoric acid to extract the water and hydriodic acid from the excess iodine coupled with distillation to separate the HI from the water-phosphoric acid mix (extractive distillation) and a potentially attractive process whereby the HI decomposes via reactive distillation from the $HI_X$ at high pressure. The third alternative, complete vaporization of the $HI_X$ and decomposition of the HI at very high temperatures, while demonstratably feasible, would lead to a very low overall efficiency.

Over the years there have been a number of published flowsheets. The major flowsheets have been:

1979 - Emphasis on high efficiency (hand calculations)
- countercurrent Bunsen reactor with reactive separation of $SO_2/O_2$ and Freon power bottoming cycle using waste heat
- multistage countercurrent flash evaporation concentration to azeotropic sulfuric acid
- extractive distillation with 4 stage countercurrent vapor recompression concentration of phosphoric acid.

1981 – Emphasis on reduced capital cost (hand calculations)
- Co-current Bunsen reactor with reactive separation of SO2/O2 and no low temperature power bottoming cycle
- Azeotropic sulfuric acid recovery via a combination of high-pressure flash evaporation of water from dilute acid coupled with vacuum distillation of concentrated acid
- Extractive distillation with 3 stage countercurrent vapor recompression concentration of phosphoric acid.

2002 – Goal of improving efficiency while lowering capital cost (combination of hand calculations and modern computational thermodynamic models)
- co-current Bunsen reactor with no low temperature power bottoming cycle
- azeotropic sulfuric acid recovery via a combination of high-pressure flash evaporation of water from dilute acid coupled with vacuum distillation of concentrated acid
- reactive distillation decomposition of HI

2005 – Improved efficiency and high pressure H2SO4 decomposer matched to high pressure helium loop. (combination of hand calculations and modern computational thermodynamic models)
- co-current Bunsen reactor with no low temperature power bottoming cycle
- sulfuric acid decomposer operating at 70 bar with minimal concentration of sulfuric acid
- reactive distillation decomposition of HI

2007 – Simplification and process demonstration . (combination of hand calculations and modern computational thermodynamic models)
- countercurrent Bunsen reactor with SO2/O2 separation via SO2 liquefaction and no power bottoming cycle
- single unit “bayonet” type integral sulfuric acid decomposer and concentrator
- extractive distillation with gas phase decomposition of HI

Future – Optimized flowsheet (completely using modern computational thermodynamic models

# SULPHUR-IODINE PLANT FOR LARGE SCALE HYDROGEN PRODUCTION BY NUCLEAR POWER

***Giovanni CERRI*, Coriolano SALVINI, Claudio CORGNALE, Ambra GIOVANNELLI** - *Università degli studi – Roma Tre (DIMI – I) (i) / cerri@uniroma3.it*
***Daniel de LORENZO MANZANO, Alfredo ORDEN MARTINEZ*** – *Empresarios Agrupados (EA – SP) (ii) / ddm@empre.es*
***Alain LE DUIGOU, Jean-Marc BORGARD, François WERKOFF*** - *Commissariat à l'Energie Atomique (CEA – F) (iii) / alain.le-duigou@cea.fr*

## Abstract

The objective of HYTHEC project - HYdrogen THErmochemical Cycles - is to investigate the effective potential of the Sulfur-Iodine (S_I) thermo-chemical cycle for massive hydrogen production, and to perform a comparison between the S_I cycle and the Hybrid-Sulfur (HyS) one. The project aims to establish models for flow-sheeting, industrial scale-up, safety and cost evaluations. Moreover experiments on fundamental thermochemical steps of both of the cycles have been carried out. S_I cycle reference flow-sheets have been proposed. First data and results on the coupling of S_I cycle with a Very High Temperature Nuclear Reactor (VHTR), the scale-up to industrial level and cost estimation have been obtained. In order to support a high temperature aggressive environment, the reference case took into account both well established chemical engineering methods and suitable materials and devices. A parametric analysis by varying the hydrogen production level showed that an optimized $H_2$ production cost exists. Due to relevant heat recovery processes the minimum cost is not achieved for the maximum allowable $H_2$ production level.

**Keywords:** HYTHEC, hydrogen production, thermo-chemical, Sulphur Iodine cycle, nuclear energy, costs.

## Introduction

Today, hydrogen is used for various purposes and is mainly produced from fossil resources. In the long term, given the prospect of an increasing energy demand (+20% by 2020, expected to double by 2030, with a possible threefold increase by 2050), a lack of fossil resources and limitations on the release of greenhouse gases, renewable and nuclear sources will play a more and more relevant role. $H_2$ as an intermediate artificial energy carrier can solve storage problems related to the use of nuclear and renewable energy sources. Thus it becomes more and more important with time depletion of fossil resources, which intrinsically store sun energy as chemical energy.

Biomass is a long term candidate raw material for hydrogen production by various processes (e.g. gasification, pirolysis, etc). Water contains $H_2$ (11wt%) and can represent another raw material which can be split by adding power. Energy stored in the produced hydrogen (fuel) can be released by an oxidation (combustion) process that produces water again. In this way hydrogen will play a relevant role both in storage of energy and in power production as an artificial energy carrier. The two processes that have the greatest likelihood of successful massive hydrogen production from water are electrolysis and thermo-chemical cycles. The latter are processes where water is decomposed into hydrogen and oxygen via chemical reactions using intermediate elements which are recycled. As the heat can be directly used, these cycles have the potential of a better efficiency than alkaline electrolysis. The required heat can be either provided by nuclear plants or by solar ones. HYTHEC project takes into consideration nuclear energy as the primary heat source to feed the S_I cycle.

## S_I Thermo-Chemical Cycle background

The Sulfur-Iodine (S_I) cycle which is schematically sketched in fig. 1, was extensively studied by the General Atomics Company [1]. Japan has recently built a small pilot plant of this process [2]. Accordingly, the S_I cycle seems to be the best known, internationally leading candidate, as a very promising thermo-chemical option. The major chemical reaction stages involved in the S_I cycle may be summarized as follows:

R1 - $9\,I_2 + SO_2 + 16\,H_2O \rightarrow (2HI + 10H_2O + 8I_2) + (H_2SO_4 + 4H_2O)$ [120°C] (1)

which represents the iodidric and sulphuric acid production step;

R2 - $2\,HI \rightarrow H_2 + I_2$ [220-330°C] (2)

where the iodidric acid is split into $H_2$ and $I_2$;

R3 - $H_2SO_4 \rightarrow SO_2 + H_2O + \frac{1}{2}\,O_2$ [850°C] (3)

which represents the $H_2SO_4$ decomposition step.

Globally, the sum gives the decomposition of water into hydrogen and oxygen:

R4 $H_2O \rightarrow H_2 + \frac{1}{2}\,O_2$ (4)

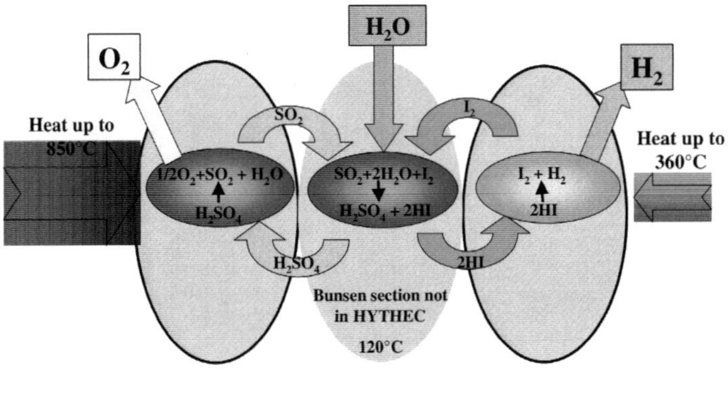


**Figure 1**. The Sulphur_Iodine (S_I) cycle

The first reaction, called the Bunsen-reaction, proceeds exothermically in the liquid phase and produces two immiscible aqueous acid phases whose compositions are aqueous sulfuric acid and a mixture of hydrogen iodide, iodine and water named $HI_x$. These acids are concentrated and the excesses of water and iodine are recycled in R1. HI and $H_2SO_4$ are then decomposed according to reactions R2 and R3. Only high temperature heat sources, such as nuclear VHTR or solar energy concentrators, may be relevant for the reaction R3. Reaction R2 is the HI decomposition step with a little endothermic heat of reaction. Reaction R3 is the major endothermic reaction releasing water, oxygen and sulfur dioxide. It takes place in the vapor phase in a catalytic reactor at about 900°C. The concentration by distillation of the two acids HI and $H_2SO_4$ involves significant heat consumption, which has a direct influence on the efficiency of the cycle. Among all options available for the $HI_x$ section (extractive distillation using phosphoric acid, electrodialysis and so on), HYTHEC is being focused on the reactive distillation concept as proposed by Knoche et al. [3]. Their approach allows this step to be done in one reactor so it seems to have the highest efficiency potential. An improved version has already been proposed in [4].

The scale up study for the feasibility of an industrial SI is of great relevance, for the assessments of safety aspects of the process, the manufacturing of the main components at industrial scale, and $H_2$ production costs. Specific R&D needs may arise. For this purpose, the cycle coupling scheme has been modeled, and taken as a basis for heat and electricity duty calculations. After that plant components have been sized and a first, but nevertheless accurate, overall $H_2$ production cost analysis has been accomplished.

Aims and first results of the overall HYTHEC project are described in a IJHE recent publication [5].

## Industrial plant design

One of the goals of HYTHEC project is to assess the potentialities of S_I thermo-chemical cycle to achieve a cost effective massive hydrogen production by nuclear thermal source. In order to select suitable plant design solutions the following main aspects have to be taken into consideration:

- definition of the process in terms of required unit operations and related process quantities (pressures, temperatures, compositions, heat and work requirements and so on);
- coupling between the nuclear thermal source and the S_I process;
- set up of the industrial plant lay out in order to find the most effective arrangement to perform the required operations;
- selection of appropriate technology and materials for each plant component;
- sizing and cost accounting of components;
- estimation of $H_2$ production overall specific costs.

The plant design is achieved when the values of a set of quantities (degrees of freedom, DOF) are established. All the other quantities are obtained by solving the plant model, which can be also split into modules, each concerning a single component or a group of components. Conceptually, a plant model containing all the above aspects can be established and the design problem solved by adopting a global optimization procedure aimed at the evaluation of DOF's which minimize an objective function (i.e hydrogen production costs). In practice such an approach is really challenging due to the complexity of the plant and the great number of DOF's involved. To simplify the design problem, DOF's related to the different aspects have been separately taken into consideration according to the following.

### *S_I thermo-chemical cycle reference flowsheet*

A review of previous flowsheets published in the literature has been undertaken. The heat transfer optimization has been performed using ProSim code. Internal heat recovery has been studied taking a minimum temperature difference between hot and cold streams equal to 10°C.

The $H_2SO_4$ section is handled with a multiple flash distillation, followed by $H_2SO_4$ evaporation and $SO_3$ decomposition. $SO_3$ is recombined in the late stages of the distillation process to optimize the global efficiency. The $HI_x$ section is handled by a reactive distillation process as proposed in [4]. To enhance the heat recovery a water heat pump has been adopted to transfer the heat rejected by the column condenser to the reboiler. The flash at the exit of the column is needed to recover some of the latent heat of water steam. The flow-sheets of the $H_2SO_4$ and $HI_x$ sections are given in Figs 2 and 3. The Bunsen section flowsheet has been taken according to General Atomics report [6]. The external source thermal power required by the plant results in 477 kW/mol$H_2$.

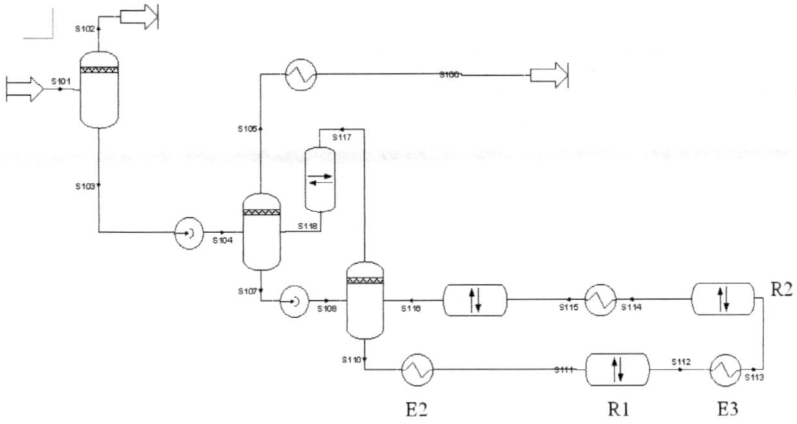


**Figure 2**. $H_2SO_4$ section flow-sheet

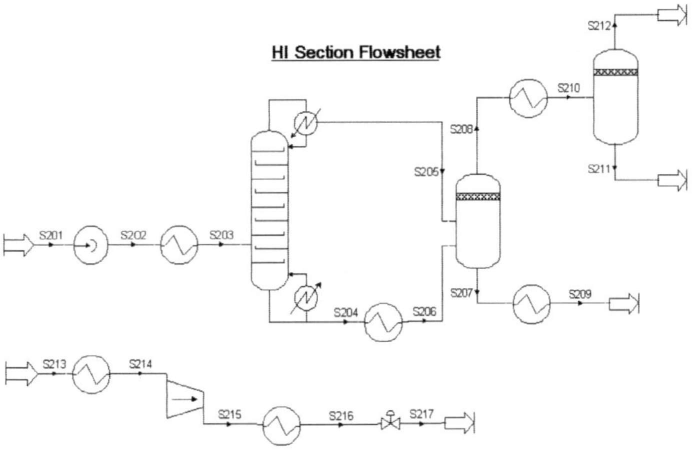


**Figure 3.** HIx section flow-sheet

***S_I cycle and nuclear reactor coupling***

The coupling of the S_I cycle to a Nuclear Reactor has been studied. Due to the high temperatures needed for the $H_2SO_4$ decomposition, the best connection option is with a VHTR. Fig. 4 shows a connection scheme as an example, between HYTHEC and the European Project RAPHAEL. This scheme represents a self-sustainable plant concept, in which, in addition to the heat supply to the S_I cycle, the electrical demand of the internal consumers is provided by the nuclear reactor. Then, the high temperature flow coming out of the reactor is derived to an Intermediate Heat eXchanger (IHX). This IHX provides heat to a secondary loop that interacts with the S_I cycle components, improving heat recovery. The flow goes partially to the S_I cycle and another part goes to a Brayton cycle for an electricity production that equals the S_I cycle consumption.

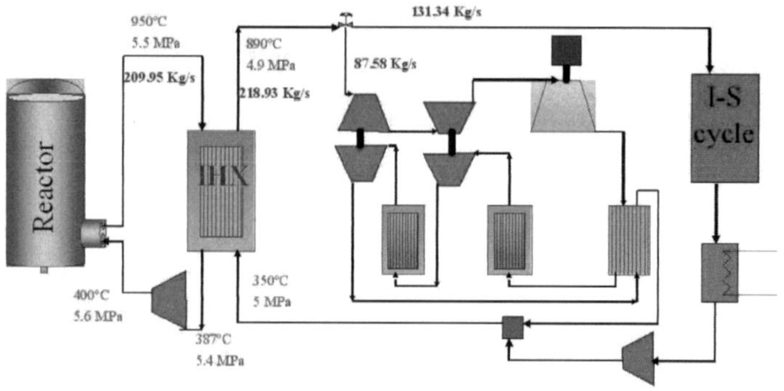


| | |
|---|---|
| Nuclear Power reactor [MW] | 600 |
| H2 production | 100% |
| IHX Inlet / Outlet Helium temperature [°C] | 350/890 |
| IHX Inlet / Outlet S_I temperature [°C] | 890/442 |
| Helium mass flow rate [kg/s] | 218.93 |
| Helium pressure [MPa] | 5 |

**Figure 4.** S_I Cycle coupling to a Nuclear Reactor

***Industrial plant lay out, technology and materials***

The $H_2SO_4$ section plant layout (fig. 5) has been established according to the corresponding flowsheet shown in fig. 2. The first two flash units (Flash1 and Flash2) are arranged as kettle pool boilers. The third flash (Flash3) is an adiabatic reactive vapour-liquid separator tank. The heat needed to vaporise the concentrated $H_2SO_4$ solution is supplied by helium and by the stream exiting the high temperature reactor R2, in two shell & tubes heat exchangers in parallel arrangement (E2). A large amount of $H_2SO_4$ is then dehydrated in the R1 reactor. The remaining is decomposed and heated in the E3 reactive heat exchanger. $H_2SO_4$ is practically all decomposed into $SO_3$ before entering R2. The reactor R1 as well as the heat exchanger E3 are sized as shell & tubes heat exchangers, as the de-hydration reaction is reasonably assumed instantaneous, without catalyst. The $SO_3$ decomposition reactor was assumed to be tubular with reactants flowing inside the tubes internally coated by catalysts. Both tubes and shell of this reactor are made of Incoloy. Heat exchanger tubes were assumed to be made of ceramic materials (SiC), as SiC tubes show an excellent thermal conductivity and seem to have a good behaviour also in highly aggressive environments. Tanks, as well as heat exchanger shells, are made of CS, internally covered by acid brick liners (ABL) that act as corrosion resistant materials.

The $HI_x$ section layout has been set up according to the reference flow-sheet and is schematically depicted in fig. 6.

The entering flow is preliminarily heated by streams exiting the first flash unit (Flash1) and is further heated up by the liquid exiting the distillation column. To reach suitable thermo-chemical conditions for the distillation process, a further amount of heat is supplied by external thermal source (Helium). All the $HI_x$ section heat exchangers are arranged as shell & tubes ones. Flash1 has been assumed to be a shell & tubes condenser followed by an accumulator-separator tank and Flash2 is a vapour-liquid separator tank, with gaseous hydrogen separation from liquid mixture. Distillation column sieve plates were assumed to be made of Monel, coated by catalysts on the surface. Tanks and shell&tube heat exchangers materials have been selected following the same criteria adopted for $H_2SO_4$ section.

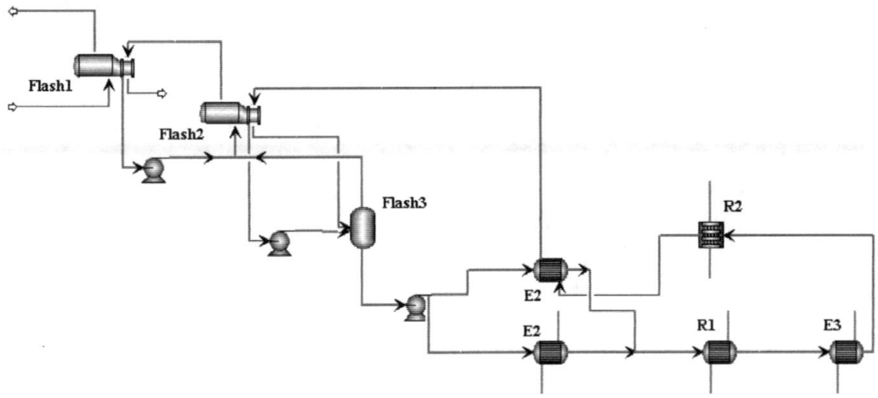


**Figure 5**. Simplified industrial $H_2SO_4$ section plant layout (helium circuit is in red line)

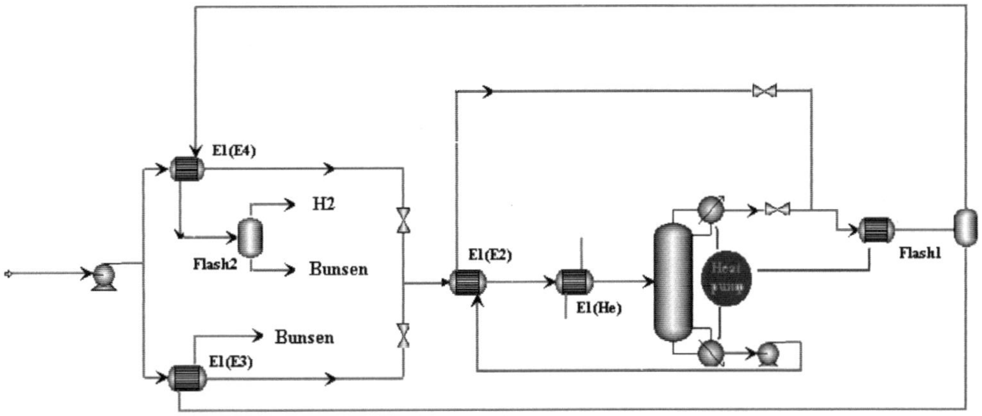


**Figure 6**. Simplified industrial HIx section plant layout (helium circuit is in red line)

***Sizing of components and cost evaluation***

Flow-sheet data and the selected nuclear reactor size (600 MW) allow a hydrogen production level of 633 mol/s (the reference case). The sizing of components has been carried out by adopting the methods briefly described in the following.

Tanks (accumulator, vapour-liquid separators and so on) have been sized using the residence time method (also called liquid phase surge time, $t_s$ method) [7]. Shell & Tubes heat exchangers have been sized by the LMTD method according to TEMA standards. Various kinds of issues (e.g. pressure drops, vibration and resonance and so on) have been considered and solved by adopting suitable kinds of baffles, inlet and outlet nozzles sizes, tubes diameters or tube pitches and so on. Distillation column has been sized, by adopting the nested design method and Fair flooding calculation method [8]. The high temperature reactor R2, for the $SO_3$ decomposition into $SO_2$, has been preliminary sized as tubular reactor, using the kinetics law reported in [9] by CHEMKIN code. Pumps and compressors have been sized according to standard engineering methods [10].

Constraints on sizes of heat transfer devices and the adoption of good engineering practices have led to the following results. The $H_2SO_4$ concentration subsection is arranged on one line, while the dehydration subsection and the $SO_3$ decomposition one are arranged on two parallel lines. Six parallel lines are required for the $HI_x$ section and each heat pump is equipped with two compressors.

More than 850MW have to be internally recovered within the $HI_x$ section with low temperature differences (10 °C) between hot fluids and cold ones. This leads to increased heat exchanger surfaces.

The most challenging component is the E1/E3 heat exchanger due to its remarkable heat transfer surfaces (i.e. relevant unit number). Relatively to the $H_2SO_4$ section the most important component is the high temperature $SO_3$ decomposition reactor R2. Another prominent result of sizing activity refers to the needed electric power being, for the reference case, approximately 86 MW, more than 85% of which is due to heat pump compressor units.

To evaluate costs of the S_I plant equipment a factored approach has been adopted. The equipment base costs (FOB costs) are evaluated, taking temperature, pressure, material and specific design suitable factors into consideration. The method also modifies basic equipment costs through the use of "adders" accounting for piping, concrete, instruments and so on. Further additional costs (labour and indirect costs) are evaluated by introducing appropriate cost factors and, finally, total equipment installed costs are estimated.

To evaluate the total investment cost, standard factors have been used to take further aspects (e.g. contingency, fees, auxiliary facilities, inventory, interest during construction) into account. These evaluations have been performed taking values available in literature ([10]) and by means of "ad hoc" developed databases.

The most expensive section (85% of total installed costs) is the $HI_x$ one. The internal recovery heat exchangers costs represent approximately 65% of the overall $HI_x$ section. The cost of chemicals (almost exclusively iodine) deserves its prominence. The major amount of the iodine inventory is within internal recovery heat exchangers and within the distillation column. This cost reaches almost 20% of the overall $HI_x$ section installed cost. The total investment cost of the whole S_I plant was evaluated as approximately 970 M€.

To evaluate the $H_2$ production cost, an annual operation scenario has been assumed as reference. Among various cost items taken into consideration, maintenance & repairs cost and taxes have been taken as 4% and 2% of the fixed capital investment respectively. A plant availability of 90% has been assumed, along with a plant lifetime of 30 years, a discount rate of 0.06 and a nuclear power cost of 1.2 c€/kWh$_{th}$. At the end a specific hydrogen cost $C_s$ of approximately 5.3 €/kg$_{H2}$ has been obtained. Total capital investment influence on the overall costs is approximately 37% and is the most relevant voice, although cost related to the power generated by the VHTR are significant (30%) as well. The remaining is substantially due to maintenance and taxes.

## Parametric techno-economic analysis

In order to explore how lowering the hydrogen production cost, a parametric analysis has been carried out by varying the production level keeping the overall available thermal power $Q_{TH}$ at the reference value (600 MW) and maintaining the self sustaining plant concept. The plant thermal power balance is expressed by:

$$Q_{TH} = Q_{TH,S} + Q_{TH,HI} + Q_{TH,EL} \qquad (5)$$

being $Q_{TH,S}$ the thermal power required by the $H_2SO_4$ section, $Q_{TH,HI}$ that required by the $HI_x$ section and $Q_{TH,EL}$ the quote absorbed to generate the electric power. By reducing the hydrogen production level respect to the reference case $Q_{TH,S}$ decreases, being fixed the heat from helium required to produce 1 mol of hydrogen. $Q_{TH,EL}$ decreases too, being the electrical power demand related to the hydrogen production level. As a result a higher amount of thermal power is available to feed the HI section. An augmented $Q_{TH,HI}$ implies the reduction of the thermal power internally recovered and the raise of the mean temperature difference between hot and cold fluid in heat recovery devices. As stated in the previous paragraph, such devices represent the most important installed cost items. Thus a reduction of the required heat transfer surfaces is expected to bring relevant cost benefits. Conversely, by lowering the production level the specific heat consumption raises with an increase of costs related to the power demand per kg of produced $H_2$.

Following up with the considerations above, several plants for different hydrogen production levels have been analysed. These plants have been sized in detail and hydrogen production costs have been evaluated following rules and assumptions previously established.

Results are given in fig. 7. The minimum cost doesn't correspond to the maximum allowable $H_2$ production level (corresponding to the maximum plant efficiency) due to the relevance of capital investment costs. The minimum cost is observed for a production level of 540 $molH_2/s$, with a cost reduction of more than 20% relative to the reference plant.

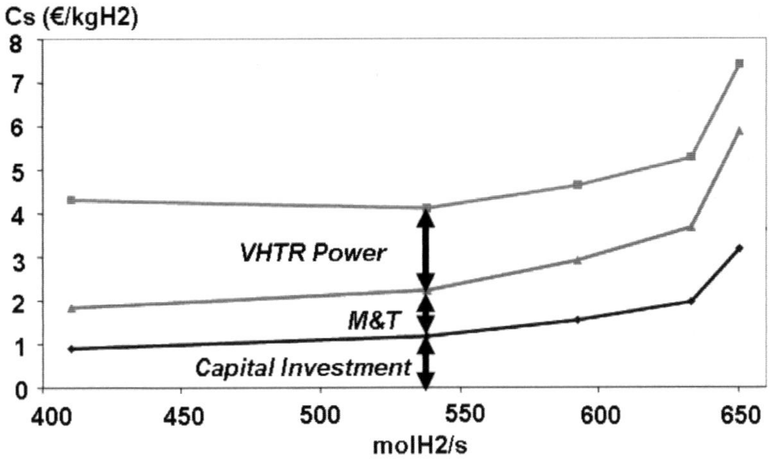


**Figure 7**. H2 specific costs vs. H2 production

For low $H_2$ production levels a flat cost curve is observed (fig. 7): the increasing costs related to the needed power are balanced by the decreasing of capital, maintenance and taxes costs.

Tab. 1 shows a comparison on installed costs as well as total capital investment, between the reference plant and the minimum cost one. The 540mol/s plant installed costs are dramatically reduced if compared with the reference plant ones, especially for the $HI_x$ section. $HI_x$ installed costs at 540 mol/s represent nearly the 75% of the total installed cost and their influence on production cost is still relevant. Therefore to achieve further reduction, technology improvements concerning the $HI_x$ section recovery heat exchangers have been introduced. To enhance heat transfer coefficients bare tubes have been replaced by corrugated ones and costs evaluated at different hydrogen production levels.

**Table1.** Cost comparison between the reference S_I plant and the minimum cost one

| | 633 mol/s | 540mol/s |
|---|---|---|
| $H_2SO_4$ installed costs [M€] | 84 | 66 |
| Bunsen installed costs [M€] | 20 | 18 |
| HIx installed costs [M€] | 624 | 268 |
| Total installed costs [M€] | 728 | 352 |
| Total capital investment [M€] | 970 | 470 |

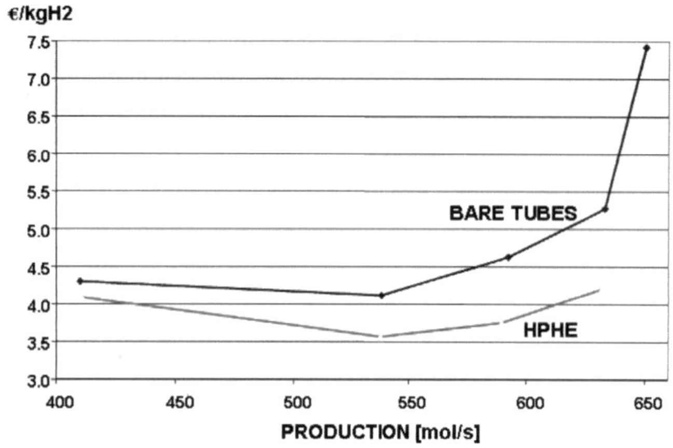


**Figure 8**. Hydrogen production costs: HPHE and Bare Tubes comparison

In fig. 8 results achieved by adopting corrugated tubes (HPHE, High Performance Heat Exchangers) are reported together with those already given in fig. 7 (Bare Tubes).

A reduction of costs is observed along all the production range, especially at high $H_2$ production levels, and the minimum is still achieved at 540 mol/s with a noticeable 40% cost reduction relative to the reference case.

Finally a further cost improvement has been achieved by searching an optimised arrangement of the $HI_x$ section heat exchangers network by adopting the pinch technique.

Results achieved for a hydrogen production level of 540 mol/s are given in fig. 9 and compared with those obtained at the same production level for the reference plant lay out with bare tubes and HPHE. The adoption of an optimum lay out lead to a production cost of some 3.4 €/kg$H_2$.

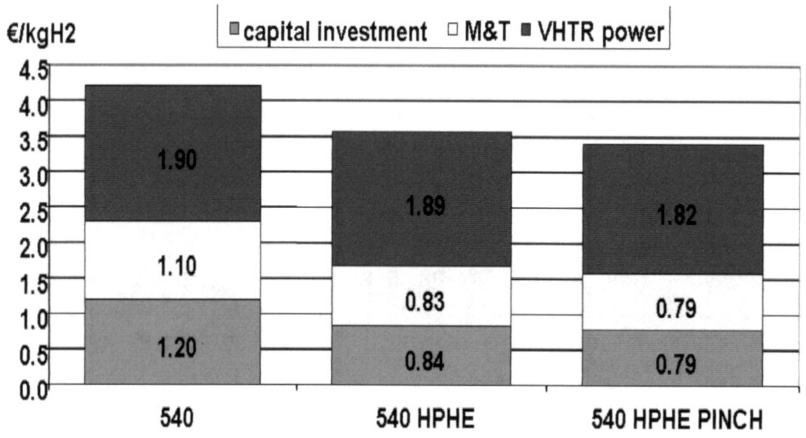


**Figure 9**. Contributions to hydrogen production costs at 540 mol/s: bare tubes, HPHE, HPHE and optimized plant lay out

## Conclusions

A preliminary techno-economic analysis has been carried out to investigate the potential for massive hydrogen production by S_I thermo-chemical cycle coupled with a 600 MW nuclear plant. A flow-sheet aimed at the achievement of the maximum efficiency has been developed and assumed as reference. The coupling with the nuclear reactor allowed a hydrogen production of 633 mol/s. After plant component sizing, production costs have been estimated. A parametric analysis, carried out by varying the hydrogen production level, has shown that the minimum cost does not correspond to the maximum achievable efficiency. Further cost decreases have been obtained by introducing other technologies about heat transfer devices, replacing traditional ones, and by optimizing the plant lay out for selected $H_2$ production levels. Cost analysis results are particularly useful to identify which aspects most influence the whole production cost. Future activities are going to be focused on the possibility of using less expensive materials, adopting compact heat exchangers and improving flow-sheet arrangement to reduce thermal as well as electrical power demand.

## References


1. G.E.Besenbruch and coll – GA-A18257 (1982)
2. H. Nakajima, K. Ikenoya, K. Onuki et al. – "Closed-Cycle Continuous Hydrogen Production Test by Thermochemical S_I Process", Kagaku Kogaku Ronbunshu, 24 (1998) 352 (in Japanese)
3. M.Roth, K.F. Knoche – Thermo-chemical water splitting through direct HI decomposition from $H_2O$ -HI- $I_2$ solutions - Int. J. Hydrogen Energy, 14, N°8, 545-549 (1989).
4. Stephen Goldstein, Jean-Marc Borgard and Xavier Vitart - Upper bound and best estimate of the efficiency of the iodine sulphur cycle - International Journal of Hydrogen Energy, Volume 30, Issue 6, May 2005, Pages 619-626
5. A.Le Duigou et.al. – "HYTHEC: An EC funded search for a long term massive hydrogen production route using solar and nucelar technologies", Int. J. Hydrogen Energy, 10-11, N°32, 1516-1529 (2007).
6. L.C. Brown, et al. – « High efficiency Generation of Hydrogen Fuels using nuclear power » Final technical Report, 1999-2002, Grant N.DE-FG03-99SF21888 US.DOE
7. R.N. Watkins – "Sizing separators and accumulators" Hydrocarbon Proceedings, 46, 11, 252, 1967
8. J.R. Fair – "Distillation, Handbook of Separation Process Technology", John Wiley, NY, 1987
9. S. Spewock, L.E. Brecher, F.Talko – "The thermal catalytic decomposition of sulfur trioxide to sulfur dioxide and oxygen" First WHEC, Miami Beach, March, 1976
10. E.Douglas – "Industrial Chemical Process Design" McGraw-Hill Professional Engineering, 2003

# Flow sheet Evaluations For The Thermochemical Water-Splitting Iodine-Sulfur Process (II)

*Shinji Kubo, Yoshiyuki Imai, Hirofumi Ohashi, Seiji Kasahara, Nobuyuki Tanaka, Hiroyuki Okuda and Kaoru Onuki*
*Japan Atomic Energy Agency*
*Narita-cho 4002, Oarai-machi, Higashi Ibaraki-gun, Ibaraki-ken, Japan*

## INTRODUCTION

Concerning a heat application system for the high temperature gas-cooled reactors (HTGRs), to achieve high level of thermal efficiency is main issue for IS process [1], so that we has proposed a preconcentration method using a membrane technology, electro-electrodialysis [2], and mass and heat balances using the electro-electrodialysis (EED) was reported [3]. In this system, while almost pure HI gas can be obtained from the top of the distillation column, to maintain operating conditions for a HI decomposition unit became relatively difficult because vapor pressure of gaseous HI is so high that $H_2$ in the undecomposed gaseous HI is not easy to separate; and the product of $I_2$ can be lead to solidification on heat transfer surfaces in low-temperature separation procedure. In this study, we explored measures to remove the $I_2$ before the separation procedure with minimum amount of energy losses.

## HI VAPOR PRESSURE

Figure 1 shows view of HI vapor pressure on HI-$H_2O$ solution. In high concentration region, the partial pressures of HI gas are high. In case that $H_2$ separation is performed by the low-temperature separation, measurable HI is entrained. Accordingly, $H_2$ in the undecomposed gaseous HI is not easy to separate. To perform good separation, low HI concentration of solution, low temperature and high pressure operation are required. This

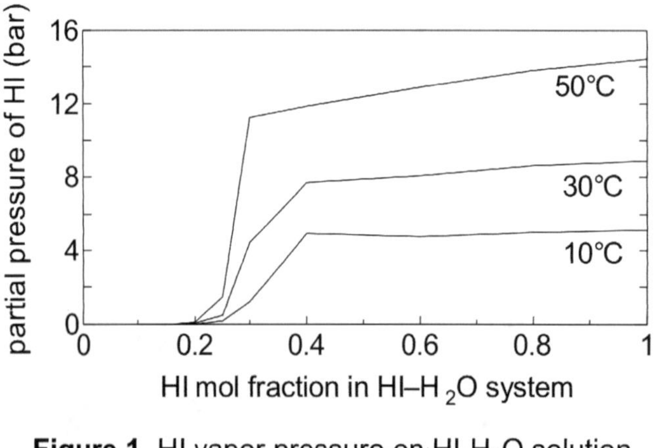


**Figure 1**. HI vapor pressure on HI-$H_2O$ solution.

indispensable low temperature operations lead to solidification of $I_2$, which disturb continuous plant operations.

## FLOW SHEET DESCRIPTION

Figure 2 shows setup of the HI processing section. This study focuses on the HI decomposition unit. Figure 3 shows a new flow sheet of HI decomposition unit and a case of

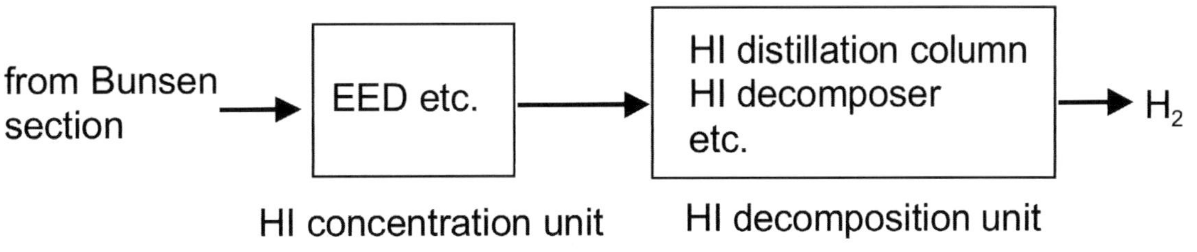


**Figure 2**. Setup of HI processing section.

mass and heat balance. For good $H_2$ separation from unreacted gaseous HI and mach HI gain by the HI distillation column due to the shift of pseudo-azeotropic composition [4], operating pressure of 5MPa, which is a prospective Helium gas pressure in a HTGRs'

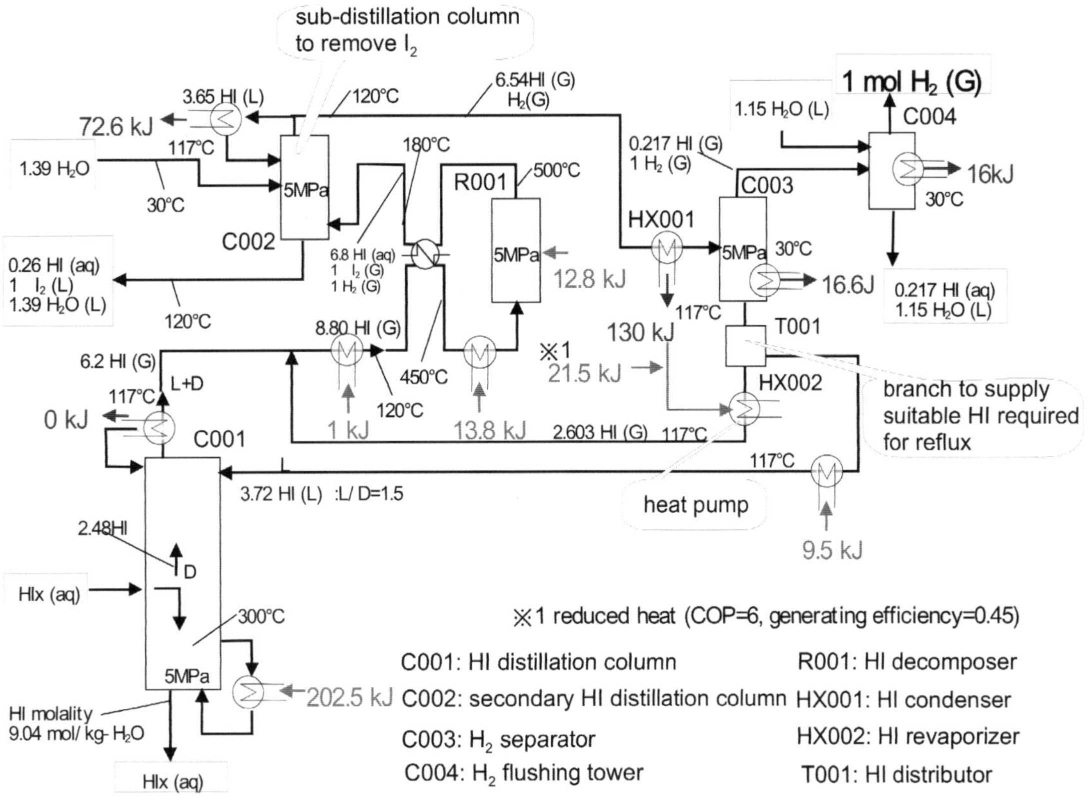


**Figure 3**. Flow sheet of HI decomposition unit, a case of mass and heat balance to gain 1 mol of hydrogen, operating conditions.

design [5], is adapted as an upper limit. The HIx solution concentrated to over pseudo-azeotropic by EED cell is fed to the HI distillation column. HI-rich vapor is distilled from the top, and HIx solution at pseudo-azeotropic composition is obtained at the bottom. The HI-rich vapor is fed to the HI decomposer; from viewpoint of the kinetics, reaction temperature of 500°C is adapted. By providing $H_2O$, the product $I_2$ in unreacted HI gas can be flashed away in the secondary HI distillation column. The remaining HI and $H_2$ are cooled down to make condensation to separate $H_2$, then recovered liquid HI is recycled to the top of HI distillation column. To save energy, suitable amount of HI required for the reflux to distillation, is fed in liquid phase; residual HI is vaporized by the heat pump, then recycled in gaseous phase, which recover heat from the HI condenser. Because of no $I_2$ in recycling gaseous HI, this gas dose not discourages the HI decomposition reaction. The tops of the $H_2$ separator are supplied to the $H_2$ flushing tower. To flush gaseous $H_2$ and HI with externally provided $H_2O$, the product of $H_2$ is obtained by producing lean HI solution at the bottom.

The equilibrium conversion in gas phase thermal decomposition of HI is so low that to improve the conversion ratio can lead to increase of thermal efficiency. That four HI decomposer with total condensers sandwiched between the decomposers to remove $H_2$ is able to rise a one-pass conversion up to 54 % [6]. Figure 4 shows a flow diagram of a similar multi-stage HI decomposition system in this study. To raise the one-pass conversion with minimum amount of energy loses, instead of $H_2$, $I_2$ is removed with HI partial condensation.

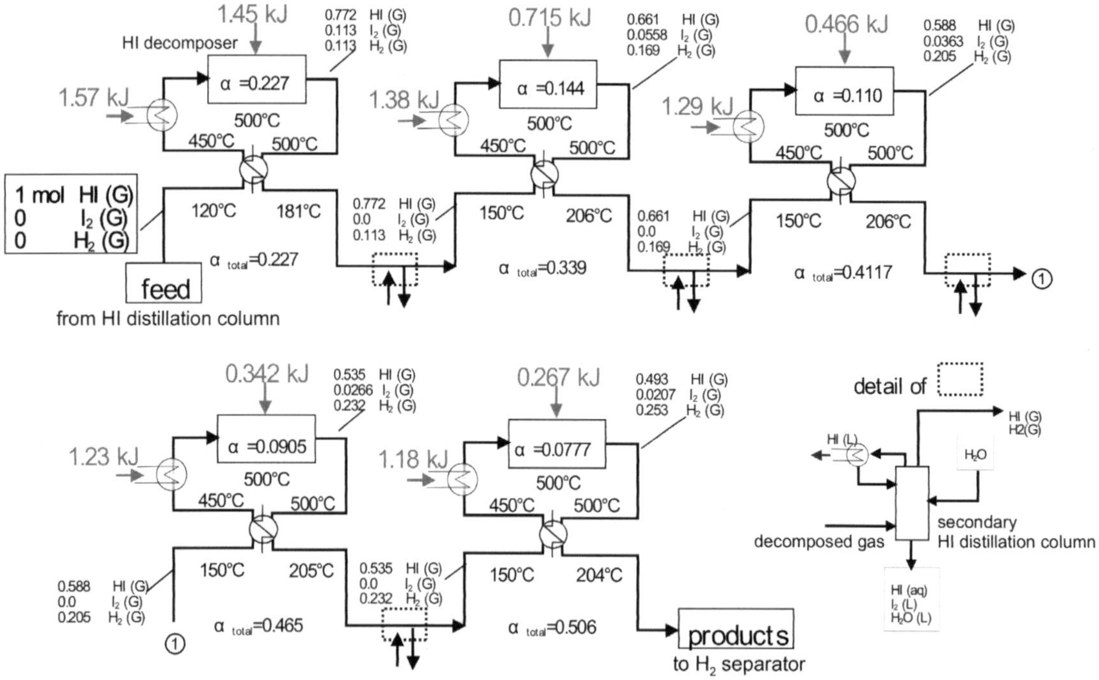


**Figure 4**. Flow diagram of five-stage HI decomposition system.

This system consists of regenerated heat exchangers (minimum $\Delta T=50^{\circ}C$), preheaters, decomposers and sub-distillation column. HI gas processing atlantes between the decomposition to produce $H_2$ and the distillation to remove $I_2$. While the required heat increase as to stage repetitions, overall conversion ratio attains 0.506 in consequence of 5 stages processing.

## REQUIED HEAT TO DRIVE HI DECOMPOSITION UNIT

To examine effect of this HI decomposition system, heat balances were evaluated focusing on influences of HI concentration at the HI distillation feed and the overall conversion of HI decomposition. As for heat balance, the enthalpies at a certain state were calculated [7] from the thermodynamic data [8, 9]. For the enthalpy changes calculated from the difference in enthalpies, reaction heat, latent heat, sensible heat, mixing heat and separating heat were counted except for the dilution enthalpies of $I_2$ solutions, and the effects of pressure were not considered. The reflux ratios of the HI distillation column were assumed 1.5 times of the minimum reflux ratios that were evaluated from the graphic solution using the x-y diagram for HI-$H_2O$ system. On this occasion, the pseudo-azeotrope concentration at the bottom of the HI distillation column was fixed at HI molality of 9.04 mol/kg-$H_2O$.

Figure 5 shows required heat to drive the HI decomposition unit depending on feed concentration of HI distillation and number of decomposition stages. The required heat decay in function of the HI molality of feed solution of the HI distillation column. The reason is that the reflux ratio lower as to HI concentration of the feeds. As for system without the

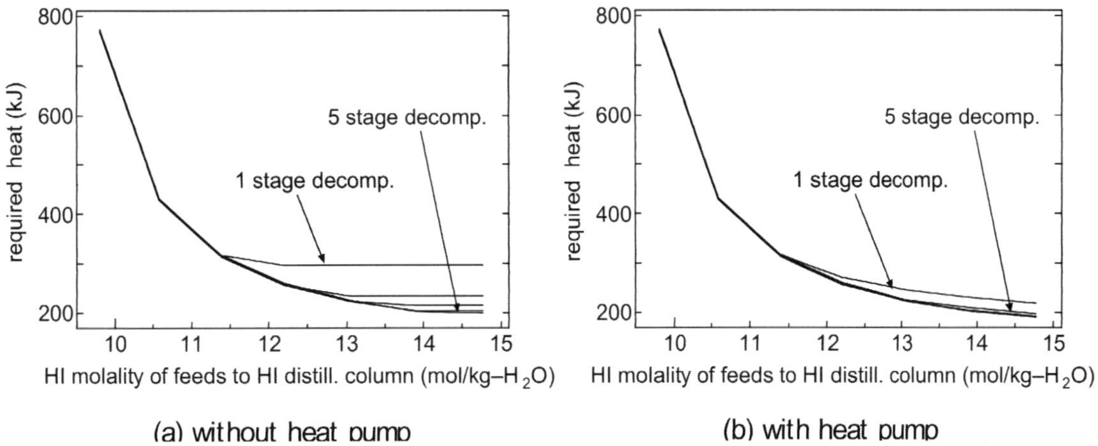


**Figure 5**. Required heat to drive HI decomposition unit depending on feeds of HI distillation and number of decomposition stages.

heat pump, the reason for no effect over 5 stages is that optimum amount of undecomposed HI recycle is determined the reflux ratio depending on the feed HI concentration and the bottom of pseudo-azeotrope concentration. As for system with the heat pump, 2 stages is sufficient to minimize required heat.

## SUMMARY

The flow sheet of the HI decomposition unit to remove the $I_2$ before the separation procedure was proposed. HI concentration fed to the HI distillation column strongly influence the energy budget. The total required heat reaches minimum value at optimum amount of undecompose HI recycle that correspond to necessary amount of the reflux. The required heat for the HI decomposition unit, except for the HI concentration unit, can be reduced around 200 kJ/mol-$H_2$ by adapting the multi-stage HI decomposition in the HI molality of HI distillation feed over 14 mol/kg-$H_2O$.

## REFERENCES

[1] J.H.Norman, G.E.Besenbruch, L.C.Brown, D.R.O'Keefe, C.L.Allen, Thermochemical water-splitting cycle, bench-scale investigations, and process engineering, Final report for the period February 1977 through December 31, 1981, GA-A16713 (1982).

[2] K. Onuki, G.-J. Hwang, Arifal, S. Shimizu: Electro-electrodialysis of hydriodic acid in the presence of iodine at elevated temperature, *J. Membr. Sci.*, 192 (2001) 193-199.

[3] Flow sheet Evaluations For The Thermochemical Water-Splitting Iodine-Sulfur Process (I), Shinji Kubo, Ijichi Masanori, Masatoshi Hodotsuka, Mitsunori Yoshida, Seiji Kasahara, Kazuyoshi Isaka, Nobuyuki Tanaka, Yoshiyuki Imai and Kaoru Onuki, 2007 AIChE Annual Meeting, Salt Lake City, USA.

[4] H. Engels, K. F. Knoche, Vapor pressure of the system HI/H2O/I2 and H2, Int. J. Hydorgen Energy, Vol 11, No. 11, pp. 703-707, 1986.

[5] X. Yan et al., GTHTR300 design variants for production of electricity, hydrogen or both, Proc. of 3$^{rd}$ Information exchange meeting on the nuclear production of hydrogen, 9, Nuclear Energy Agency, Oarai, Japan (2005).

[6] K. Onuki et al., Studies on an Iodine-Sulfur process for thermochemical hydrogen production, Proc. of the 8$^{th}$ WHEC, Vol. 2, pp. 547-556, Honolulu, U.S.A. (1990).

[7] S. Kasahara et al., Proc. of 15th World Hydrogen Energy Conf., 29J-03, Yokohama, Japan (2004).

[8] Chase, M. W.; NIST-JANAF Thermochemical Tables, 4$^{th}$ ed., Am. Chem. Soc. And Am. Inst. of Physics, Washington, D. C., USA (1998).

[9] Nihonkagakukai, ed.; Kagaku Binran Kisohen, 4$^{th}$ ed., Maruzen, Tokyo, Japan (1993).

# Flow sheet Evaluations For The Thermochemical Water-Splitting Iodine-Sulfur Process (I)

*Shinji Kubo, Ijichi Masanori, Masatoshi Hodotsuka, Mitsunori Yoshida, Seiji Kasahara, Kazuyoshi Isaka, Nobuyuki Tanaka, Yoshiyuki Imai and Kaoru Onuki*
*Japan Atomic Energy Agency*
*Narita-cho 4002, Oarai-machi, Higashi Ibaraki-gun, Ibaraki-ken, Japan*

## INTRODUCTION

To achieve high level of thermal efficiency is main issue for IS process [1]. The presence of azeotrope and pseudo-azeotrope in $HI$-$I_2$-$H_2O$ system imposes large thermal burden on conventional distillation method, and the gas phase thermal decomposition of hydrogen iodide suffers from low equilibrium conversion ratio, so that efforts to improve thermal efficiency have been concentrated on how to separate the hydrogen iodide produced in the Bunsen reaction and on how to carry out the hydrogen iodide processing to produce hydrogen. To solve this issue, we have proposed a preconcentration method using a membrane technology, electro-electrodialysis [2], and implemented steady state process simulations [3, 4]. This study provides a mass and heat balance of the HI decomposition section using the electro-electrodialysis (EED).

## PROCESS SIMULATION

Mass and heat balance of the process were evaluated by using the PRO-II and the mixed-solvent electrolyte model (OLI systems Inc.). Circulating fluids within IS process are mixtures of multi-component electrolytes. In particular, $HI$-$I_2$-$H_2O$ and $HI$-$I_2$-$H_2O$-$H_2SO_4$ have very complicated system. In addition, they have many phases, such as gaseous, aqueous, solid, liquid, and secondary liquid, and temperature range and concentration range are so wide. To deal such thermodynamic data in process simulation, we employed MSE model developed by OLI systems. For gas phase in the range of 500 to 850 °C in $SO_3$ and HI

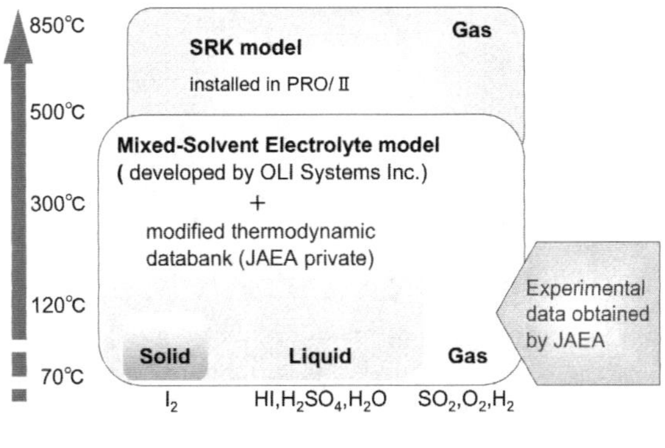


**Figure 1**. Thermodynamic model for mass and heat analysis of IS process.

decomposition, the SRK model installed as default in PRO-Ⅱ is adapted. While for liquid phase below 500 °C, MSE model with the modified thermodynamic databank are adapted for the vapor-liquid equilibrium, the liquid-liquid equilibrium and enthalpy, in which experimental data obtained from our experiments were considered.

**ARTIFICE CONFIGURATION FOR HIx CONCENTRATION UNIT**

A Role of HI decomposition section including HIx concentration unit is to produce hydrogen through following reaction.

$$2HI = H_2 + I_2 \qquad (1)$$

A view of HI processing using EED for efficient hydrogen production is shown in Figure 2.

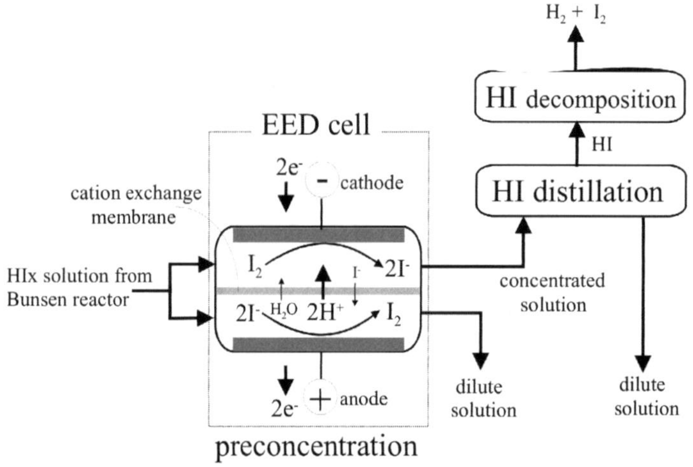


**Figure 2**. Schematic of HI processing by electro-electrodialysis (EED) device.

The HIx solution from the Bunsen reaction is introduced into the electro-electrodialysis (EED) cell. EED cells have two flow channels, an anode and a cathode, which are separated by the cation exchange membrane. The HIx solution is supplied to both sides, and the redox reactions given by following occur on the electrodes.

$$2I^- \rightarrow I_2 + 2e^- \qquad : \text{anode} \qquad (2)$$

$$I_2 + 2e^- \rightarrow 2I^- \qquad : \text{cathode} \qquad (3)$$

On the cathode side, the HI molality of the HIx ($HI$-$I_2$-$H_2O$) solution increases to over-azeotropic composition to distill only HI from the HIx solution. By separating HI without water, the thermal burden for HI processing is gradual reduced.

Figure 3 shows a variety of configuration for HIx concentration unit. Bunsen reaction solution is fed into EED device, then concentrated HIx solution, rich in HI, is supplied to a HI distillation column in which almost pure gaseous HI can be obtained for next HI decomposer.

Figure 3 (a) is a brief flow sheet of closed system. From eqs. (1) - (3), the HI decomposition section may separate completely from the Bunsen section by the membrane. In practice, as shown in Figure 2, mass transfers except $H^+$ through the membrane take place, so that operation to satisfy mass balance using the closed system is difficult. For that reason, we employed Figure 3 (b) of a semi-closed system and Figure 3 (c) of a open system.

Advantage of the semi-closed system is that $I_2$ concentration in circulating solution in HI distillation side can be adjusted at given condition. Obviously, quite low concentration of

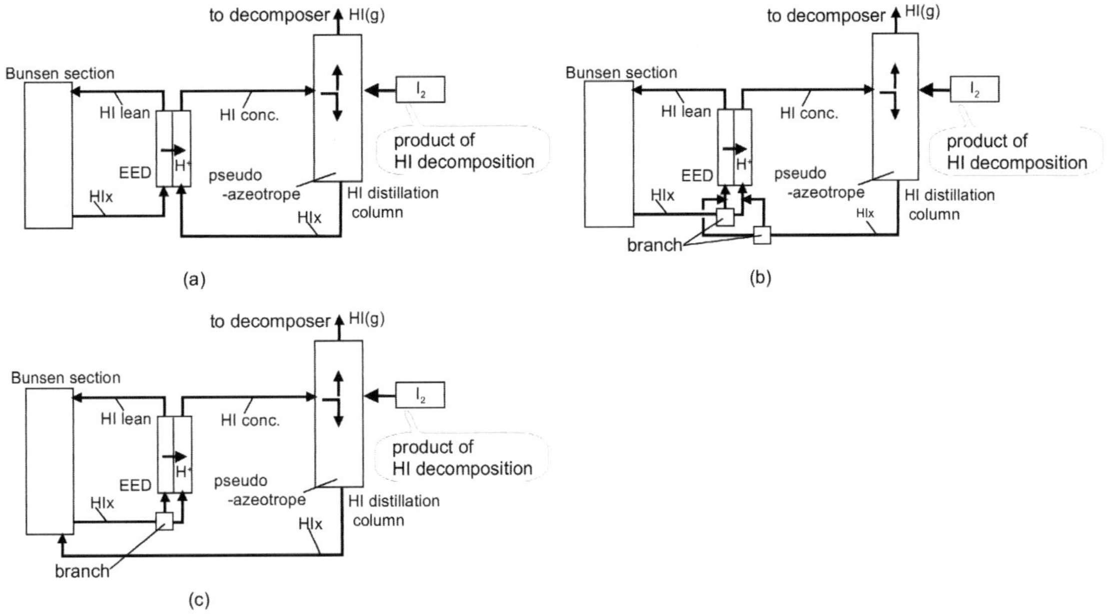


**Figure 3**. Variety of configuration for HIx concentration unit by use of EED.
(a) closed system, (b) semi-closed system, (c) open system

expensive $I_2$ can be adapted. While relatively complicated configuration, there are a lot of space to optimize by using two of the branch which distribute the solution at any ratio of flow rate. Figure 3 (c) is a brief flow sheet of closed system, and Figure 4 shows a view of HI molality and locations of existence. This can draw a portion of HI from Bunsen solution to the

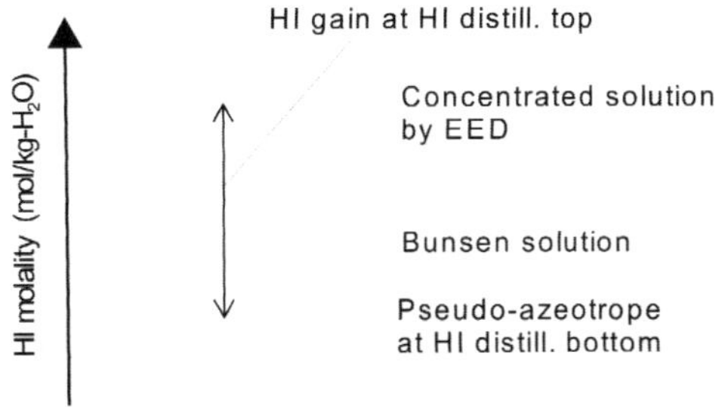


**Figure 4**. HI drawing from Bunsen solution.

top of HI distillation column with a assistance of shift of pseudo-azeotropic composition [5] which save on electricity to drive EED.

## EVALUATION OF MASS AND HEAT BALANCE ON HI PROCESSING SECTION

Figure 5 shows the schematic flow sheet of the HI processing section used for process analyses. Membrane equipment, that is, EED cell for the HI concentration is featured. This system can be operated on the each configuration in Figure 3. The HIx solution from the Bunsen section (S001) is, at first, concentrated to over pseudo-azeotropic by EED cell (S004). The concentrated HIx solution is then fed to the HI distillation column (S007). HI-rich vapor is distilled from the top (S010) and HIx solution of pseudo-azeotropic composition is obtained at the bottom (S011). The bottom solution is returned to the EED or the Bunsen section through the HI distributors. The HI-rich vapor is fed to the HI

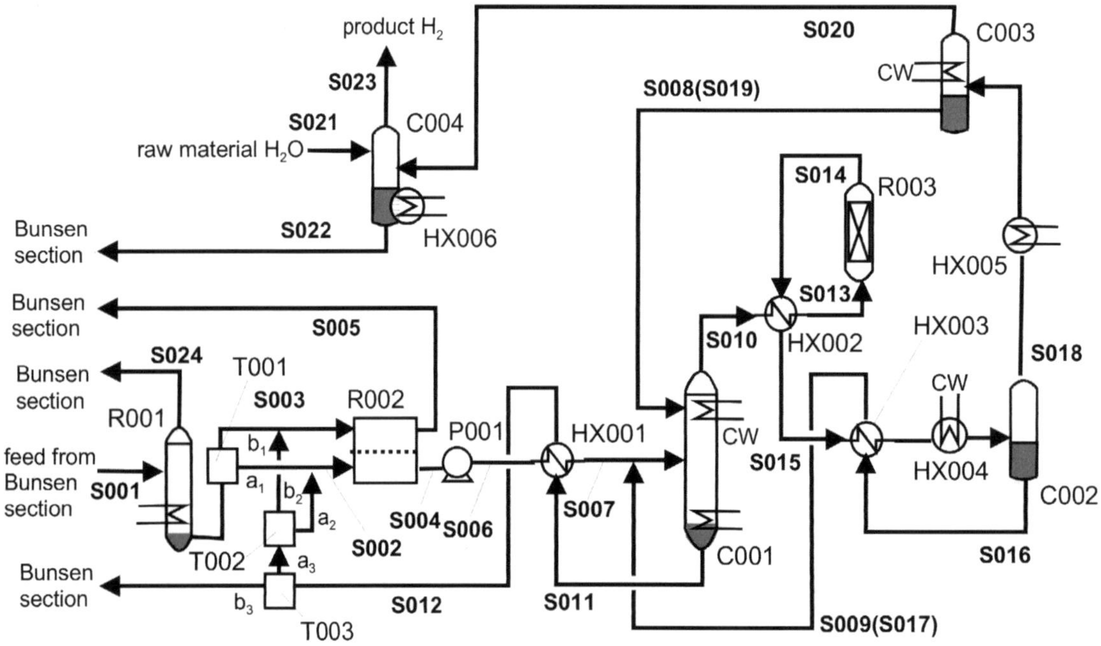


| | | |
|---|---|---|
| R001: HIx purifier | C001: HI distillation column | T001: HIx distributor 1 |
| R002: EED | C002: $I_2$ recovery column | T002: HIx distributor 2 |
| R003: HI decomposer | C003: HI recovery column | T003: HIx distributor 3 |
| | C004: $H_2$ separator | HX005: refrigerator |

**Figure 5**. Flow sheet of HI processing section featuring EED device.

decomposer (S013). Unreacted HI, remaining $H_2$ and $I_2$ are then cooled down to make partial condensation of mixture of $I_2$ and HI. The liquid of HI and $I_2$ is separated at the $I_2$ recovery column (S016) and returned to the feed stream of the HI distillation column. The rest is supplied to the HI recovery column (S018). The bottoms are recycled into the HI distillation column (S008), and the tops are fed to the $H_2$ separator (S020). To flush gaseous $H_2$ and HI with $H_2O$, the product of $H_2$ is obtained (S023).

| | | | | middle-efficiency case | example of parameter set for η of over 40% |
|---|---|---|---|---|---|
| key parameters | Pressure in HI section | | MPa[abs] | 4 | 5 |
| | EED | apparent electro-osmosis coefficient β | mol/mol-e- | 2.0 | 1.0 |
| | | apparent transport number of proton, t | mol/mol-e- | 0.9 | 0.9 |
| | | molality difference at EED outlet (rich side-lean side) | mol/kg-H2O | 5.0 | 5.0 |
| | | average voltage | V | 0.25 | 0.20 |
| | duty for HIx purifier | | kJ/mol-$H_2$ | 0.00 | 0.00 |
| | flow distribution | HIx distributor 1 | a1:b1 | 0.05 : 0.95 | 0.1: 0.9 |
| | | HIx distributor 2 | a2:b2 | 0.6 : 0.4 | - |
| | | HIx distributor 3 | a3:b3 | 1 : 0 (semi-closed system) | 0 : 1 (open system) |
| | duty for H2SO4 section | | kJ/mol-$H_2$ | 373.0 ** | 373.0 ** |
| HI decomposition unit | HI decomposer | | kJ/mol-$H_2$ | 27.2 | 26.9 |
| | refrigerator | | kJ/mol-$H_2$ | 19.3 * | 14.5 * |
| HI concentration unit | EED | | kJ/mol-$H_2$ | 155.0 * | 93.7 * |
| | pump | | kJ/mol-$H_2$ | 7.7 * | 15.1 * |
| | HI distillation column | | kJ/mol-$H_2$ | 193.0 | 154.8 |
| cooling water system | pumps | | kJ/mol-$H_2$ | 8.6 * | 6.1 * |
| total of required heat | | | kJ/mol-$H_2$ | 783.8 | 684.1 |
| thermal efficiency (HHV), η | | | % | 36.5 | 41.8 |

* : converted heat, generating efficiency=45.8%

** : Ozturk, I.T et al., Energy Conversion and Management, 36(1), p. 11-21(1995)

**Table 1**. Key parameter sets and thermal efficiencies.

Table 1 shows key parameter sets and thermal efficiencies, middle-efficiency case and the high-efficiency case, of this flow sheet analyses. Through conversion of the electricity to heat, the generating efficiency of 45.8 % [6] was applied. To employ a potentially improved key parameter set, the thermal efficiency can reach to 40 % (HHV). Table 2 details the operating condition and mass flow rate for the high efficiency case ( $\eta$

| Stream | | S001 | S002 | S003 | S004 | S005 | S006 | S007 | S008 | S009 | S010 |
|---|---|---|---|---|---|---|---|---|---|---|---|
| Phase | | Liquid | Liquid | Liquid | Liquid | Liquid | Liquid | Liquid | Liquid | Liquid | Vapor |
| Temperature | °C | 100 | 100 | 100 | 98.5 | 98.5 | 98.6 | 266.7 | -15 | 194.5 | 116.0 |
| Pressure | kPA [abs] | 200 | 200 | 200 | 200 | 200 | 5000 | 5000 | 5000 | 5000 | 5000 |
| Flow rate | mol | | | | | | | | | | |
| H2O | | 183.556 | 18.356 | 165.201 | 20.366 | 163.190 | 20.366 | 20.366 | 0.000 | 0.000 | 0.000 |
| I2 | | 124.887 | 12.489 | 112.398 | 11.372 | 113.515 | 11.372 | 11.372 | 0.000 | 0.999 | 0.000 |
| HI | | 34.653 | 3.465 | 31.187 | 5.476 | 29.177 | 5.476 | 5.476 | 0.526 | 5.898 | 8.488 |
| H2 | | 0.000 | 0.000 | 0.000 | 0.000 | 0.000 | 0.000 | 0.000 | 0.000 | 0.000 | 0.000 |
| | | total flow rate = 343 mol | | | | | | molality =14.94 mol/kg-H2O | | | |

| Stream | | S010 | S013 | S014 | S015 | S016 | S017 | S018 | S019 | S020 | S021 |
|---|---|---|---|---|---|---|---|---|---|---|---|
| Phase | | Vap | Vap | Vap | Mix | Liq | Mix | Vap | Liq | Vap | Liq |
| Temperature | °C | 116 | 450 | 500 | 214.5 | 60 | 194.5 | 60 | -15 | -15 | 25 |
| Pressure | MPa[abs] | 5 | 5 | 5 | 5 | 5 | 5 | 5 | 5 | 5 | 5 |
| Flow rate | mol/s | | | | | | | | | | |
| H2O | | 0.000 | 0.000 | 0.000 | 0.000 | 0.000 | 0.000 | 0.000 | 0.000 | 0.000 | 1.001 |
| I2 | | 0.000 | 0.000 | 1.000 | 1.000 | 1.000 | 1.000 | 0.000 | 0.000 | 0.000 | 0.000 |
| HI | | 8.487 | 8.487 | 6.487 | 6.487 | 5.906 | 5.906 | 0.581 | 0.526 | 0.055 | 0.000 |
| H2 | | 0.000 | 0.000 | 1.000 | 1.000 | 0.000 | 0.000 | 1.000 | 0.000 | 1.000 | 0.000 |
| | | | | decomposition rate= 23.6 % | | | | | | | |

| Heat exchanger | | HX001 | HX002 | HX003 | HX004 | HX006 |
|---|---|---|---|---|---|---|
| duty | kJ/mol-H2 | 652.0 | 108.0 | 120.0 | -25.0 | -2.0 |
| ΔTmin. | °C | 14.0 | 50.0 | 20.0 | - | - |

**Table 2**. Operating conditions and flow rates of HI processing section in case of using parameter set for $\eta$ over 40%.

=41.8 %). To achieve high level of thermal efficiency, we will have conducted experimental activities and the process simulation approaches. The parameters relating to the EED cell has large influence on thermal efficiency, so that we have implemented to develop a radiation-grafted membranes (RGM) [7] for the cation exchange membrane. On primary object, RGMs are for fuel cell application. RGM has high proton conductivity and low water permeation. Now, R&D for RGM are underway, to modify membrane characteristics for EED application and to conduct HI concentrating test [8] by using test fabricated RGMs. For purification procedure, the duty to remove the impurities, $H_2SO_4$ and $SO_2$, in HIx of Bunsen solution is so high that reduction of heat duty for purification is also important. Accordingly, to reduce contaminations of Bunsen solution will have been examined by a semi-batch reaction test [9]. As for the process simulation approach, to pick out a superior configuration and to reduce the total flow rate provided to the HI processing section, further optimization on the flow rate distributions will be conducted with the aid of the branches. Also, for the $H_2SO_4$ processing section, to prove low heat duty such as the Ozturk's flow sheet [10] is indispensable.

## SUMMARY

Flow sheet analyses on the HI processing section featuring the electro-electrodialysis is carried out. The thermal efficiency to produce hydrogen can reach to 40 % (HHV) by adapting the potentially improved key parameter set. To prove such high level of thermal efficiency, hereafter, we will have conducted experimental and the process simulation studies.

## REFERENCES

[1] J.H.Norman, G.E.Besenbruch, L.C.Brown, D.R.O'Keefe, C.L.Allen, Thermochemical water-splitting cycle, bench-scale investigations, and process engineering, Final report for the period February 1977 through December 31, 1981, GA-A16713 (1982).

[2] K. Onuki, G.-J. Hwang, Arifal, S. Shimizu: Electro-electrodialysis of hydriodic acid in the presence of iodine at elevated temperature, *J. Membr. Sci.*, 192 (2001) 193-199.

[3] M. Ijichi et al., Hydrogen production with high-temperature gas-cooled reactors –(1) steady state process evaluation for thermochemical IS process, 2007 Annual Meeting of the Atomic Energy Society of Japan, Nagoya, Japan.

[4] M. Ijichi et al., Hydrogen production with high-temperature gas-cooled reactors –(10) Development of process simulation code for thermochemical IS process, 2006 Annual Meeting of the Atomic Energy Society of Japan, Oarai, Japan.

[5] H. Engels, K. F. Knoche, Vapor pressure of the system $HI/H_2O/I_2$ and $H_2$, Int. J. Hydorgen Energy, Vol 11, No. 11, pp. 703-707, 1986.

[6] K. Kunitomi et al,. Reactor core design of gas turbine high temperature reactor 300, Nucler Engineering and Design, 230, pp 349-366 (2004).

[7] T. Yamaki et al., Preparation of Proton Exchange Membranes Based on Crosslinked Polytetrafluoroethylene for Fuel Cell Applications, Polymer, 45 (2004) 6539-6573.

[8] H. Okuda et al., Improvement of the thermal efficiency of hydrogen iodine concentration in I-S process by using radiation-induced polymerized membrane in electrodialysis system, AIChE spring meeting (2006).

[9] H. Nakajima et al., Effect of sulfur dioxide partial pressure on the reaction of iodine, sulfur dioxide and water, Kagaku Kogaku Ronbunshu, 33(3), pp. 257-260 (2007).

[10] Ozturk, I.T., Hammache, A., Bilgen, E. (1995). An Improved Process for H2so4 Decomposition Step of the Sulfur Iodine Cycle. Energy Conversion and Management, 36(1), p. 11-21.

## Control Techniques For Bunsen Reaction Solution To Regulate Process Condition

*Shinji Kubo, Hayato Nakajima, Yoshiyuki Imai, Seiji Kasahara and Kaoru Onuki*
*Japan Atomic Energy Agency*
*Narita-cho 4002, Oarai-machi, Higashi Ibaraki-gun, Ibaraki-ken, Japan*

### INTRODUCTION

A stable hydrogen production via the IS process [1] is relatively difficult because of multiple components and corrosive solution. This issue is therefore a high targeted priority when industrializing the process as feasible in a chemical plant. For stable operations by IS process, to regulate the composition of Bunsen reaction solution is one of key techniques. This paper discusses a specific method to maintain the composition of Bunsen reaction solution.

### FUNDAMENTAL CONSIDERATION

At the setout, we demonstrate output composition in consequence of Bunsen reaction can be controlled at given value by two feed flow rate under particular circumstances. Bunsen reaction (eq.1) produces two acids, hydriodic acid and sulfuric acid from water, sulfur dioxide and iodine as reactants in an aqueous solution.

$$SO_2 + I_2 + 2H_2O = 2HI + H_2SO_4 \quad (1)$$

Figure 1 shows brief mass balance around Bunsen reactor. $F$ indicates flow rate of each stream, $x$ is molar fraction of each component of HI, $H_2SO_4$, $I_2$, $H_2O$, $b$ represents rates about production or consumption by the reaction. Supposing $SO_2$ is the limiting reactant, reaction yield is 1 and dissolved $SO_2$ is zero, mass balance is given by:

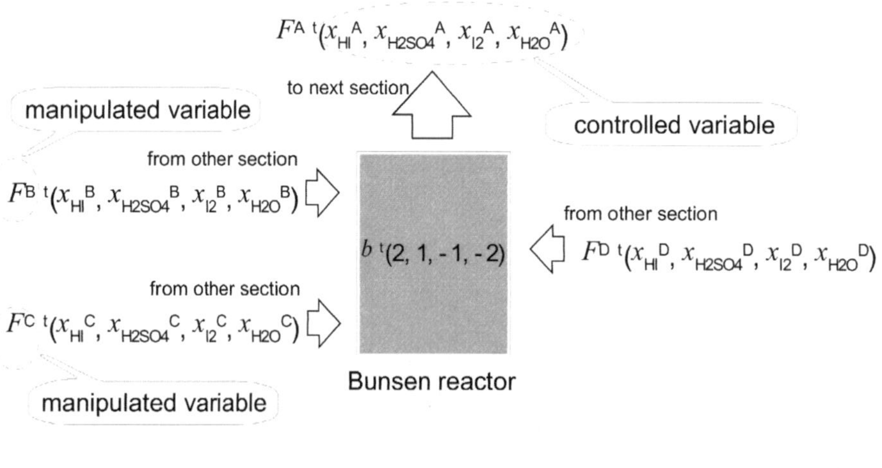


**Figure 1**. Mass balance of Bunsen reactor.

$$\boldsymbol{A} = \begin{pmatrix} x_{\mathrm{HI}}^{A} & x_{\mathrm{HI}}^{B} & x_{\mathrm{HI}}^{C} & 2b + F^{D}x_{\mathrm{HI}}^{D} \\ x_{\mathrm{H_2SO_4}}^{A} & x_{\mathrm{H_2SO_4}}^{B} & x_{\mathrm{H_2SO_4}}^{C} & b + F^{D}x_{\mathrm{H_2SO_4}}^{D} \\ x_{\mathrm{I_2}}^{A} & x_{\mathrm{I_2}}^{B} & x_{\mathrm{I_2}}^{C} & -b + F^{D}x_{\mathrm{I_2}}^{D} \\ x_{\mathrm{H_2O}}^{A} & x_{\mathrm{H_2O}}^{B} & x_{\mathrm{H_2O}}^{C} & -2b + F^{D}x_{\mathrm{H_2O}}^{D} \end{pmatrix}, \quad \boldsymbol{B} = \begin{pmatrix} F^{A} \\ F^{B} \\ F^{C} \\ 1 \end{pmatrix}, \quad \boldsymbol{AB} = \begin{pmatrix} 0 \\ 0 \\ 0 \\ 0 \end{pmatrix}.$$

In case of this homogeneous equation has nontrivial solution, determinant of matrix $\boldsymbol{A}$:

$$\det \boldsymbol{A} = 0$$

should be satisfied. In case that one row vector of $\boldsymbol{A}$ is in proportion to another row vector, $\det \boldsymbol{A}$=0 can be satisfied in all settings. Specifically, matrix of

$$\mathbf{A}' = \begin{pmatrix} 2x_{\mathrm{H_2SO_4}}^{A} & 0 & 0 & 2b \\ x_{\mathrm{H_2SO_4}}^{A} & 0 & 0 & b \\ x_{\mathrm{I_2}}^{A} & x_{\mathrm{I_2}}^{B} & x_{\mathrm{I_2}}^{C} & -b + F^{D}x_{\mathrm{I_2}}^{D} \\ x_{\mathrm{H_2O}}^{A} & x_{\mathrm{H_2O}}^{B} & x_{\mathrm{H_2O}}^{C} & -2b + F^{D}x_{\mathrm{H_2O}}^{D} \end{pmatrix}$$

is employed in order to make the determinate vanish. The matrix $\boldsymbol{A}'$ means the feed streams contain no components of HI and $H_2SO_4$. Assuming $\boldsymbol{A}'$, particular circumstances, is formed, $\boldsymbol{A}'\boldsymbol{B}$=0 has a nontrivial solution. In short, the output composition in consequence of Bunsen reaction is able to be controlled at given value by manipulating two flow rate of feed stream; on this occasion, $x^{B}$ should not disagree with $x^{C}$.

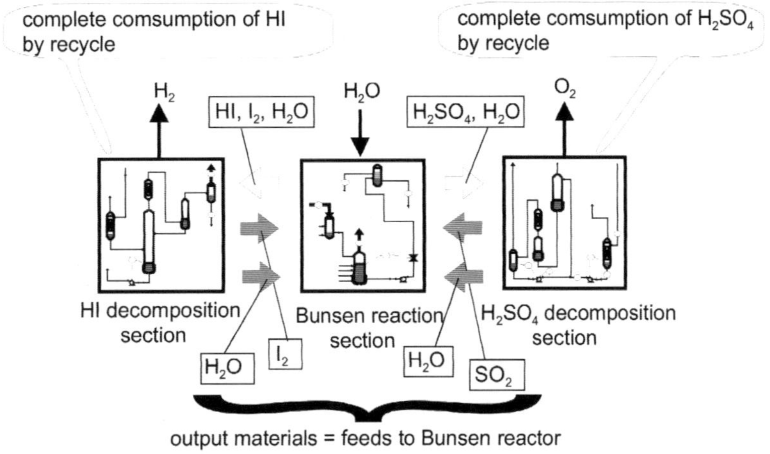


**Figure 2**. Compositions of feed streams to Bunsen reactor.

## METHOD TO MAINTAIN COMPOSITION OF BUNSEN SOLUTION

Figure 2 shows a schematic of composition of feed stream to Bunsen reactor. IS process is operated on the closed-cycle condition, so that the feeds to Bunsen reactor correspond to the output materials from upstream sections. In case that the $H_2SO_4$ and HI sections consume the fed $H_2SO_4$ and HI completely to produce $O_2$ and $H_2$, feed streams of Bunsen reactor contain no HI and $H_2SO_4$. The det $\boldsymbol{A'} = 0$ is able to be satisfied in this way.

To measure Bunsen solution by the in situ methods is very difficult due to multiple components and corrosive solution. With this in mind, we devised a simple and easy method [2] for the Bunsen section control. Figure 3 (a) shows the liquid-liquid phase separation

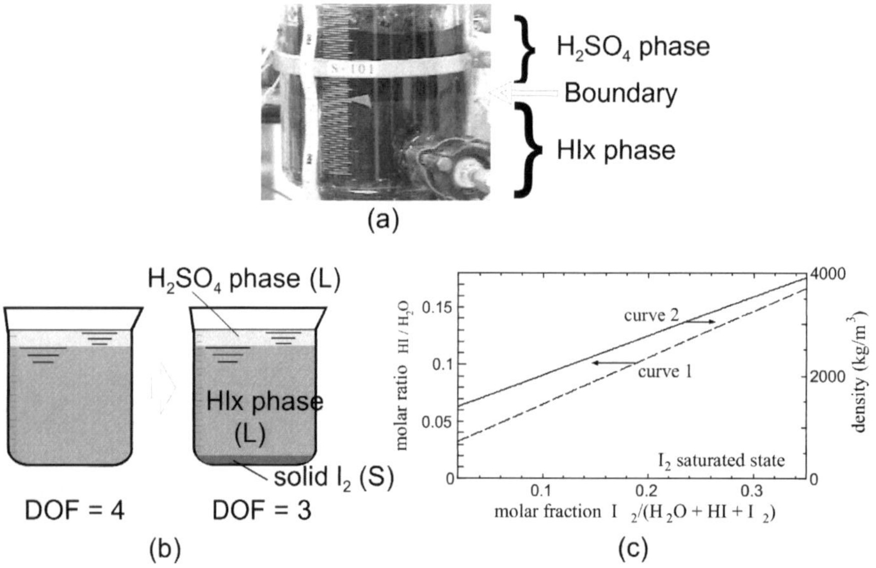


**Figure 4**. Method to maintain composition of Bunsen reaction solution.

phenomenon in a glass vessel. The two produced acids divide into $H_2SO_4$ and HIx phase with a clear boundary. Figure 3 (b) shows the phases and Gibbs' degree of freedom (DOF) in the Bunsen reaction at equilibrium. Here, the number of reactions is 1, the number of components of the reaction (eq. 1) is 5, and the number of phases is 2 or 3 where the gas phase is not at equilibrium, because as reactants flow, the gas mixture flows from the reaction field via such a flow reactor. To purposely make solid $I_2$, the DOF can be reduced. Figure 3 (c) shows the molar ratio ($HI/H_2O$) and the density of the HIx phase, excluding $H_2SO_4$ as a minor component, with the molar fraction ($I_2/H_2O+HI+I_2$) on $I_2$ saturated state. With constant temperature and pressure, the composition varies on the curve 1. Because the curve 1 corresponds one-on-one with the curve 2, the composition of the HIx phase is able to fix by its density. Therefore to make HIx phase $I_2$ saturated state and regular density stabilizes the composition of the Bunsen reaction solution at given value on the curve 1.

## TECHNIQUES USED TO MAINTAIN COMPOSITION ON PLANT OPERATION

Figure 4 (a) shows the Bunsen reactor system. The vessel of the reactor equipped with an ultrasonic sensor (US sensor) externally and on the bottom. The sensor releases an outgoing wave, which is reflected via the reflector in the solution, and then reaches the sensor. Where solid substances exist between the sensor and reflector, the intensity of the reflected wave signal drastically decreases (Figure 4 (b)). Herewith the US sensor detects information to confirm the solid $I_2$ existence or not. The pump used to manipulate the flow

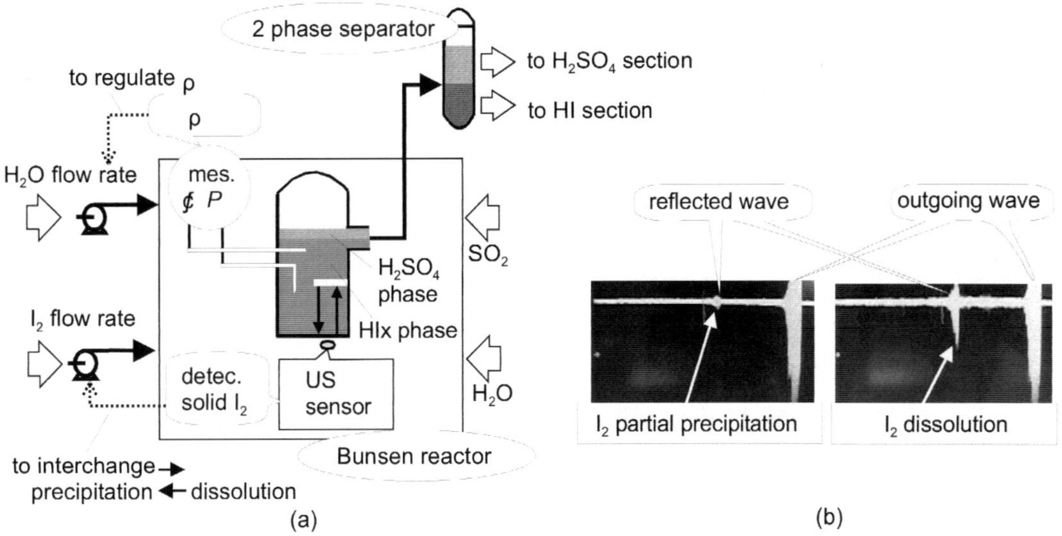


**Figure 4**. Techniques to maintain Bunsen composition on plant operation.

rate of $I_2$ is controlled by acting on interchange $I_2$ precipitation with dissolution, while the differential pressure instrument is also equipped to measure the HIx phase density and the pump to manipulate the flow rate of $H_2O$ is controlled to regulate the HIx phase density. Therefore, the composition of the Bunsen reaction solution can be maintained under a stable condition by applying the techniques.

## COMPOSITION CONTROL ON HYDROGEN PRODUCTION TEST

A bench-scaled closed-cycle hydrogen production test [3] was carried out to confirm usefulness of the control methods maintaining process conditions stably. The test lasted for 175 hours in the closed-cycle condition. The hydrogen production is almost stable at a rate of 31 L/h, and the production ratio of oxygen to hydrogen almost agrees to 0.5:1, which showed the stable and stoichiometric water-splitting was accomplished.

In order to investigate stability of compositions of the circulatory solutions, the solutions were sampled periodically at outlets of the two phase separator to obtain the HI phase and the $H_2SO_4$ phase. Figures 5 (a) and (b) show the changes in the composition of the two phases with time. The interval between sampling was longer than the holding time in the separator. The sample compositions for both phases were analyzed by chemical titration methods. The figures highlight that the compositions of both phases were retained at around regular molar fractions, which confirmed the suitability of this control method.

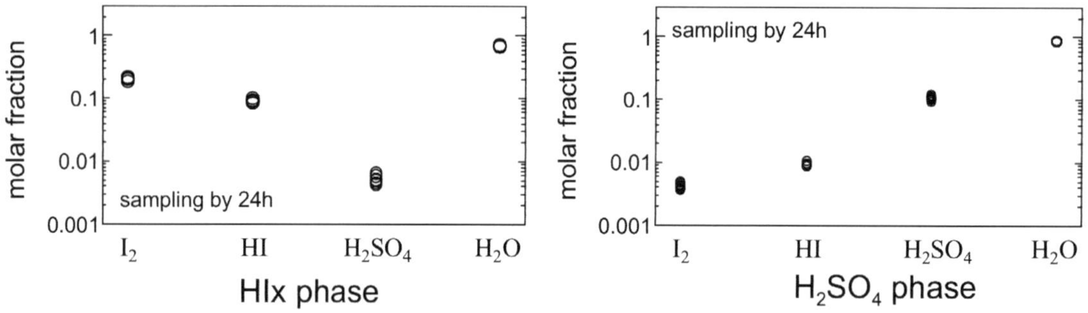


**Figure 5**. Composition changes of HIx and $H_2SO_4$ phase in two phase separator.

## SUMMARY

For stable operations by IS process, to regulate the composition of Bunsen reaction solution is one of key techniques. We devised a simple and easy method for the composition control on plant operation. In the bench-scaled hydrogen production test, the compositions were retained at around regular molar fraction. Eventually suitability of this control method was confirmed.

## REFERENCES

[1] J.H.Norman, G.E.Besenbruch, L.C.Brown, D.R.O'Keefe, C.L.Allen, Thermochemical water-splitting cycle, bench-scale investigations, and process engineering, Final report for the period February 1977 through December 31, 1981, GA-A16713 (1982).

[2] S. Kubo et al., Closed cycle and continuous operation by a thermo-chemical water-splitting IS process, Int. Conf. On Non-Electric Applications of Nuclear Power, April 16-18, 2007, Oarai, Japan.

[3] S. Kubo et al., A bench scale hydrogen production test by the thermochemical water-splitting Iodine-Sulfur process, Proc. GLOBAL 2005, No. 474, Oct 9-13, Tsukuba, JAPAN (2005).

# Recent Progress in High Temperature Electrolysis

*J. Stephen Herring[1], Carl M. Stoots[1], James E. O'Brien[1], Joseph J. Hartvigsen[2] and Gregory Housley[1]*
*(1) Idaho National Laboratory, P.O. Box 1625, MS 3860, Idaho Falls, ID 83415-3860,*
*(2) Ceramatec, Inc., 2425 South 900 West, Salt Lake City, UT 84119*

***Abstract***

Hydrogen production technologies will provide a means through which nuclear energy can be used to reduce the nation's critical vulnerability on imported petroleum. Eventually, hydrogen may be a major transportation fuel through the use of fuel cells in automobiles. However, even today, the annual demand for hydrogen in the US is about 13 million tons, primarily for the production of fertilizers, for the upgrading of low quality crude oils and for the removal of sulfur to meet increasingly stringent environmental requirements. As we come to rely on poorer grades of crude oil and on the Athabasca oil sands, the demand for hydrogen in the petroleum industry will grow significantly. Average crude oils require 1 to 2 kg of $H_2$ per barrel, heavy crudes 2-3 kg per barrel and the oil sands 3-5 kg per barrel. Today, 95% of our hydrogen is produced through the steam reforming of natural gas. Both the increasing price of natural gas and the need to reduce $CO_2$ emissions will promote the nuclear production of hydrogen.

The development of High Temperature Electrolysis (HTE) of steam for the production of hydrogen using nuclear energy has progressed through a series of experiments and analyses during the past year. The purpose of these experiments was to demonstrate the feasibility of using solid-oxide electrolytic cells to efficiently split steam into hydrogen and oxygen, while developing the economies of scale and long operating life needed by successful commercial devices. These tests have increased in scale from button cells through 10 and 25 cell single stacks and, most recently, to the testing of a dual stack containing 2 x 60 cells. The long-duration test of the "Half-Module" demonstrated the viability of the technology planned for the HTE Integrated Laboratory-scale (ILS) experiment. The ILS will eventually have three modules of four 60-cell stacks each. The Half-Module thus consisted of half of one of the ILS modules. The Half Module was fabricated at Ceramatec, Inc. in Salt Lake City and tested at their facility from June through September of 2006. The unit ran for 2040 hours, mostly at steady state except for minor interruptions due to power outages and testing of the stacks under a variety of current and voltage conditions. Hydrogen production was 1.21 normal $m^3$/hour at the beginning of the test, dropping to 0.65 normal $m^3$/hour at the end of the test.

Construction of the ILS experiment took place during the summer of 2007, with initial shakedown testing beginning on August 22 and the production of its first hydrogen at the end of September. Initially the ILS will have only one module (four stacks of 60 cells each) and will not have integrated heat recovery and hydrogen separation/recycle components. Eventually the ILS will include heat exchangers to recover the 800° C heat of the hydrogen and oxygen for use in heating the incoming water to steam. In addition, the ILS will incorporate condensation tanks or high temperature porous membranes for the separation of hydrogen from the residual steam. The residual steam and a small amount of hydrogen will be re-circulated to the electrolytic stacks where the hydrogen will maintain reducing conditions at the nickel-cermet electrode.

Another aspect of HTE research is the analysis of cell degradation mechanisms. The lifetime of the electrolysis cells is being limited by corrosion, transport of impurities into the electrodes and delaminations of the electrodes from the electrolyte. The HTE project is conducting progressively longer tests followed by post-test analyses to determine the causes and to find better formulations of the cell materials. Computational Fluid Dynamics (CFD) analysis and flowsheet modeling provide further insights into the operation of electrolytic cells. CFD is being used to determine the best ways to reduce thermal stresses over the area of the cells and to assure that the steam is properly distributed in our progressively larger stacks and the tubular cells. Flowsheet modeling is being used to model complete hydrogen production plants as driven by a variety of reactors.

For about 800 hours during the Half-Module test (summer 2006), the stacks were operated in a co-electrolysis mode, which we are calling *Syntrolysis.* In the *Syntrolysis* mode, the stacks are supplied with steam and carbon dioxide, producing carbon monoxide and hydrogen (synthesis gas, or "syngas"), which can be used in the Fischer-Tropsch process for the production of synthetic hydrocarbon liquids, such as diesel fuel, jet fuel, and synthetic lubricating oils. The synthesis of fuels from a variety of carbon sources in forward locations has the potential for easing fuel logistics' requirements for the military. In addition, the production of synthetic fuels through nuclear hydrogen is a means for reducing our civilian dependence on imported oil while continuing to use the existing liquid transportation fuel infrastructure.

## *I. Introduction*

Over the last five years there has been a growing interest in the use of hydrogen as an energy carrier, particularly to augment transportation fuels and thus reduce our dependence on imported petroleum. Hydrogen is now produced primarily via steam reforming of methane. However, in the long term, methane reforming is not a viable process for large-scale hydrogen production since such fossil fuel conversion processes consume non-renewable resources and emit greenhouse gases.

High Temperature Electrolysis (HTE) is built on the technology of solid oxide fuel cells (SOFCs), which were invented over a century ago, though they have been most vigorously developed during the last twenty years. SOFCs consume hydrogen and oxygen and produce heat and electricity. Solid Oxide Electrolytic Cells (SOECs) consume electricity and steam and produce hydrogen and oxygen. The purpose of our HTE research is to solve those problems unique to the electrolytic mode of operation, while building further on continuing fuel cell development.

A hydrogen production plant using the heat and electricity of a high temperature nuclear reactor is the basis for this concept. About 80 % of the thermal energy of the reactor, at ~900° C, is sent to the Brayton of the efficient (~50%) generation of electricity. The remaining 20% of the reactor thermal output is used to heat the reagent water to steam at about 850 °C. A mixture of 90 vol % steam and 10 vol % hydrogen is fed to the electrolytic cells, where the oxygen migrates through the electrolyte at $O^{=}$ ions because of the imposed voltage. The voltage necessary is about 0.3 V lower than in conventional electrolyzers because of the high operating temperature. In addition, the kinetics of the electrolytic reactions are much faster than in room temperature electrolysers, thus avoiding polarization losses. The mixture exiting the cells, about 25 vol % steam and 75 vol % hydrogen, enters a separation device where the two components are separated either be condensation of the steam or diffusion of the hydrogen. Most of the steam and some of the hydrogen is mixed with additional steam, reheated and returns to the electrolytic cells. The remaining steam can be removed from the hydrogen product stream through condensation. Like the hydrogen stream, the oxygen stream exiting the cells has a temperature of about 830° C. An integrated HTE plant would contain heat exchangers to transfer the heat of the hydrogen to in the coming water or steam. The oxygen might be cooled or it might be used in high temperature combustion processes.

The integrated operation of a high-temperature electrolytic plant offers the utility the potential to vary the output from hydrogen to electricity within a few minutes. We have found in experiments with stacks of electrolytic cells that the output can be changed from full hydrogen production to zero by simply controlling the voltage applied to the cells. By operating at either the thermal neutral voltage or the open cell voltage, the output can be varied without imposing thermal transients on the cells. When the cell voltage is reduced, the steam continues to flow through the stacks, maintaining the 830° C operating temperature, but no electrical current flows and no hydrogen is produced. Thus the hydrogen plant operator can respond to variations in the grid demand by producing less hydrogen or by accepting power from the grid and producing hydrogen beyond the capability of the reactor alone.

## *II. The High Temperature Electrolysis Integrated Laboratory Scale Experiment*

The HTE version of the ILS is now under development at the INL. In accordance with the Research and Development Plan for the NHI, the HTE research track is progressing from small-scale bench testing to large-scale demonstration. For HTE, the experimental research objectives have been organized as follows with increasing scale:

- Button cell fabrication/testing (~1 W) -- cell material development and performance
- Stack development/testing (200 W - 5 kW) -- stack design (electrode and electrolyte materials, interconnect and flow field materials and fabrication, inter-cell electrical contact, cell and manifold sealing issues, cell durability)
- Integral operation issues (15 kW) -- feed-stock heating, high-temperature gas handling, multiple-stack thermal management, heat recuperation
- Facility testing (200/500 kW and 5 MW) --production issues including energy management, utility requirements, high pressure operation, product purification.

Figure 1 shows the progression of the HTE experiments over the last four years, from a 3.2 $cm^2$ button cell to the ILS, which will have an active area of 46,080 $cm^2$ when all three modules are installed during 2008. This series of experiments represents a growth in active area (and in hydrogen production rate) by a factor of 15,000.

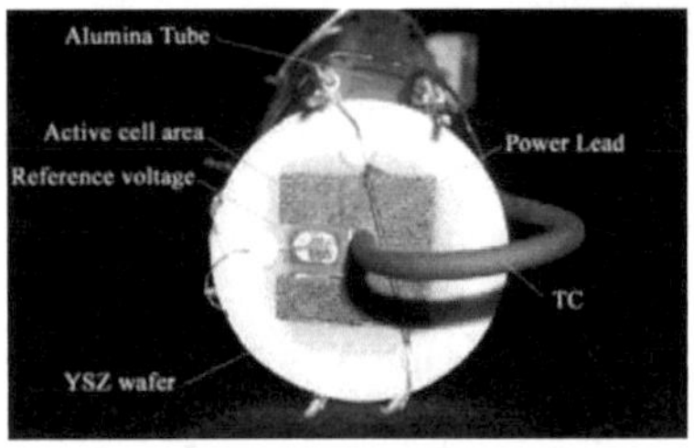


**Button cell (2003) 3.2 cm²**

**10-cell stack (2004) 640 cm²**

**120-cell half-module (2006) 7,680 cm²**

**Integrated Laboratory Scale (operational 8-22-07)**
**720 cells, 3 modules (2008) 46,080 cm²**

Figure 1. Progression of HTE Experiments

The HTE ILS facility is designed to study large-scale hydrogen-production issues such as thermal management, hot-zone design, multiple-stack electrical configuration, pre-heating of process gases, and heat recuperation. In the ILS, the inlet steam/hydrogen flow will be preheated to the full 830°C electrolyzer operating temperature prior to entering the hot zone using superheaters. This scenario more closely simulates the inlet stream heating performed in a full-scale hydrogen production plant coupled to an advanced nuclear reactor. A full-scale plant would utilize both electrical power and high-temperature process heat from the reactor system.

The configuration of the electrolysis modules in the hot zone is also noteworthy. The three modules have three separate, parallel support systems that supply power, superheated steam, preheated air, and appropriate exhaust handling. By utilizing three separate gas streams, failure of an individual module will not affect the performance or testing of the other modules. Thus the ILS system design allows each module to be tested independently.

Figure 2. HTE ILS system overview.

The HTE ILS system has been assembled on a large skid (10' x 16'). Primary skid-mounted components include electronics, power supplies, air compressor, pumps, superheaters, hot zone, condensers, and dew-point sensor vessels. Next to the skid is a computer work station to electronically control the equipment. Fig. 2 shows the ILS system configuration with important components labeled.

Table 1. Comparison of nominal and extreme design cases.

| | Nominal Case | Extreme Design Case |
|---|---|---|
| ASR (ohm cm$^2$) | 1.5 | 1.0 |
| Current Density (A/cm$^2$) | 0.25 | 0.37 |
| Per-cell Voltage, V() | 1.283 | 1.283 |
| Electrolysis Power (kW) | 14.54 | 21.8 |
| Hydrogen Production Rate (NL/hr) | 4735 | 7103 |

Downstream of the superheaters, the steam/hydrogen gas mixture will reach the three electrolysis modules incorporated into a single hot zone, as displayed in Figs. 2 and 3.

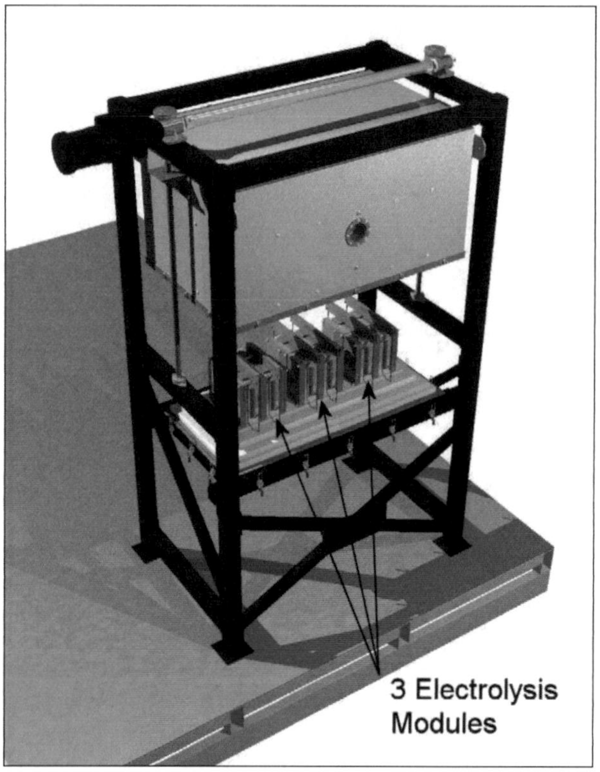


Figure 3. Hot zone.

The ILS design includes three four-stack modules, with 60 electrolysis cells in each stack, for a total of 12 stacks and 720 cells in a single hot zone. This configuration utilizes the solid oxide cell technology similar to the stacks previously tested at INL. The electrolysis module is configured so that the steam/hydrogen gas mixture enters through the bottom of the steam inlet manifold (shown on the module's ends in Fig. 4) and exits through the bottom of the outlet manifold in the module's center. Airflow enters through the bottom of the air inlet manifold, flows through the stack perpendicular to the steam/hydrogen gas mixture, and exits at the front of the stacks directly into the hot zone.

The three electrolysis modules will require separate and parallel support systems supplying electrical power for electrolysis, a feedstock gas mixture of hydrogen and steam, a sweep gas, and appropriate exhaust handling. In particular, this system must include components for controlled steam generation, mixing hydrogen with the steam, inlet and outlet dewpoint measurements, heating the feedstock and sweep gas to the appropriate electrolysis temperature (via a superheater), cooling the electrolysis product stream, condensing any residual steam out of the product stream, and venting the hydrogen product. Furthermore, each electrolysis module has its own independent instrumentation system. Because the feedstock and product gas streams for each electrolysis module are to be monitored and controlled separately; the inlet and outlet gas composition and flow rate will be known specifically for each module. This allows for independent testing of each of the three HTE ILS modules under different operating conditions (gas mixture, current density, etc.). In this way, a test matrix covering a range of operating conditions can be covered quickly. Additionally, if a module fails, that particular module can be shut down without affecting the performance or testing of the other modules.

To measure fuel cell performance, the area specific resistance (ASR) value is closely monitored. Testing of fuel cells shows that the resistance of the cell gradually increases over time, reducing the cells' hydrogen production rate. This degradation is evidenced by an increase of the calculated ASR values. As the technology of cell fabrication improves, the cell's ability to efficiently produce large quantities or hydrogen over long time-periods should improve.

Figure 4. Electrolysis module with four stack of 60 cells.

### *III. Syntrolysis, the Co-Electrolysis of Steam and Carbon Dioxide*

The INL, in conjunction with Ceramatec Inc. (Salt Lake City, USA), has initiated an experimental / computational research program investigating a novel technique for producing syngas which does not consume non-renewables or emit greenhouse gases (Refs. 1, 2). This program entails using nuclear energy to power reversible solid-oxide fuel cells, electrolyzing steam and carbon dioxide ($CO_2$) simultaneously (Eq. 1):

$$H_2O + CO_2 \longrightarrow H_2 + CO + O_2. \quad (1)$$

Co-electrolysis, however, is significantly more complex than simple steam electrolysis. This is primarily due to the multiple reactions that occur: steam electrolysis, $CO_2$ electrolysis, and the reverse shift reaction (RSR):

$$CO_2 + H_2 \leftrightarrow CO + H_2O. \quad (2)$$

Better understanding of the feasibility of producing syngas using high temperature electrolysis may initiate the systematic investigation of nuclear-powered synfuel production as a bridge to the future hydrogen economy and ultimate independence from foreign energy resources (Refs. 3, 4).

The rates of steam and $CO_2$ electrolysis are measured via three different, independent methods: 1) electrical current through the stack, 2) the measured change in inlet and outlet steam and $CO_2$ concentrations as measured by the on-line dew point and $CO_2$ sensors, and 3) an on-line microchannel GC. The GC also tests for any additional electrolysis products, such as $CH_4$, that may be produced. Some additional discussion of the test apparatus, experimental procedures, data reduction, and the stack construction may be found in Ref. 5.

Steam electrolysis, $CO_2$ electrolysis, and the RSR are endothermic reactions that tend to depress the cell temperature during electrolysis. However, the cell ohmic heating tends to increase the cell temperature, proportional to the square of the current. These two effects balance each other at the thermal neutral voltage. At operating voltages below thermal neutral, the endothermic heat of reaction dominates and the cell temperature is lower than that at open cell. At operating voltages above thermal neutral, ohmic heating dominates and the cell temperature will exceed that at open cell. For pure steam electrolysis, the thermal neutral voltage is a weak function of temperature only and is equal to 1.287 V at 800°C and 1.288 V at 830°C. For co-electrolysis, however, the thermal neutral voltage is also a function of gas composition. By curve fitting the experimental data and solving for zero temperature depression, experimental thermal neutral voltages were estimated. Experimental open cell potentials were directly measured. To predict the theoretical open cell potentials and thermal neutral voltages, the chemical equilibrium co-electrolysis model discussed above was extended to include an energy balance. This more complex model is discussed in Ref. 6.

The experiments demonstrated that even at zero current there was a drop in $CO_2$ and $H_2$ mole fractions from the cold inlet values, with CO produced. This is solely due to the RSR. As the electrolysis current was increased, the yield of syngas increased linearly while the concentration of $CO_2$ and $H_2O$ decreased. These experiments also show overall good agreement between experimental GC data and results from the chemical equilibrium co-electrolysis model for the range of testing performed in this study.

## Acknowledgment

This work was supported by the U.S. Department of Energy, Office of Nuclear Energy, under the Nuclear Hydrogen Initiative Program.

## References

1. C.M. Stoots, J.E. O'Brien, G.L. Hawkes, J.S. Herring, and J.J. Hartvigsen, "High Temperature Co-Electrolysis of H2O and CO2 for Syngas Production," *2006 Fuel Cell Seminar*, Honolulu, Hawaii, Nov. 13-17, 2006, paper no. 418.
2. G.L. Hawkes, J.E. O'Brien, C.M. Stoots, R. Jones, "Three Dimensional CFD Model of a Planar Solid Oxide Electrolysis Cell for Co-Electrolysis of Steam and Carbon-Dioxide," *2006 Fuel Cell Seminar*, Honolulu, Hawaii, Nov. 13-17, 2006, paper no. 298.
3. S.H. Jensen, J.V.T. Høgh, R. Barfod, M. Mogensen, "High temperature electrolysis of steam and carbon dioxide," In *Energy technologies for Post Kyoto targets in the medium term,* Proceedings of *Risø International Energy Conference*, Risø, Denmark, May 19-21, 2003.
4. S. H. Jensen, and M. Mogensen, "Perspectives of High Temperature Electrolysis Using SOEC," *19th World Energy Congress 2004*, Sydney, Australia, September 5-9, 2004.
5. J.E. O'Brien, C.M. Stoots, and J.Hartvigsen, "Hydrogen Production Performance of a 10-Cell Planar Solid-Oxide Electrolysis Stack," *Journal of Fuel Cell Science and Technology*, **3**, 1 (May 2006).
6. J.E. O'Brien, C.M. Stoots, G.L. Hawkes, J.S. Herring, and J.Hartvigsen, "High-Temperature Co-electrolysis of Carbon Dioxide and Steam for the Production of Syngas; Equilibrium Model and Single-Cell Tests," *Safety and Technology of Nuclear Hydrogen Production, Control and Management (ST-NH2), Embedded Topical Meeting, ANS Annual Meeting*, June 24-28, 2007.

# RESULTS OF RECENT HIGH TEMPERATURE COELECTROLYSIS STUDIES AT THE IDAHO NATIONAL LABORATORY

*Carl Stoots, Idaho National Laboratory, Idaho Falls, ID, USA*
*James O'Brien, Idaho National Laboratory, Idaho Falls, ID, USA*
*Joseph Hartvigsen, Ceramatec Inc., Salt Lake City, UT, USA*

## Introduction

For the past several years, the Idaho National Laboratory (INL) and subcontractor Ceramatec, Inc. have been studying the feasibility of high temperature solid oxide electrolysis for large-scale, nuclear-powered hydrogen production. Parallel to this effort, the INL and Ceramatec have been researching high temperature coelectrolysis of steam/$CO_2$ mixtures to produce syngas:

$$H_2O + CO_2 \longrightarrow H_2 + CO + O_2 \tag{1}$$

the raw material for synthetic fuels production. The INL and Ceramatec have been conducting experiments to characterize the electrochemical performance of coelectrolysis, as well as validate INL-developed computer models. An inline methanation reactor has also been tested to study direct methane production from coelectrolysis products. These activities include bench scale experimentation, modeling, and flow sheet analysis. When linked to nuclear power, high temperature coelectrolysis can provide a carbon neutral means of producing syngas while consuming $CO_2$.

Coelectrolysis is significantly more complex than simple steam electrolysis. This is primarily due to the multiple reactions that occur: steam electrolysis, $CO_2$ electrolysis, and the reverse shift reaction (RSR):

$$CO_2 + H_2 \leftrightarrow CO + H_2O. \tag{2}$$

Reaction kinetics govern the relative contributions of these three reactions. It is also important to note that the electrolysis reactions are not equilibrium reactions. The electrolyte separates the products from the reactants. However, the RSR is a kinetically fast, equilibrium reaction at high temperature in the presence of a Ni catalyst. Also, if the cell potential is high enough, CO can be further electrolyzed to elemental C:

$$CO \rightarrow C + 1/2\,O_2 \tag{3}$$

producing solid particulates that can then deposit on cell surfaces and reduce cell performance. Finally, there could be material compatibility issues related to corrosion and seal leakage.

It is also feasible to produce syngas by separately electrolyzing steam and $CO_2$. There are, however, significant advantages to electrolyzing steam and $CO_2$ simultaneously, the primary one of which is electrical efficiency. For a given solid oxide electrolysis cell, pure $CO_2$ electrolysis will exhibit a higher area specific resistance (ASR) than steam electrolysis. This is due to the slower overall kinetics of $CO_2$ electrolysis and the higher overpotentials required. In coelectrolysis, the reverse gas shift reaction is relied upon for most of the CO production and therefore the overall electrical requirement is less.

Some results of $CO_2$ / $H_2O$ electrolysis experiments performed to date using button cells and three different 10-cell planar solid oxide stacks are presented and discussed. These results include electrolysis performance at various temperatures, gas mixtures, and electrical settings. Product gas compositions, as measured via an online micro gas chromatograph (GC), are compared to predictions

**Fig. 1**. Photograph of the INL high-temperature electrolysis laboratory.

obtained from an INL-developed chemical equilibrium coelectrolysis model (CECM). Better understanding of the feasibility of producing syngas using high temperature electrolysis may initiate the systematic investigation of nuclear-powered synfuel production as a bridge to the future hydrogen economy and ultimate independence from foreign energy resources.

## Experimental Test Facility

A comprehensive discussion of the INL high temperature solid oxide electrolysis test facility is presented elsewhere [1,2,3]. This same facility is used for button cell testing as well as stack testing. A photograph of the test hardware is in Fig. 1. Primary components include gas supply cylinders, mass-flow controllers, a humidifier, dewpoint measurement stations, carbon dioxide concentration measurement stations, microchannel gas chromatograph, temperature and pressure measurement, high temperature furnaces, and a solid oxide electrolysis cell/stack.

For single-cell testing, an electrolysis button cell is bonded to the bottom of a zirconia tube, as shown in Fig. 2. During testing, the tube is suspended in the smaller furnace. The cell is an electrolyte-supported single button cell with a scandia-stabilized zirconia electrolyte, about 150 μm thick. The outside electrode, which acts as the cathode in fuel cell mode and the anode in electrolysis mode, is a doped manganite. The inside electrode (electrolysis cathode) material is a nickel cermet. Both button-cell electrodes incorporate a platinum wire mesh for current distribution. The button

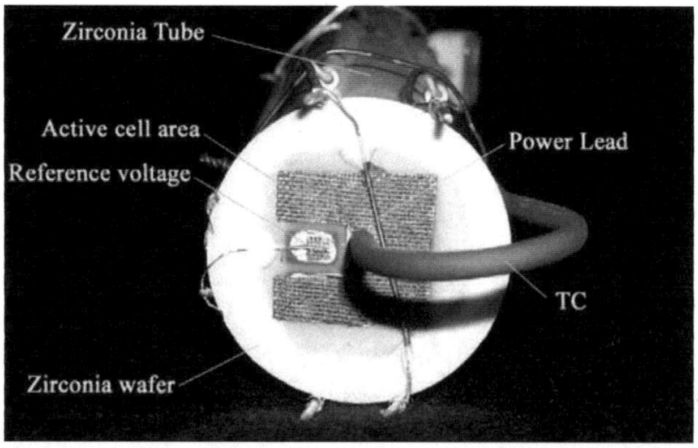


**Fig. 2.** Detail of button cell.

**Fig. 3.** 10-cell stack mounted on test fixture on furnace base, ready to test.

cell includes both an active cell area (2.5 $cm^2$ for the cell shown) and a reference cell area. A type-K stainless-steel sheathed thermocouple, is mounted on the manifold tube and bent around in front of the button cell in order to allow for continuous monitoring of the button-cell temperature.

For stack testing, the inlet gas mixture is directed to the larger high temperature furnace (Skutt Model KS818-3), capable of producing temperatures up to 1250°C, which heats and maintains the electrolyzer at the appropriate test temperature via computer-based feedback control. The furnace also preheats the inlet gas mixture and the air sweep gas. A photograph of the stack, mounted on its inconel test fixture and resting on the furnace base, is shown in Fig. 3. The button cells and stacks were fabricated by Ceramatec, Inc., of Salt Lake City, UT. The stacks have a per-cell active area of 64 $cm^2$, for a total active area of 640 $cm^2$ each.

## Single-Cell Experimental Results

To understand the impact of the electrolysis reactions and the RSR discussed above, and to assist with interpretation of experimentally measured data, a CECM was developed. This model also served to help determine the necessary inlet conditions for the range of experiments that were conducted. A discussion of the INL-developed model can be found in [4,5].

Coelectrolysis performance was characterized through a series of stepwise DC potential sweeps. Results of three sweeps are presented in Fig. 4. The furnace temperature for all three sweeps was 800°C. Inlet gas flow rates and inlet dewpoint values for each of these three sweeps are provided in Table 1. The corresponding inlet volume flow rate of steam is also provided in the table. Sweep 1 had the highest steam flow rate; sweep 3 had the lowest. Note that the flow rates used for these test were quite small. Low flow rates were required in order to achieve reasonable steam and $CO_2$ utilization values with low values of total cell current. The single cell, with an active area of 2.5 $cm^2$, could only support a maximum total current of about 0.75 A.

Fig. 4 plots the mole percent of $H_2$, CO, and $CO_2$ as a function of cell current on a dry basis. The data symbols represent measurements obtained from the gas chromatograph. The lines represent predictions based on our CECM. Two lines are shown for each case. The dashed lines represent CECM predictions based on an effective equilibrium temperature of 700°C. The dotted lines represent CECM predictions based on an effective equilibrium temperature of 650°C. The RSR equilibrium constant is a function of temperature. Since product gases cool to room temperature before

**Table 1.** Inlet conditions for DC potential sweeps.

| | Sweep 1 | Sweep 2 | Sweep 3 |
|---|---|---|---|
| $Q_{N2}$ (sccm) | 35 | 50 | 40 |
| $Q_{H2}$ (sccm) | 4 | 3 | 8 |
| $Q_{CO2}$ (sccm) | 8 | 6 | 8 |
| $T_{dp,i}$ (°C) | 55.1 | 44.6 | 30.1 |
| $Q_{H2O}$ (sccm) | 10.67 | 7.27 | 2.93 |
| $T_f$ (°C) | 800 | 800 | 800 |

analysis in the micro GC, it was not certain what value to use for an "apparent" equilibrium temperature for the products. Therefore, the chemical equilibrium coelectrolysis model was run for several different equilibrium temperatures.

During coelectrolysis, the mole fractions of $CO_2$ and steam (not shown in Fig. 4) decrease with current, while the mole fractions of $H_2$ and CO increase. For the conditions chosen for these tests, the ratio of $H_2$ to CO is close to the desired 2-to-1 value for syngas production. Measured compositions of $CO_2$ and CO agree best with predictions based on an effective equilibrium temperature of 700°C. Measured compositions of $H_2$ agree best with predictions based on an effective equilibrium temperature of 650°C.

## Discussion of 10-Cell Stack Test Results

Table 2 lists the coelectrolysis stack testing conducted. Cell area-specific resistance, ASR, quantifies the loss mechanisms associated with cell operation, including ohmic resistances, contact resistances, and ionic resistances, activation and concentration overpotentials. The ASR value is dependent on the type of electrolysis being conducted, with pure $CO_2$ electrolysis exhibiting a significantly higher ASR than pure steam electrolysis [4,5]. However, in coelectrolysis the RSR is relied upon for most $CO_2$-to-CO conversion, and steam electrolysis is the primary electrolytic reaction. Therefore, there is little change in ASR from steam electrolysis to coelectrolysis. To demonstrate this, polarization curves were generated for stack #3 for steam electrolysis, $H_2O/CO_2$ coelectrolysis, and $CO_2$ electrolysis. Once the stack was at the operating temperature of 800°C, a steam electrolysis polarization curve was generated by performing a voltage sweep for the conditions labeled "Test #20" in Table 2. This same voltage sweep was repeated for

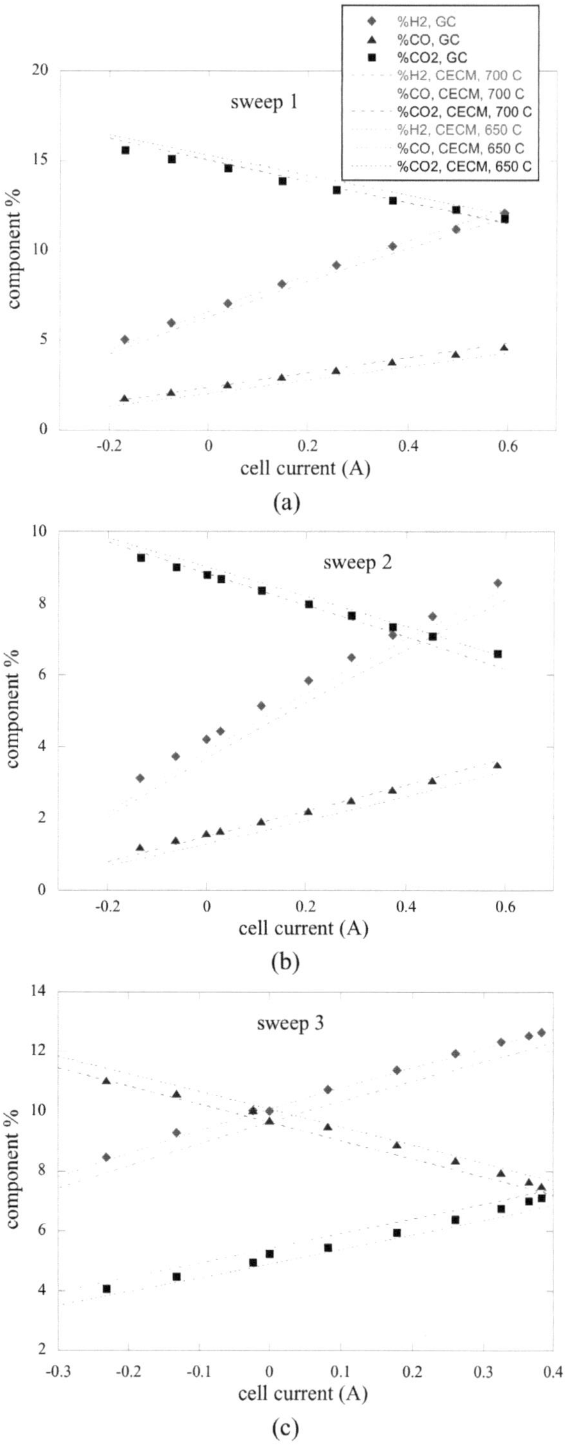


**Fig. 4.** Measured outlet gas compositions, with comparisons to predictions from the chemical equilibrium coelectrolysis model, sweeps 1 – 3.

**Table 2.** Summary of stack testing conditions.

| Test # | Stack # | Sweep Type | Stack Age | $T_{furnace}$ (C) | Flow Rates | | | Inlet Dew Point (C) | Molar Composition | | | |
|---|---|---|---|---|---|---|---|---|---|---|---|---|
| | | | | | $H_2$ (sccm) | $CO_2$ (sccm) | $N_2$ (sccm) | | $H_2$ (mol %) | $CO_2$ (mol %) | $N_2$ (mol %) | $H_2O$ (mol %) |
| 0 | 1 | Fast | Fresh | 800 | 497 | 0 | 3010 | 51.5 | 12.0 | 0 | 72.6 | 15.4 |
| 1 | 1 | Fast | Fresh | 800 | 497 | 605 | 3010 | 51.5 | 10.2 | 12.4 | 61.9 | 15.5 |
| 2 | 1 | Slow | Fresh | 800 | 497 | 605 | 3010 | 51.5 | 10.2 | 12.4 | 61.9 | 15.5 |
| 3 | 1 | Slow | Fresh | 800 | 497 | 505 | 2510 | 45.5 | 12.6 | 12.7 | 63.3 | 11.4 |
| 4 | 1 | Slow | Fresh | 800 | 497 | 705 | 1010 | 66.0 | 15.6 | 22.2 | 31.8 | 30.4 |
| 5 | 1 | Slow | Fresh | 800 | 497 | 756 | 3010 | 74.0 | 6.7 | 10.1 | 40.2 | 43.0 |
| 6 | 1 | Slow | Fresh | 828 | 497 | 605 | 3011 | 51.5 | 10.2 | 12.4 | 61.9 | 15.5 |
| 7 | 1 | Slow | Fresh | 828 | 497 | 756 | 3513 | 65.3 | 7.3 | 11.2 | 52.0 | 29.5 |
| 8 | 2 | Fast | Fresh | 800 | 450 | 0 | 1413 | 75.3 | 13.2 | 0 | 41.4 | 45.4 |
| 9 | 2 | Fast | Fresh | 800 | 449 | 753 | 1414 | 59.8 | 13.2 | 22.2 | 41.6 | 23.0 |
| 10 | 2 | Fast | Fresh | 800 | 334 | 564 | 1064 | 59.7 | 13.1 | 22.2 | 41.8 | 22.9 |
| 11 | 2 | Fast | Fresh | 800 | 213 | 378 | 710 | 60.7 | 12.4 | 22.1 | 41.5 | 24.0 |
| 12 | 2 | Slow | 1 Week | 800 | 449 | 753 | 1414 | 59.8 | 13.2 | 22.2 | 41.6 | 23.0 |
| 13 | 2 | Slow | 1 Week | 800 | 334 | 564 | 1064 | 59.7 | 13.1 | 22.2 | 41.8 | 22.9 |
| 14 | 2 | Slow | 1 Week | 800 | 213 | 378 | 710 | 60.7 | 12.4 | 22.1 | 41.5 | 24.0 |
| 15 | 2 | Slow | 1 Week | 830 | 497 | 756 | 3513 | 65.3 | 7.4 | 11.2 | 52.0 | 29.4 |
| 16 | 2 | Slow | 2 Weeks | 800 | 450 | 752 | 1410 | 60.3 | 13.2 | 22.0 | 41.3 | 23.5 |
| 17 | 2 | Slow | 2 Weeks | 800 | 334 | 564 | 1064 | 60.2 | 13.0 | 22.0 | 41.5 | 23.5 |
| 18 | 2 | Slow | 2 Weeks | 800 | 213 | 378 | 710 | 60.7 | 12.4 | 22.1 | 41.5 | 24.0 |
| 19 | 2 | Slow | 2 Weeks | 830 | 497 | 756 | 3513 | 65.3 | 7.4 | 11.2 | 52.0 | 29.4 |
| 20 | 3 | Fast | Fresh | 800 | 996 | 0 | 1009 | 80 | 22.5 | 0 | 22.7 | 54.8 |
| 21 | 3 | Fast | Fresh | 800 | 996 | 1003 | 0.0 | 80 | 22.5 | 22.6 | 0 | 54.9 |
| 22 | 3 | Fast | Fresh | 800 | 0 | 1500 | 0 | 0 | 0 | 100.0 | 0 | 0 |

the coelectrolysis conditions "Test #21" and for the $CO_2$ electrolysis conditions "Test #22". These results are shown in Fig. 5.

There was almost no change in apparent ASR for coelectrolysis versus steam electrolysis. However, the ASR for $CO_2$ electrolysis was significantly higher, reinforcing the hypothesis that steam electrolysis is the principal electrolysis reaction and that the RSR is mostly responsible for CO production.

Tests 1 through 6 comprised very slow voltage sweeps where for each power supply voltage setting the stack was allowed to reach thermal and chemical equilibrium. Thermal equilibrium was determined by watching the response of the stack internal thermocouples. Chemical equilibrium was recognized when the downstream dew point reading reached steady state. These two conditions were usually met after approximately 10 minutes at each setting.

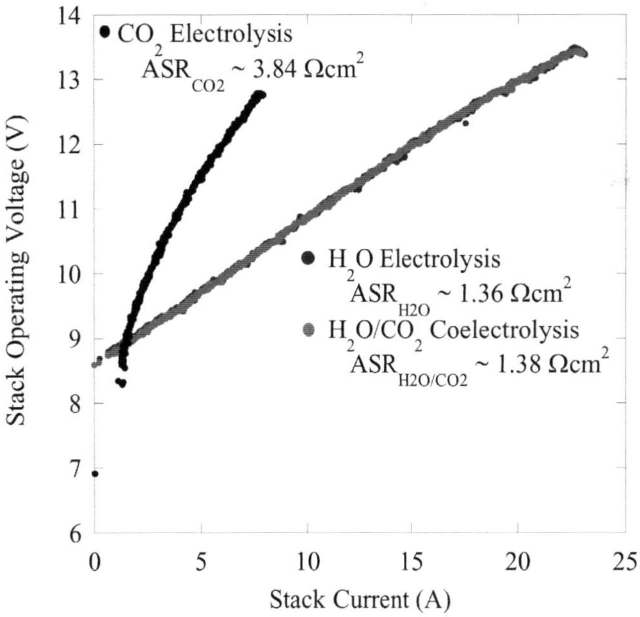


**Fig. 5.** Polarization curves for $H_2O$ electrolysis, $H_2O/CO_2$ coelectrolysis versus $CO_2$ coelectrolysis, with mean ASR values.

Fig. 6 presents internal stack temperature depression (the difference between the temperature measured during the sweep and the temperature at open cell conditions) as a function of stack operating voltage for the electrolysis conditions listed in Table 2. When a solid oxide cell is operated as a fuel cell, the exothermic heat of reaction as well as ohmic heating tends to increase the cell temperature and excess air is typically required for cooling. Steam electrolysis, $CO_2$ electrolysis, and the RSR, however, are endothermic reactions which tend to reduce the cell temperature in proportion to the electrolysis current. Ohmic heating is proportional to the square of the electrolysis current. These two effects balance each other at the thermal neutral voltage. At operating voltages below thermal neutral, the endothermic heat of reaction dominates and the cell temperature is lower than that at open cell. At operating voltages above thermal neutral, ohmic heating dominates and the cell temperature will exceed that at open cell. For pure steam electrolysis, the thermal neutral voltage is a weak function of temperature only and is equal to 1.287 V at 800°C and 1.288 V at 830°C. For coelectrolysis the thermal neutral voltage is also a function of gas composition.

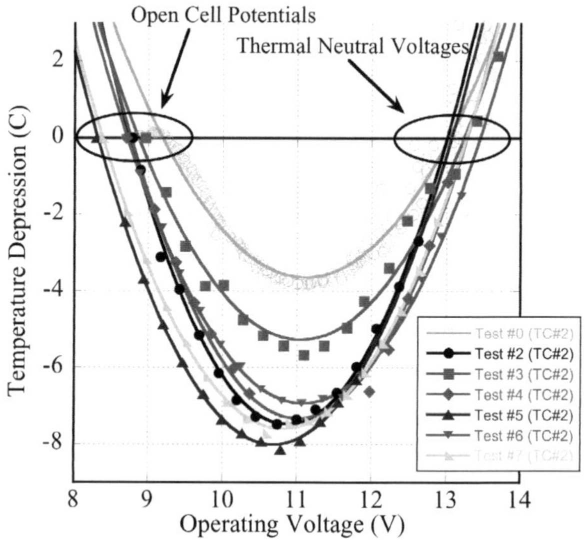


**Fig 6.** Internal stack temperature (thermocouple #2) for various test conditions.

Figs. 7-12 present the steady state outlet compositions of steam, $CO_2$, hydrogen, and CO as a function of electrolysis current on a dry basis for tests 1-6. Lines represent various model predictions and symbols represent experimental measurements. An in-line gas chromatograph was used to measure outlet compositions from the stack. The chemical equilibrium coelectrolysis model was run for several different equilibrium temperatures ranging from 650°C to 800°C (Fig. 7). It was found that setting the chemical equilibrium coelectrolysis model equilibrium temperature equal to the furnace temperature produced the best comparisons, indicating that the products are kinetically frozen after they leave the hot zone. This is different than the button cell case and is due to the much higher gas flow rates and correspondingly shorter residence times, lack of any significant catalyst surface, and the rapid cool-down. Predicted compositions were therefore evaluated at the electrolyzer temperature for all subsequent evaluations (Fig. 8-12).

Figs. 7-12 demonstrate that even at zero current there was a drop in $CO_2$ and $H_2$ mole fractions from the cold inlet values, with CO produced. This is solely due to the RSR. As the electrolysis current was increased, the yield of syngas increased linearly while the concentration of $CO_2$ (and $H_2O$, not shown in the figures) decreased. These figures also show overall good agreement between experimental GC data and results from the CECM for the range of testing performed in this study. Finally, in the case of Test #6, at the maximum current studied the product $H_2$ concentration was doubled and product $CO_2$ concentration was reduced to half that of the process inlet mixture. Coelectrolysis significantly increases the yield of syngas over the reverse water gas shift reaction equilibrium composition.

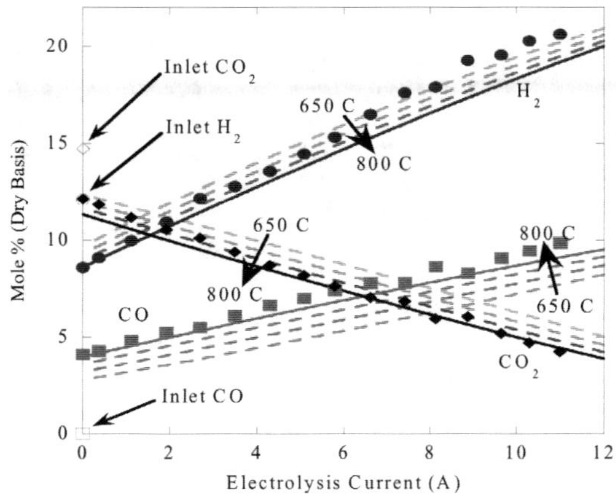


**Fig. 7.** Effect of varying chemical equilibrium coelectrolysis model equilibrium temperature with comparison to Test 1 experimental data.

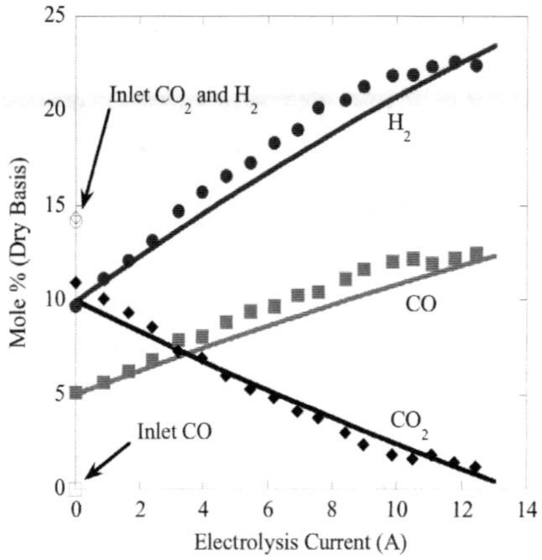


**Fig 8.** Test 2 experimental and chemical equilibrium coelectrolysis model results, $T_{eq} = 800°C$.

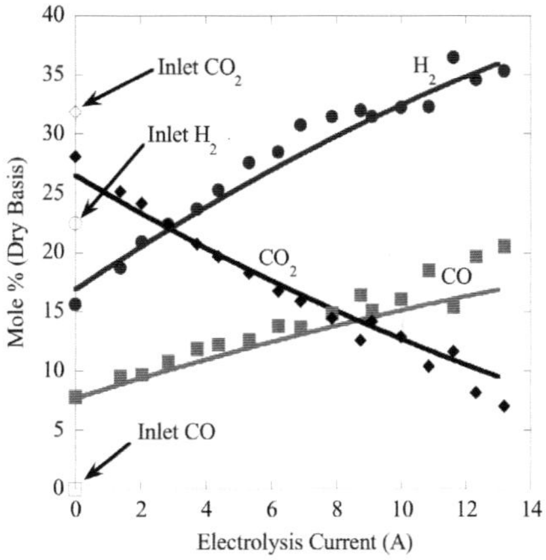


**Fig. 9.** Test 3 experimental and chemical equilibrium coelectrolysis model results, $T_{eq} = 800°C$.

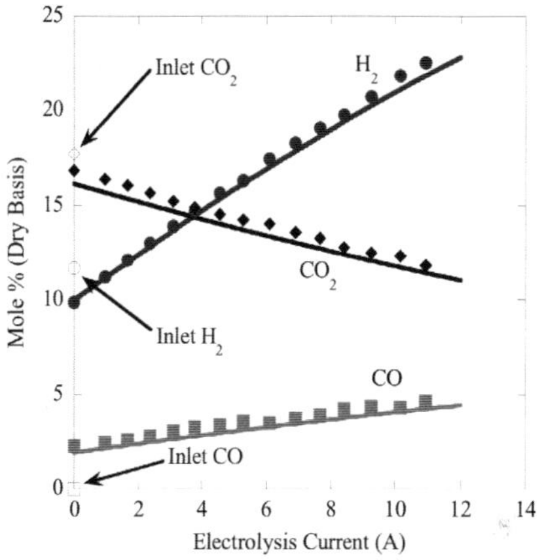


**Fig. 10.** Test 4 experimental and chemical equilibrium coelectrolysis model results, $T_{eq} = 800°C$.

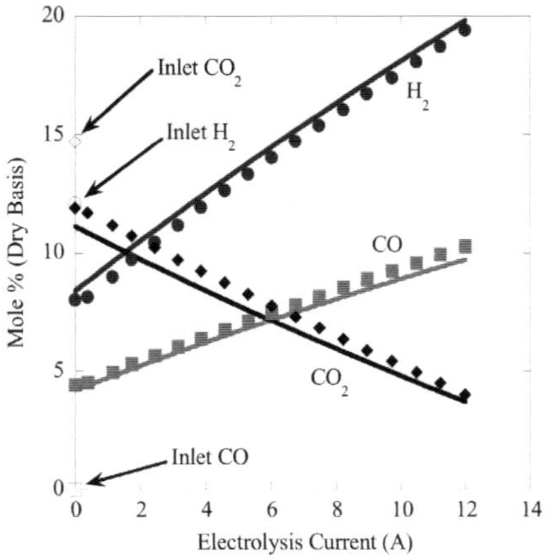


**Fig. 11.** Test 5 experimental and chemical equilibrium coelectrolysis model results, $T_{eq} = 828°C$.

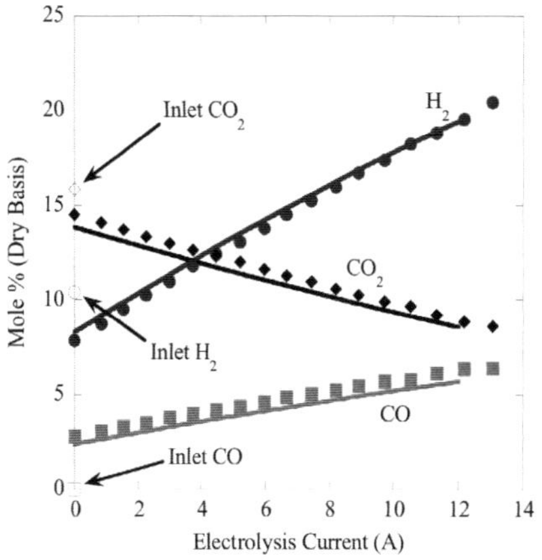


**Fig. 12.** Test 6 experimental and chemical equilibrium coelectrolysis model results, $T_{eq} = 828°C$.

## Methanation Test Results

Ceramatec Inc. extended their 10-cell stack testing apparatus by addition of a methanation reactor downstream of the stack. The methanation reactor consisted of a 18mm inner diameter stainless steel tube, approximately 1.5 m in length. Within this tube was placed a commercial steam reforming catalyst (R-67R from Haldor Topsoe). This is a nickel catalyst on magnesium aluminate, a ceramic inert oxide of the spinel family. The reactor tube was placed within a zinc-aluminum sleeve to homogenize the axial temperature gradient. The reactor was maintained at approximately 300 C for testing. Testing consisted of high temperature coelectrolysis, with the coelectrolysis products immediately fed to the methanation reactor. Fig. 13 summarizes the stack inlet, stack outlet, and methanation outlet stream compositions (volume %) for 5 tests. Between 40 and 50 volume % methane product was produced.

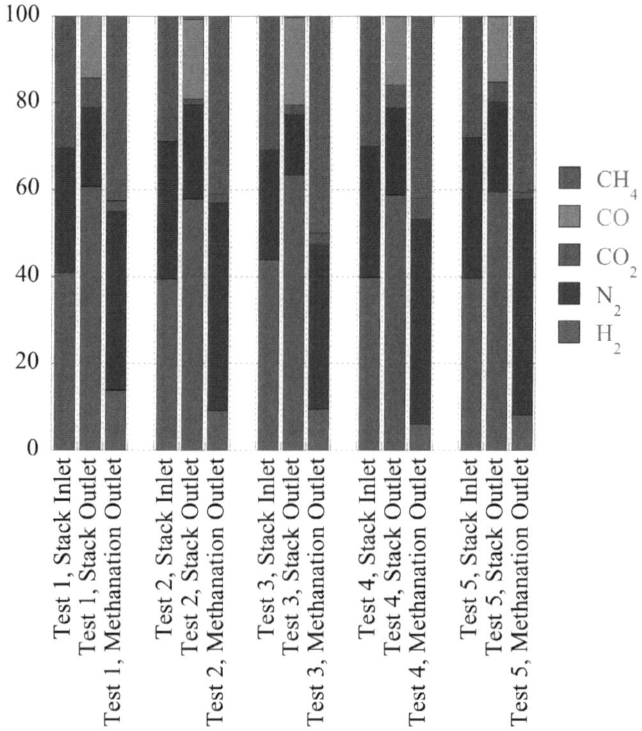


**Fig 13.** Coelectrolysis with subsequent methanation.

## Summary

Synthetically-derived hydrocarbon fuels can offer an interim solution for domestic energy dependence and a bridge to the hydrogen economy. The raw material for synfuel production is syngas, a mixture of $H_2$ and CO. The Idaho National Laboratory has demonstrated the feasibility of using high temperature solid oxide cells to coelectrolyze $H_2O$ and $CO_2$ simultaneously to produce syngas. ASRs did not vary substantially between steam electrolysis and steam/$CO_2$ coelectrolysis, but increased dramatically with $CO_2$ electrolysis. There was overall good agreement between experimental gas chromatograph data for outlet compositions and results from the chemical equilibrium coelectrolysis model. These measurements and model results indicate that coelectrolysis effectively increases the syngas yield for a given steam/$CO_2$ process stream. The concept of directing coelectrolysis products through a methanation reactor was tested, with yields of 40-50 volume % methane being produced. Overall, the coelectrolysis process appears to be a promising technique for large-scale syngas production.

## Acknowledgements

This work was supported by the Idaho National Laboratory, Laboratory Directed Research and Development program and the U.S. Department of Energy, Office of Nuclear Energy, Nuclear Hydrogen Initiative Program.

# References


1. Stoots, C.M., O'Brien, J.E., Hawkes, G.L., Herring, J.S., Hartvigsen, J.J, "High Temperature Co-Electrolysis of H2O and CO2 for Syngas Production," 2006 Fuel Cell Seminar, Honolulu, Hawaii, Nov. 13-17, 2006, paper no. 418.
2. O'Brien, J.E., Stoots, C.M., Herring, J.S., Hawkes, G.L., Hartvigsen, J.J., "Thermal And Electrochemical Performance Of A High-Temperature Steam Electrolysis Stack," 2006 Fuel Cell Seminar, Honolulu, Hawaii, Nov. 13-17, 2006, paper no. 417.
3. O'Brien, J. E., Stoots, C. M., Herring, J. S., and Hartvigsen, J. J.,"Hydrogen Production Performance of a 10-Cell Planar Solid-Oxide Electrolysis Stack," *Journal of Fuel Cell Science and Technology*, Vol. 3, pp. 213-219, May, 2006.
4. Stoots, C.M., O'Brien, J.E., Hartvigsen, J.J., "Syngas Production Via High-Temperature Coelectrolysis Of Steam And Carbon Dioxide In A Solid-Oxide Stack," ASME 5$^{th}$ International Fuel Cell Science, Engineering & Technology Conference, New York, June 18-20, 2007, paper no. 25208.
5. Stoots, C.M., O'Brien, J.E., Hartvigsen, J.J., "Test Results Of High Temperature Steam/$CO_2$ Coelectrolysis In A 10-Cell Stack," American Nuclear Society Embedded Topical Meeting: Safety and Technology of Nuclear Hydrogen Production (ST-$NH_2$), Boston, June 24-28, 2007.

# PARAMETRIC STUDY OF LARGE-SCALE PRODUCTION OF SYNGAS VIA HIGH TEMPERATURE CO-ELECTROLYSIS

*J. E. O'Brien, M. G. McKellar, C. M. Stoots, J. S. Herring and G. L. Hawkes*
*Idaho National Laboratory*
*Idaho Falls, ID 83415, USA*

## ABSTRACT

A process model has been developed to evaluate the potential performance of a large-scale high-temperature co-electrolysis plant for the production of syngas from steam and carbon dioxide. The co-electrolysis process allows for direct electrochemical reduction of the steam – carbon dioxide gas mixture, yielding hydrogen and carbon monoxide, or syngas. The process model has been developed using the Honeywell UniSim systems analysis code. Using this code, a detailed process flow sheet has been defined that includes all the components that would be present in an actual plant such as pumps, compressors, heat exchangers, turbines, and the electrolyzer. Since the electrolyzer is not a standard UniSim component, a custom one-dimensional co-electrolysis model was developed for incorporation into the overall UniSim process flow sheet. The one dimensional co-electrolysis model assumes local chemical equilibrium among the four process-gas species via the gas shift reaction. The electrolyzer model allows for the determination of co-electrolysis outlet temperature, composition (anode and cathode sides); mean Nernst potential, operating voltage and electrolyzer power based on specified inlet gas flow rates, heat loss or gain, current density, and cell area-specific resistance. The one-dimensional electrolyzer model was validated by comparison with results obtained from a fully three dimensional computational fluid dynamics model developed using FLUENT, and by comparison to experimental data. This paper provides representative results obtained from the UniSim flow sheet model for a 300 MW co-electrolysis plant, coupled to a high-temperature gas-cooled nuclear reactor. The co-electrolysis process, coupled to a nuclear reactor, provides a means of recycling carbon dioxide back into a useful liquid fuel. If the carbon dioxide source is based on biomass, the overall process, from production through utilization, would be climate neutral.

## NOMENCLATURE

| | |
|---|---|
| $A_{cell}$ | electrolysis cell active area, $cm^2$ |
| $ASR$ | area-specific resistance, $Ohm \cdot cm^2$ |
| $F$ | Faraday number, 96487 C/mol |
| $I_e$ | total ionic current, Ampere |
| $i$ | current density, $Ampere/cm^2$ |
| $LHV_i$ | lower heating value of component i, J/mol |
| $\dot{N}_i$ | molar flow rate of component i, mol/s |
| $N_{cells}$ | number of cells in electrolysis stack |
| $\dot{Q}$ | heat transfer rate, W |
| $T$ | temperature, K |
| $\eta_{syn}$ | overall thermal-to-syngas process efficiency |

## INTRODUCTION

Large-scale production of synthetic liquid fuels represents one possible path toward greater energy independence. Primary advantages of synthetic liquid fuels, as compared to hydrogen, are that the infrastructure for liquid fuel distribution is already in place and on-board-vehicle storage is not an issue. However, these fuels will release carbon dioxide to the atmosphere when burned. Therefore, climate-neutral methods of synfuel production are most desirable. If the energy input to the synfuel production process is based on nuclear energy, and if the carbon source is based on biomass, a large-scale climate-neutral synthetic fuel production strategy could be achieved. Nuclear-powered high-temperature electrolysis of steam and carbon dioxide to produce syngas, with subsequent Fisher-Tropsch conversion to liquid fuel represents such a strategy.

This research is an outgrowth of ongoing work at the Idaho National Laboratory (INL), on high-temperature steam electrolysis for hydrogen production [1], funded by the US Department of Energy under the Nuclear Hydrogen Initiative. High-temperature electrolysis (HTE) is one of two technologies under evaluation for large-scale hydrogen production based on nuclear energy. Advanced high-temperature nuclear reactors have the potential to enable efficient, large-scale, carbon-free hydrogen production [2]. Large-scale nuclear hydrogen production based on water-splitting is already under serious consideration in the short term to supply hydrogen for upgrading of low-quality petroleum resources such as the Athabasca Oil Sands [3]. In the intermediate term, large-scale hydrogen production will be required for the production of synthetic liquid hydrocarbon fuels. In the long term, large-scale hydrogen production may fuel the hydrogen economy.

High-temperature electrolysis is based on solid oxide fuel cell (SOFC) technology and materials. The zirconia electrolytes used for SOFCs conduct oxygen ions, so they can be used to electrolyze steam ($H_2O$) to hydrogen ($H_2$), and/or carbon dioxide ($CO_2$) to carbon monoxide (CO). When both steam and carbon dioxide are present simultaneously in the feed stream, the total amounts of hydrogen and carbon monoxide that are produced depend on the electrolysis current. The relative amount of hydrogen produced versus carbon monoxide is determined by the relative amounts of steam, hydrogen (included in the feed stream as a reducing agent) and carbon dioxide included in the feed stream and by the effect of the gas shift reaction:

$$CO_2 + H_2 <=> H_2O + CO \tag{1}$$

The desired molar ratio of hydrogen to carbon monoxide in the gaseous product depends on the particular liquid fuel to be produced as a final product, but a 2-to-1 ratio of $H_2$ to CO is typical.

In order to evaluate the potential syngas-production performance of large-scale high-temperature co-electrolysis operations, we have developed an engineering process model at INL using the commercial system-analysis code UniSim. Using this code, several detailed process flow sheets have been defined that include all of the components that would be present in an actual high-temperature co-electrolysis (HTCE) plant such as pumps, compressors, heat exchangers, turbines, and the electrolyzer. However, since the actual electrolyzer is not a standard UniSim component, custom one-dimensional co-electrolysis models have been developed for both steam [4, 5] and steam/$CO_2$ electrolysis for incorporation into the overall process flow sheet. Details of this one-dimensional co-electrolysis electrolyzer model were provided in [5]. This paper will provide details of an overall co-electrolysis system process model, with representative results, over a range of operating conditions. Results of similar process simulations for pure steam electrolysis were provided in reference [6].

## PROCESS DESCRIPTION

A graphical representation of the process model developed for this study is presented in Fig. 1. The primary process feedstock streams are liquid water and carbon dioxide. The inlet water stream is compressed in the liquid phase to the process operating pressure of 3.5 MPa using a pump. This operating pressure was selected because it is approximately equal to the desired operating pressure for a Fisher-Tropsch process using a cobalt catalyst. Downstream of the pump, condensate from the water knockout tank is recycled back into the inlet stream at M3. The water stream is then vaporized and pre-heated in the electrolysis recuperator, which recovers heat from the post-electrolyzer process and sweep-gas outlet streams. Downstream of the recuperator, at M2, the steam is mixed with carbon dioxide plus recycled hydrogen and carbon monoxide product gas. A fraction of the product gas is recycled in this way in order to assure that reducing conditions are maintained on the steam/hydrogen electrode.

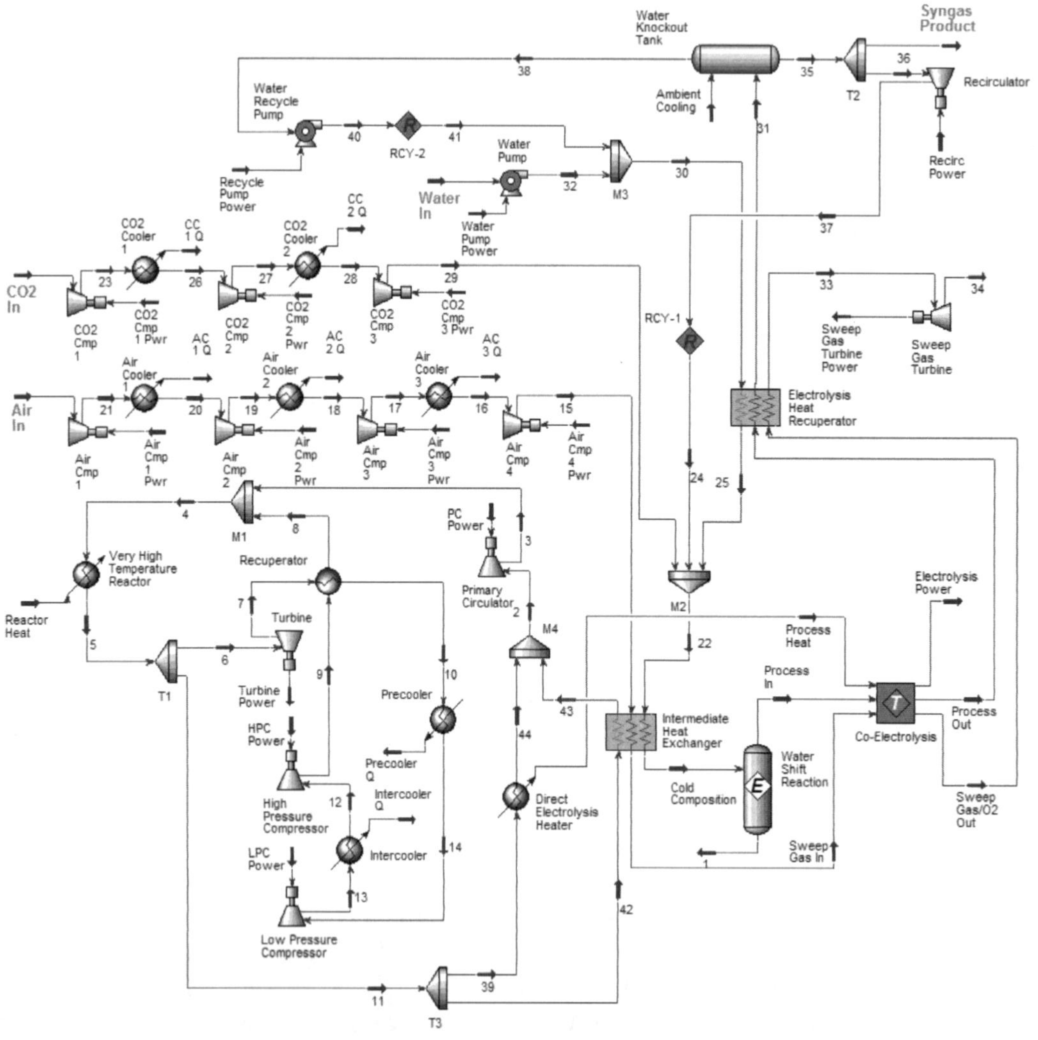


**Figure 1.** Process flow diagram for co-electrolysis plant.

Downstream of the mixer, the process gas mixture enters the intermediate heat exchanger (IHX), where final heating to the electrolysis operating temperature occurs, using high-temperature process heat from the nuclear reactor. A gas shift reaction occurs with heated gas mixture represented by an equilibrium reactor in the process flow diagram, allowing chemical equilibrium to be achieved. The process stream then enters the electrolyzer, where oxygen is electrolytically removed from the system, producing hydrogen and carbon monoxide.

The UniSim version of the co-electrolysis model utilizes built-in features of UniSim as much as possible. UniSim inherently ensures mass and energy balances across all components, includes thermodynamic data for all chemical species, calculates chemical equilibrium states for the gas shift reaction, and calculates the heats of reaction for $CO_2$ and $H_2O$ electrolysis. The rate of electrolytic oxygen removal from the process stream, the mean Nernst potential and the electrolyzer operating voltage are calculated by means of an embedded spreadsheet. An expanded process flow diagram of the electrolyzer module is shown in Fig. 2. This diagram represents the UniSim implementation of the one-dimensional chemical equilibrium co-electrolysis model discussed in detail in [5]. Within this module, the hot shifted process stream enters a conversion reactor where the steam and/or carbon dioxide are electrolytically reduced. The conversion reactor unit includes both the steam and carbon dioxide reduction reactions. Based on the percent conversion of the steam and $CO_2$, the reactor will calculate the associated heat of reaction. The percent conversion of steam and/or $CO_2$ is determined by the total electrolysis current, which is the product of the current and the number of cells. The molar oxygen removal rate is therefore given by Faraday's law:

$$\Delta \dot{N}_O = \frac{I_e}{2F}; \quad where \quad I_e = i A_{cell} N_{cells} \tag{2}$$

To avoid oxygen starvation, the minimum required inlet steam and $CO_2$ molar flow rates must satisfy the following constraint:

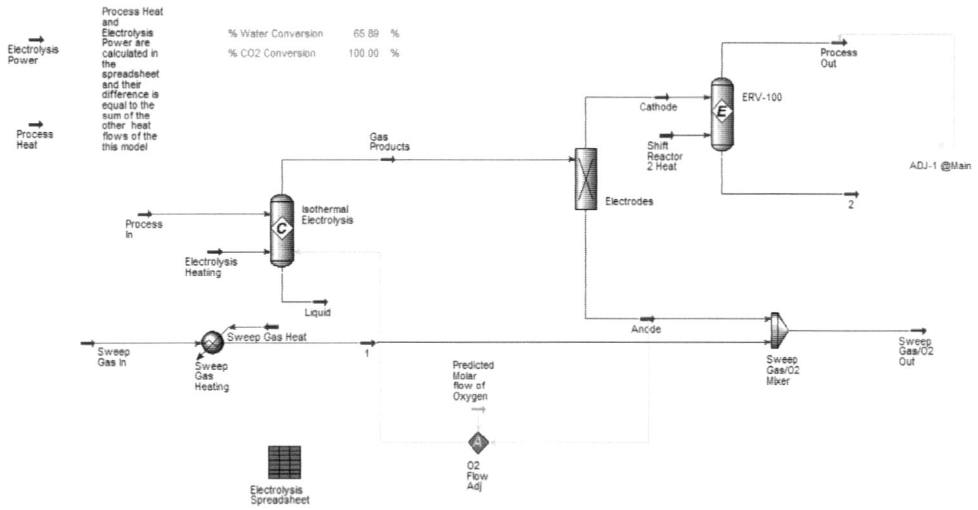


**Figure 2.** Process flow diagram for the electrolysis module within UniSim.

$$\dot{N}_{H_2O} + \dot{N}_{CO_2} \geq \frac{I_e}{2F} \quad (3)$$

Note that the oxygen contribution from the $CO_2$ is only counted once, since we want to avoid creation of carbon soot, which could foul the cells.

This value of the molar flow rate of produced oxygen is stored in a dummy stream. A logical adjust is used to change the percent conversion of steam and carbon dioxide until the oxygen molar flow rate leaving the conversion reactor is the same as the calculated value. The oxygen is split from the rest of the reacted process-gas components by means of a component splitter unit (labeled Electrodes). The split oxygen combines with the sweep gas. The remaining components are passed through a second equilibrium gas shift reactor to determine the outlet equilibrium composition.

The UniSim implementation of the one-dimensional chemical equilibrium co-electrolysis (CEC) model was validated by comparison with an analytical version that was implemented in MathCad and with a full three-dimensional (3-D) computational fluid dynamic (CFD) model developed using FLUENT. Representative results are presented in Fig. 3. A more comprehensive comparison is provided in reference [5]. Electrolyzer-outlet mole fractions of the four process-gas species are presented in Fig. 3 (a) as a function of current density for the per-cell inlet molar flow rates, inlet temperature, area-specific resistance, and thermal boundary condition indicated in the figure. The cold inlet mole fractions corresponding to the indicated molar flow rates are also plotted in the figure for each species. Outlet mole fraction results obtained from the MathCad and UniSim implementations of the CEC are plotted, along with results obtained from a 3-D CFD co-electrolysis simulation obtained using FLUENT. Details of the FLUENT co-electrolysis model are provided in reference [7]. The FLUENT model includes reaction kinetics via separate forward and backward reaction rate constants for the shift reaction. Outlet mole fraction predictions based on the three independent methods are virtually identical. Mole fractions of steam and carbon dioxide decrease with current density, while the mole fractions of hydrogen and carbon monoxide increase. For the inlet

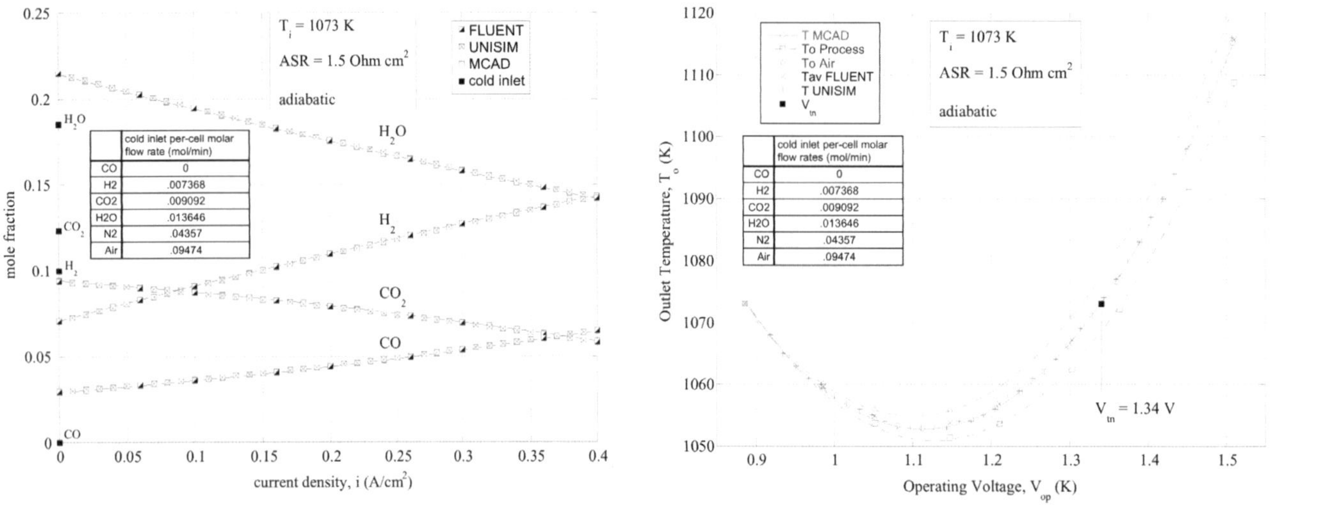


(a) (b)

**Figure 3.** Electrolyzer model results and comparisons, (a) Variation of species mole fraction as a function of current density; (b) Electrolyzer outlet temperature as a function of operating voltage.

composition chosen for this particular run, the ratio of produced hydrogen to carbon monoxide is slightly higher than 2-to-1.

Electrolyzer outlet temperature predictions from the CEC and FLUENT models are presented in Fig. 3 (b) for the same conditions as the results presented in Fig. 3 (a). For low values of operating voltage (and current density), outlet gas temperatures for adiabatic operation are lower than the inlet temperature. The magnitude of this adiabatic temperature depression is dependent on the voltage and the flow rates of the process gases and sweep gas. For operating voltages higher than the thermal neutral voltage (1.34 V in this case), the gas outlet temperatures are higher than the inlet temperature due to the dominance of ohmic heating. The figure shows predictions from the MathCad and UniSim versions of the CEC model as well as predictions from the FLUENT model. The 3-D FLUENT model simulates a single electrolysis cell as it would exist in a planar stack, with a cross-flow arrangement of process and sweep gases. Consequently, the average process-gas and sweep-gas outlet temperatures are not necessarily the same. For this particular case, the outlet temperature of the air sweep was higher than the outlet temperature of the process gas, as shown in the figure. An overall heat-capacity-rate-weighted mean outlet gas temperature was also calculated based on the process-gas and sweep-gas flow rates and temperatures. This result is also shown in Fig. 3 (b). The weighted mean outlet temperature agrees very well with predictions obtained from the CEC model.

Returning to Fig. 1, downstream of the electrolyzer, the hydrogen and carbon monoxide – rich product stream flows through the electrolysis recuperator where the product stream is cooled and the inlet process stream is preheated. The cooled product stream is split at T2 and a fraction of the product gas is recycled into the inlet process stream, as discussed previously. A recirculating blower is required to repressurize the recycle stream to the upstream pressure at M2. The remainder of the product stream is cooled further at the water knockout tank, where the majority of any residual steam is condensed and separated, yielding dry syngas product.

The process flow diagram shows air in use as a sweep gas, to remove the excess oxygen that is evolved on the anode side of the electrolyzer. For the air-sweep cases, inlet air is compressed to the system operating pressure of 3.5 MPa in a four-stage compressor with intercooling. The final compression stage is not followed by a cooler, so the air enters the IHX at about 105°C. The sweep gas is heated to the electrolyzer operating temperature of 800°C via the IHX which supplies high-temperature nuclear process heat directly to the system. The sweep gas then enters the electrolyzer, where it is combined with product oxygen. Finally, it passes through the electrolysis recuperator to help preheat the incoming process gas. Some of the sweep gas compression work is recovered using a sweep-gas turbine located at the sweep-gas exit.

In order to avoid the work requirement associated with compression of the sweep gas, it is possible to operate with no sweep gas, and to allow the system to produce pure oxygen, which could potentially be supplied to another collocated process such as an oxygen-blown gasifier. For this mode of operation, the four-stage air compressor would not be included in the process flow diagram and there would be no air flow through the intermediate heat exchanger. Oxygen would simply be evolved from the anode side of the electrolyzer at the electrolysis operating pressure and temperature. It would flow through the electrolysis heat recuperator and the outlet turbine. The results of the system analysis will show that this concept is desirable from the standpoint of overall process efficiency, but there are significant technical issues associated with handling high-temperature pure oxygen that would have to be addressed.

The final portion of the process flow diagram in the lower left represents the high temperature gas reactor power cycle and nuclear process heat supply. The reactor is assumed to have a capacity of 600 MW thermal. The outer flow loop downstream of splitter T1 supplies high-temperature nuclear process heat to the IHX. The remainder of the high-temperature gas flows through a recuperated Brayton cycle. The cycle includes a gas turbine, recuperator, and two-stage compressor with intercooling. For the baseline case, the reactor inlet and outlet temperatures were assumed to be 490 and 900°C, respectively, consistent with the direct Brayton cycle concept proposed by General Atomics [8]. The turbine inlet pressure is 7 MPa. A 20°C temperature difference was assumed across the recuperator. The thermal efficiency of this baseline power cycle is 48.3%.

Two sets of system simulations were performed. The first set examined the overall syngas production efficiency variation as a function of current density for two *ASR* values and for both isothermal and adiabatic electrolyzer operation. The second set examined the variation in system performance as a function of reactor outlet temperature.

The methodology followed in performing the first set of system simulations will now be discussed. To assure that the results generated by the model were consistent for all cases, the following constraints were imposed:

- The sum of the inlet hydrogen and carbon monoxide mole fractions before electrolysis and before any water shift reaction was set to 0.1 by adjusting the fraction of syngas product recycled back to the electrolysis process.
- The sum of the electrolysis outlet hydrogen and carbon monoxide mole fractions was maintained at 0.9 by adjusting the total mass flow rate of the process gas into the electrolysis process as the current density was varied. This is a fixed oxygen utilization condition.
- For the air-sweep cases, the oxygen mole fraction exiting the electrolysis process on the air side was fixed at 0.5 by adjusting the inlet air flow rate as the current density was varied. Obviously, for the no-sweep cases, the air flow was zero and the outlet oxygen mole fraction was 1.0.
- The ratio of hydrogen to carbon dioxide in the product stream was fixed at 2.12 by adjusting the carbon dioxide inlet flow rate. This ratio is optimal for the production of synthetic fuel using a Fischer Tropsch process with a cobalt catalyst.
- The minimum approach temperature of the Electrolysis Heat Recuperator was set at 50°C by adjusting the flow split between the power cycle and the intermediate heat exchanger.
- The temperatures of the helium entering and exiting the reactor were fixed at 490°C and 900°C respectively.
- The overall heat transfer coefficient and heat exchanger area product (UA) was kept constant for the recuperator of the Brayton power cycle.

Power generated by the direct Brayton cycle was supplied to the high temperature electrolysis process and its support equipment. For each case, care was taken to ensure that the net power remaining (the power from the power cycle not used for electrolysis) was positive.

The per-cell active area for electrolysis was assumed to be 225 $cm^2$ for all cases considered in this study. This cell size is well within the limits of current technology for planar cells. The total number of cells used in the process simulations was determined by specifying a maximum current density for each *ASR* value considered that was large enough to ensure that the operating voltage would just exceed the thermal neutral voltage. For the higher *ASR* value of 1.25 Ohm·cm$^2$, the maximum current density was set at 0.25 A/cm$^2$ and an adiabatic thermal boundary condition was assumed. The total number of cells for this base case was adjusted until the total remaining power was zero. In other words, the full power cycle output at this operating point is dedicated to electrolysis. At lower current densities, the power cycle output exceeds the value required for electrolysis and this excess power would be supplied to the grid. For the case of *ASR* = 0.25 Ohm·cm$^2$, the maximum current density was set at 1.0 A/cm$^2$. A much higher maximum current density was required for the lower *ASR* case, again in order to assure that the thermal neutral voltage was just exceeded. This procedure resulted in $3.488 \times 10^6$ cells required for the high-*ASR* cases and $8.733 \times 10^5$ required for the low-*ASR* cases. Obviously, for specified target hydrogen production rate, a low ASR value is desirable from the standpoint of capital cost.

For the isothermal cases, heat from the reactor was directly supplied to the electrolysis process to maintain isothermal conditions. This heat exchange was parallel to the heat exchange process via the intermediate heat exchanger. For the adiabatic cases, the direct electrolyzer heater was not used.

System simulations were also performed to examine the effect of reactor outlet temperature on overall syngas production performance. As the reactor outlet temperature is varied, both the power cycle and the electrolysis processes are directly affected. Lower reactor outlet temperatures result in lower power-generation efficiencies, and poorer electrolyzer performance.

The variable reactor-outlet-temperature simulations were constrained similarly, with the following exceptions:

- The number of electrolysis cells was adjusted for each case until the total electric power remaining is less than 50 kW. In other words, the plant is totally dedicated to hydrogen production.
- The co-electrolysis process was run at the thermal neutral voltage, such that the electrolyzer outlet gas temperatures for each case were the same as the electrolyzer inlet temperatures. This constraint required adjustment of the current density until the thermal neutral condition was met.
- The cases for which the exiting temperature of the reactor coolant stream are 850°C or less, the electrolyzer was operated at 800°C. For reactor outlet temperatures of 800°C or less, two auxiliary electrical heaters (one on the process stream and one on the air stream) were used to maintain the 800°C electrolyzer operating temperature. The power for these heaters is supplied by the power cycle and is accounted for in the syngas production efficiency.

An important performance parameter for all of the cases studied is the overall thermal-to-syngas process efficiency, defined as follows:

$$\eta_{syn} = \frac{\dot{N}_{H_2} LHV_{H_2} + \dot{N}_{CO} LHV_{CO}}{\sum_i \dot{Q}_i} \qquad (4)$$

This efficiency definition quantifies the ratio of the low heating value of the produced syngas to the total thermal energy required to produce it. This total thermal energy includes direct process heat inputs such as at the IHX and at the electrolyzer (isothermal cases). In addition, the thermal equivalent of any power-consuming devices such as pumps and compressors, electrical heaters, etc. in the system must be accounted. The thermal equivalent of the power-consuming devices is given by the power requirement divided by the thermal efficiency of the power cycle.

The *ASR* value used in the electrolyzer module is temperature-dependent per the following Arrhenius equation:

$$ASR(T) = ASR_{1100K} - 0.463 + 3.973 \times 10^{-5} \exp\left(\frac{10300}{T(K)}\right) \quad (5)$$

where $ASR_{1100K}$ represents the user-specified cell ASR at the temperature 1100 K. This constant allows one to shift the entire *ASR* curve to higher or lower *ASR* values, to mimic lower or higher performing cells, respectively. This equation for *ASR(T)* is based on empirical data obtained from an actual operating stack, modified to allow user specification of the *ASR* value at 1100 K. In order to show the trends that can be expected with higher or lower *ASR*, two values of $ASR_{1100K}$ have been included in this study. The $ASR_{1100K}$ value of 1.25 represents a stack-average *ASR* value at 1100 K that should be achievable in the short term with existing technology. The $ASR_{1100K}$ value of 0.25 is an optimistic value that has been observed in button cells, but will be difficult to achieve in a stack in the short term. The temperature dependence of the *ASR* is important for the adiabatic cases (since the outlet temperature in these cases is generally different than the inlet temperature) and for evaluating the effect of electrolyzer inlet temperature on overall process efficiency.

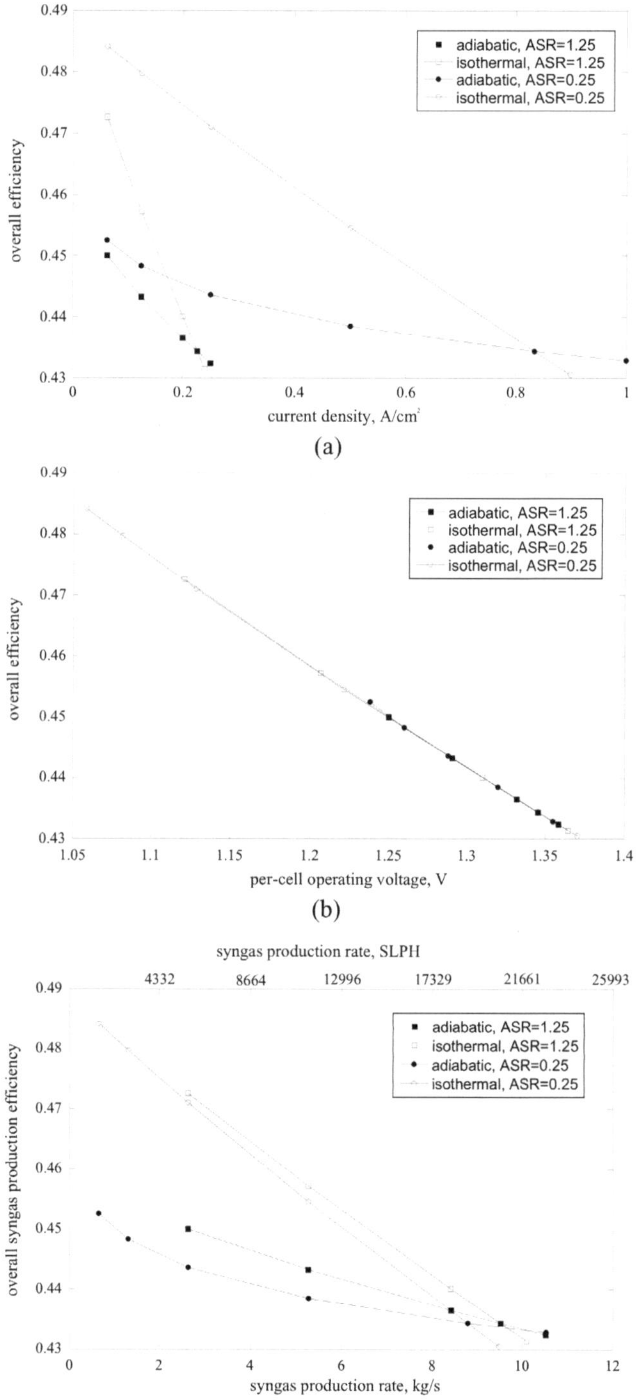


**Figure 4.** Overall syngas production efficiencies for the air-sweep cases plotted as a function of current density (a), per-cell operating voltage (b), and syngas production rate (c).

## RESULTS

Representative overall syngas production efficiency results corresponding to the process flow diagram of Fig. 1, for the fixed-utilization operating conditions described above, with air sweep, are presented in Fig. 4. Overall efficiencies are plotted against current density in Fig. 4(a), against per-cell operating voltage in Fig. 4(b), and against syngas production rate in Fig. 4(c). Results of four cases are presented: low and high *ASR*, adiabatic and isothermal electrolyzer operation. Predicted overall thermal-to-syngas efficiency values are generally within 5 percentage points of the power-cycle efficiency of 48.3%. Fig. 4(a) indicates a relatively rapid decrease in overall efficiency for the high-*ASR* cases as current density is increased. Note also that the maximum current density for the high-*ASR* cases is limited to about 0.25 A/cm$^2$. This limit necessitates deployment of many more electrolyzer cells in order to consume the available power. Electrolyzer efficiency is inversely proportional to operating voltage. Higher current densities, and correspondingly higher syngas production rates require higher operating voltages. Therefore, overall efficiencies decrease with increasing current density. The adiabatic cases show a nonlinear decrease in overall efficiency with increasing current density, as a result of the temperature-dependent *ASR* used in the simulation. For isothermal cases, the *ASR* is constant and the overall efficiencies decrease linearly with current density. For a specified current density (and syngas production rate), the isothermal cases produce higher efficiencies, at operating voltages below thermal neutral.

Interestingly, the overall process efficiencies collapse onto a single line when plotted as a function of per-cell operating voltage, as shown in Fig. 4(b). Note that the highest operating voltages shown are just above the thermal neutral voltage of 1.34 V. Note also that the highest overall efficiency

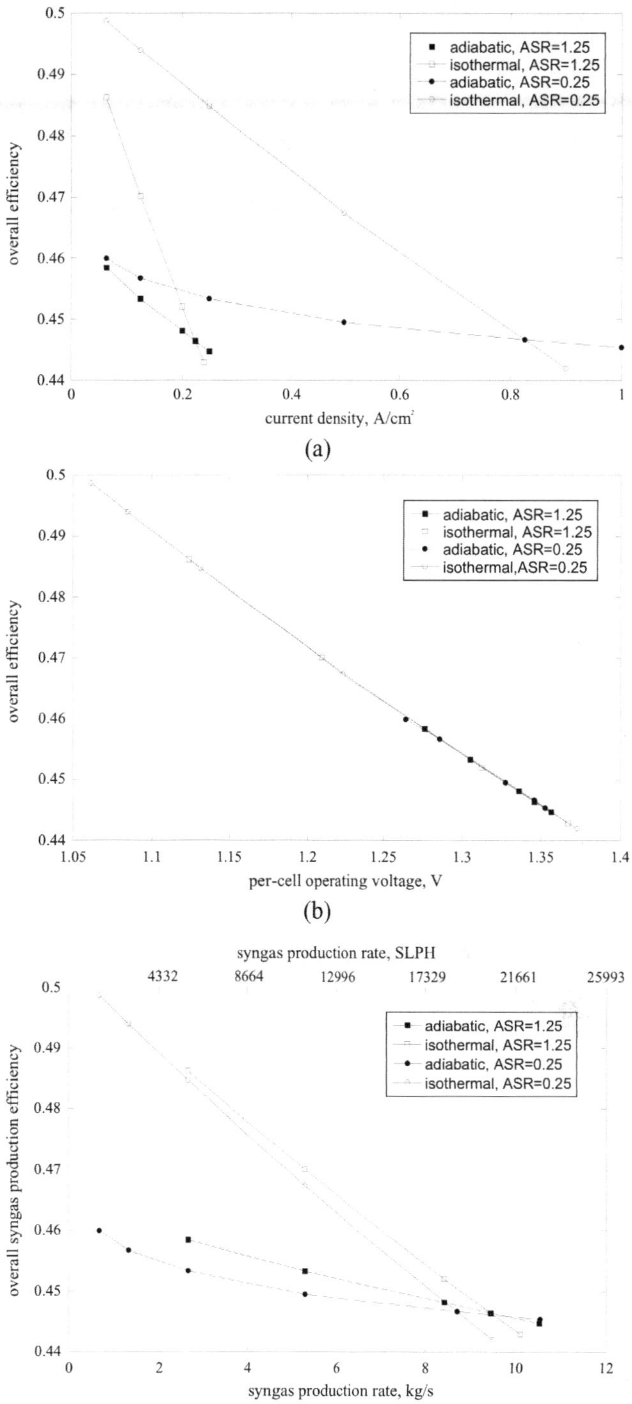


**Figure 5.** Overall syngas production efficiencies for the no-sweep cases plotted as a function of current density (a), per-cell operating voltage (b), and syngas production rate (c).

plotted in Fig. 4 (for ASR = 0.25, isothermal, i = 0.0625 A/cm$^2$) is actually slightly higher than the power cycle efficiency of 48.3%.

Fig. 4(c) shows that syngas production rates as high as 10 kg/s (25993 SLPH) could be achieved with a dedicated 600 $MW_{th}$ syngas-production plant. Fig. 4(c) also indicates similar overall efficiencies for the low-*ASR* and high-*ASR* cases at a specified electrolyzer thermal operating condition (adiabatic or isothermal) and syngas production rate. Recall, however, that the high-*ASR* plant requires four times as many cells as the low-*ASR* plant for the same syngas production rate. So the capital cost of the electrolytic plant would be significantly greater with the high-*ASR* cells compared to the low-*ASR* cells.

The results shown in Fig. 4 are qualitatively similar to results presented in reference [6] for pure steam electrolysis. However, the coelectrolysis results presented here have overall efficiencies values that are a few percentage points higher. The primary reason is that the modeled and predicted power cycle efficiency for the present study is higher (48.3%) than the assumed (not modeled) power cycle efficiency of 45% used for the pure steam electrolysis study. In addition, the operating pressure for this plant was chosen to be 3.5 MPa rather than 5 MPa used in the previous study of pure steam electrolysis. Overall plant efficiencies decrease slightly with increasing operating pressure.

Since the electrolysis cells produce oxygen, rather than consuming it as in the fuel-cell mode, a sweep gas stream is not necessarily required. There has been some discussion of the possible need for a sweeping flow of air or steam to dilute the produced oxygen in order to avoid possible materials and safety issues related to handling of pure oxygen at temperatures over 800°C. From a thermodynamic efficiency standpoint, the use of a sweep gas improves the electrolyzer efficiency, but there are also some disadvantages associated with the use of an air sweep. First, dilution of the pure oxygen that is produced in the electrolysis stack with air would be wasteful since pure oxygen is a valuable commodity that could be sold as an electrolysis by-product. Second, production of a sweeping flow of high-pressure air at even a modest flow rate requires a significant amount of compressor power, compared to the electrolysis stack power consumption, which would seriously degrade the overall process efficiency, if a corresponding outlet turbo expander is not used. Finally, our research has indicated that pure oxygen can be safely handled at high temperature, if the right materials are used.

Overall syngas production efficiencies were also calculated for the no-sweep cases. These results are presented in Fig. 5. Efficiency values are plotted against current density in Fig. 5(a), against per-cell operating voltage in Fig. 5(b), and against syngas production rate in Fig. 5(c). Again, four cases are presented: low and high ASR, adiabatic and isothermal electrolyzer operation. Results of the no-sweep cases are qualitatively similar to the air-sweep cases, but efficiency values for the no-sweep cases are about 1.0 to 1.5 percentage points higher than for the corresponding air-sweep cases. Again, the overall process efficiencies collapse onto a single line when plotted as a function of per-cell operating voltage. The highest overall efficiency plotted in Fig. 5 (for ASR = 0.25, isothermal, i = 0.0625 A/cm$^2$) is fully 1.5% higher than the power cycle efficiency of 48.3%.

The results presented in Figs. 4 and 5 show the variation in overall syngas production efficiency as a function of electrolyzer current density and operating voltage for a fixed value (88.9%) of steam/$CO_2$ utilization. In order to maintain a fixed value of steam/$CO_2$ utilization, the inlet flow rate of process gas was adjusted in proportion to the electrolyzer current density for each case. Alternately, the inlet process gas mass flow rate can be fixed at the maximum value and the effect of electrolyzer current density on overall syngas production efficiency can be calculated for the fixed-mass-flow, variable-utilization case. For this case, steam/$CO_2$ utilization is proportional to the current density. Consequently at low current density (and low utilization), there is a significant excess of process gas flowing through the system. Handling of this excess process gas results in low overall syngas production efficiencies, as shown in Fig. 6. Results of four cases are shown: low-*ASR*, isothermal and adiabatic, and high-*ASR*, isothermal and adiabatic, all with air sweep. In Fig. 6(a), overall syngas production efficiencies are plotted as a function of current density. For the low-*ASR* cases, overall efficiencies range from about 20% at the lowest current density to 43% at the highest. The corresponding range for the high-ASR cases is 37% to 43%. Again, the maximum current density for the low-ASR case is 1.0 A/cm$^2$ and 0.25 A/cm2 for the high-ASR case. Note that the overall efficiency at the highest current densities, 43%, is the same value as was plotted in Fig. 4 at the highest current densities. In fact these represent the same cases.

Fig. 6(b) shows the overall syngas production efficiencies for the variable-utilization cases plotted as a function of per-cell operating voltage. In contrast to the fixed-utilization results (Fig. 4(b)), the variable-utilization efficiencies do not collapse onto a single line when plotted versus operating voltage.

The overall efficiency results for the variable-utilization case nearly collapse onto a single curve when plotted versus utilization, as shown in Fig. 6(c). This plot indicates a strong dependence on utilization, with overall efficiencies of only 20% at the lowest utilization values shown (~5.7%), increasing to a maximum value of 43% at the highest utilization value considered (90%). Again, low utilization results in relatively high irreversibilities associated

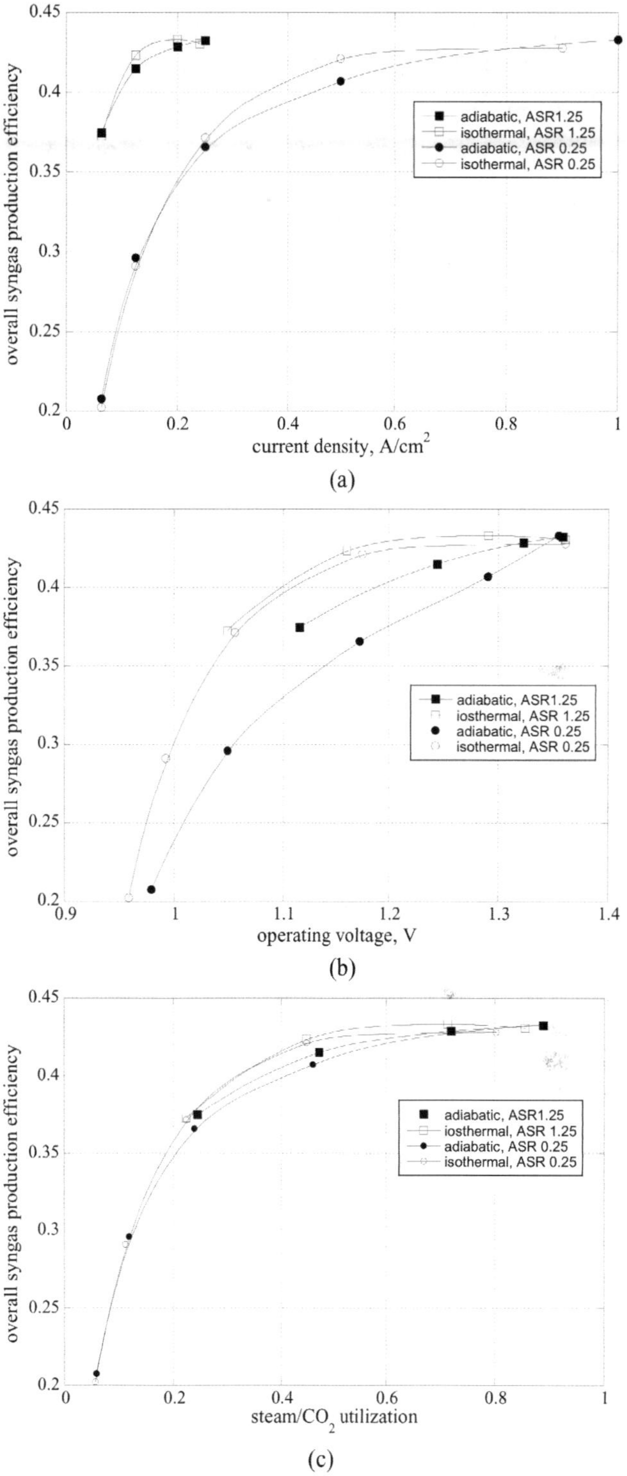


Figure 6. Overall syngas production efficiency for a fixed electrolyzer inlet flow rate with air sweep, plotted as a function of electrolyzer current density (a), as a function of electrolyzer operating voltage (b), and as a function of steam/$CO_2$ utilization (c).

with incomplete heat recuperation, pumping and compression of excess process streams, etc.

The influence of reactor outlet temperature on power cycle and overall thermal-to-syngas process efficiencies is presented in Fig. 7. The process conditions for these simulations were described in the previous section. The upper curve in the plot is the power cycle thermal efficiency and the lower curve is the overall syngas production efficiency. The results indicate the importance of utilizing a high temperature heat source for improved process efficiency. The overall syngas production efficiencies for the thermal-neutral operating point are 3 – 5% lower than the power cycle efficiencies. The results indicate a 46% improvement in overall process efficiency at 1000°C compared to at a 700°C reactor outlet temperature.

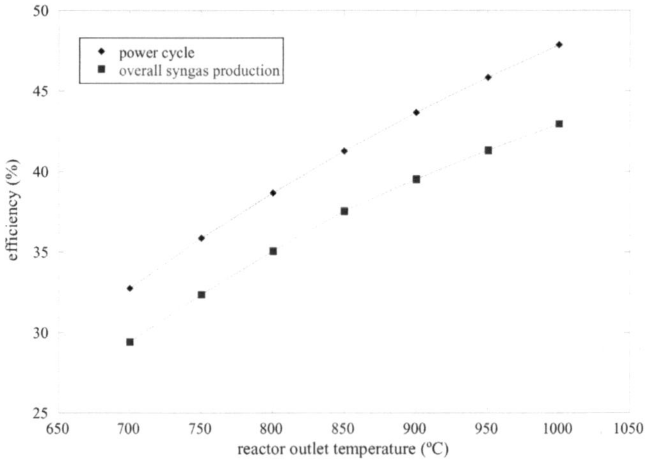


**Figure 7.** Power cycle and overall syngas production efficiencies at thermal-neutral voltage as a function of reactor outlet temperature.

## CONCLUSIONS

An engineering process model has been developed for performance evaluation of a large-scale syngas production plant based on high-temperature co-electrolysis of carbon dioxide and steam. Based on the results presented in this paper, the process appears to represent a promising technology for efficient production of syngas from nuclear energy. Overall thermal-to-syngas process efficiencies of 43 – 48 % were predicted with realistic modeling assumptions, including realistic treatment of heat exchanger performance. Higher process efficiencies result at higher reactor outlet temperatures due to increased power cycle efficiencies.

## ACKNOWLEDGMENTS

This work was supported by the Idaho National Laboratory, Laboratory Directed Research and Development program and by the U.S. Department of Energy, Office of Nuclear Energy, Nuclear Hydrogen Initiative Program.

## COPYRIGHT STATEMENT

## REFERENCES


1. Herring, J. S., O'Brien, J. E., Stoots, C. M., and Hawkes, G. L., "Progress in High-Temperature Electrolysis for Hydrogen Production using Planar SOFC Technology," *International Journal of Hydrogen Energy*, Vol. 32, Issue 4, pp. 440-450, March 2007.
2. Forsberg, C. W., "The Hydrogen Economy is Coming. The Question is Where?" *Chemical Engineering Progress*, December, 2005, pp. 20-22.
3. Finan, A. E., Miu, K., and Kadak, A. C., "Nuclear Technology and Canadian Oil Sands; Integration of Nuclear Power with In-Situ Oil Extraction," Proceedings, 2006 International Congress on Advances in Nuclear Power Plants, ICAPP06, 2006, pp. 2329-2338.
4. O'Brien, J. E., Stoots, C. M., and Hawkes, G. L., "Comparison of a One-Dimensional Model of a High-Temperature Solid-Oxide Electrolysis Stack with CFD and Experimental Results," 2005 ASME International Mechanical Engineering Congress and Exposition, Nov. 5 – 11, Orlando.
5. O'Brien, J. E., McKellar, M. G., Hawkes, G. L., and Stoots, C. M., "Development and Validation of a One-Dimensional Co-electrolysis Model for Use in Large-Scale Process Modeling Analysis," Presented at the ASME 5th International Conference on Fuel Cell Science, Engineering, and Technology, June 18 – 20, 2007, New York, NY; submitted to the *Journal of Fuel Cell Science and Technology*.
6. Stoots, C. M., O'Brien, J. E., McKellar, M. G., Hawkes, G. L., and Herring, J. S., "Engineering Process Model for High-Temperature Steam Electrolysis System Performance Evaluation," AIChE 2005 Annual Meeting, Cincinnati, Oct. 30 – Nov. 4, 2005.
7. Hawkes, G.L., O'Brien, J.E., Stoots, C.M., Herring, S.J., and Hartvigsen, J., "3D CFD Model of High Temperature H2O/CO2 Co-Electrolysis," ANS Embedded Topical: International Topical Meeting on the Safety and Technology of Nuclear Hydrogen Production, Control, and Management, June 24 - 28, 2007, Boston, Massachusetts, USA.
8. LaBar, M. P., "The Gas Turbine-Modular Helium Reactor: A Promising Option for Near Term Deployment," General Atomics Report GA-A23952, April, 2002.

**Post-Test Evaluation of the Oxygen Electrode from A Solid Oxide Electrolysis Stack and Electrode Materials Development**

*Jennifer R. Mawdsley, J. David Carter, Bilge Yildiz, Ann V. Call, A. Jeremy Kropf, Magali S. Ferrandon, Deborah J. Myers, and Victor A. Maroni, Argonne National Laboratory, Argonne, Illinois, USA.*

### *Introduction*

Steam electrolysis is a high-efficiency process that generates hydrogen from water using electricity and heat from an advanced nuclear power plant.[1,2] The process depends on multiple ceramic cells arranged in "stacks" identical to solid oxide fuel cell designs, but operating with a reverse current flow. The stack provides for a manageable voltage/current input as well as a manufacturable cell size. However, increased cell area and a multiplicity of components in the stack introduce potential sources of defects that can lead to materials, structural, and performance degradation. A recent paper from Idaho National Laboratory reports the results of a 1000-h test of a Ceramatec 25-cell HTSE stack.[2] The stack, which had generated hydrogen at a nominal rate of 200 Nl/h, experienced an overall performance degradation of ~20% over the duration of the test.

The objective of this work was to identify the causes of performance degradation of the Ceramatec stack. Our approach was to map the surface characteristics of cells and bipolar plates obtained from the disassembled stack using four-point resistivity measurements, X-ray fluorescence and Raman spectroscopies, and by analyzing cross-sections of specific regions of the cells using scanning electron microscopy (SEM). In this paper, we discuss the results of our investigation of the oxygen electrode compartment.

In addition, the oxygen electrode has been shown to be one of the major contributors to the cell overpotential.[3] Lowering the overpotential in the oxygen electrode can improve the efficiency and lower the cost of hydrogen production by the HTSE process. Here, we also present our findings in developing improved oxygen electrode materials.

### *Experimental*

Cells and bipolar plates were separated from the 25-cell Ceramatec stack[2] for analysis. The cells contained scandia-stabilized zirconia electrolytes, graded oxygen electrodes consisting of strontium-doped manganite mixed with zirconia, and cobaltite bond layers. The steam/hydrogen electrodes consisted of a nickel-zirconia cermet and a nickel bond layer. The bipolar plates were fabricated from ferritic stainless steel. As cells were separated from bipolar plates, separation usually occurred at the bond layers. These layers had been applied to either side of the cells and bipolar plates during stack assembly to provide good electrical contact.

First, we mapped the resistivity of the exposed surfaces of the cells and bipolar plates using a linear four-point probe to identify possible degraded regions by abnormal increases in resistivity. The four-point probe measures mainly in-plane resistivity with unknown depth information, and the measurements are taken at room temperature. Therefore, the measured resistivities do not relate directly to the actual resistances during high temperature stack operation. Nevertheless, regions that have degraded show up as regions of high resistivity compared to the unaffected areas. After testing the as-separated cells, the bond layers were removed with adhesive tape to expose the electrodes and the resistivity mapping procedure was repeated.

To obtain sub-surface information, X-ray fluorescence maps and radiological images were obtained at the Materials Research Collaborative Access Team beamline of Argonne's Advanced Photon Source (APS). Resolution of the maps and images was defined by a scanning step size of 1 mm. The X-ray fluorescence spectrum from each position on the sample was collected using a multi-element germanium detector (MED), consisting of 14 probes, each capable of collecting data at various take-off angles ranging from 5–35°. The fluorescence maps were taken from predetermined regions-of-interests for each atomic element. Transmitted X-rays were collected by a scintillation detector. The X-ray beam was incident along the z-axis, normal to the sample surface and the sample was moved in the x-y plane. The MED was located to the outboard side of the sample with respect to the incident beam. Since the MED probes were positioned at different angles with respect to the sample surface, it was possible to obtain a limited depth-dependent profile of elemental distribution. At low take-off angles, the detected X-rays emanate predominantly from nearer the sample surface, whereas X-rays detected at higher take-off angles emanate from locations deeper in the sample. This is due to the strong dependence of the path length on the sine of the take-off angle.

Raman micro-spectroscopy was used to identify known constituent phases and unknown impurities within the bond layers and electrodes. This near-surface technique is useful for analyzing features in a non-destructive manner with a variably focused excitation laser. Subsurface phases below the penetration depth of the laser could not be detected without scraping off the surface. Nevertheless, this method was chosen because of the relative ease of identifying known impurities and unexpected phases without needing to cut the cells or to coat them with a conductive layer.

From the analyses described above, several regions were identified where performance degradation had likely occurred. We then cross-sectioned the samples and used scanning electron microscopy to further assess the causes. It was determined that most of the cell degradation occurred along the edges of the electrodes and at the interfaces of the various layers.

To increase efficiency and lower the capital cost, we also investigated alternative materials to improve the oxygen electrode performance. Using electrochemical impedance spectroscopy on half-cells of our oxygen electrode materials, we have evaluated the area specific resistance of $Pr_2NiO_4$ as a function of temperature and time. We are currently working to optimize the performance of the nickelate materials by engineering the microstructure using functionally graded compositions and fine-grained interlayers. We have found that a thin interlayer of $Ce_{0.8}Sm_{0.2}O_{2-\delta}$ improves the performance of the nickelate electrodes.

## *Results and Discussion*

### *Four-point resistivity probe*

The four-point probe is a simple and fast method to identify areas of degradation by mapping the resistivity of the surfaces. The maps in Fig. 1 illustrate the resistivities of the cell

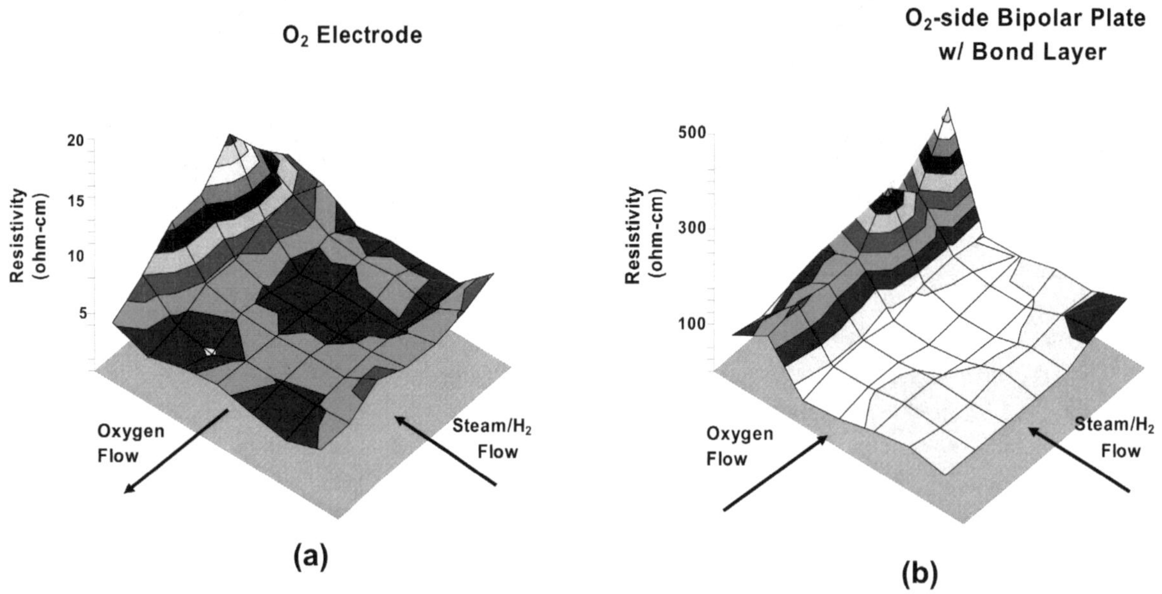


**Fig. 1.** Resistivity maps of (a) an oxygen electrode, and (b) the $O_2$-side surface of a bipolar plate with the bond layer.

and bipolar plate in the oxygen compartment. Figure 1(a) shows the behavior of the oxygen electrode surface after the bond layer was removed. Maps of the bond layer and the electrode for this cell showed similar results. The resistivity is reasonably constant over most of the electrode surface; however, it rises to a maximum along the sealed edge where the hydrogen/steam flow exited the stack (see Fig. 1a).

On the $O_2$-side of the bipolar plate, mapping of the residual bond layer (Fig. 1(b)) shows a similar ridge of high resistivity along the edge where hydrogen exits the stack. These higher resistivities are 25 times those of the unaffected regions. The stack performance, however, did not correspond to such high resistances in operation. In our tests, the particles in the bond layer were not as well interconnected as the electrode and bipolar plate surfaces, which is an artifact of the four-point measurement. Nevertheless, the increase in resistivity along the hydrogen exit edge of the stack is clearly evident and it suggests a possible leak in the seal along this edge, affecting both the bond layer and the electrode. This is a likely contributor to the observed long-term degradation.[2]

*X-ray fluorescence mapping*

Three types of information were obtained from the X-ray fluorescence experiments. Elemental maps showed the distribution of elements on the surface of the cell with undefined depth information. Rudimentary depth profiles showed abundance of elements as a function of depth into the porous electrode area. Finally, a transmission image of the cell revealed thickness and density variations.

The Mn and Co transition metals are defining elements for the electrode and bond layers, whereas Cr is an impurity that diffuses out of the stainless steel bipolar plate. Figure 2 shows maps of the distribution of these three elements in the oxygen electrode. Lighter

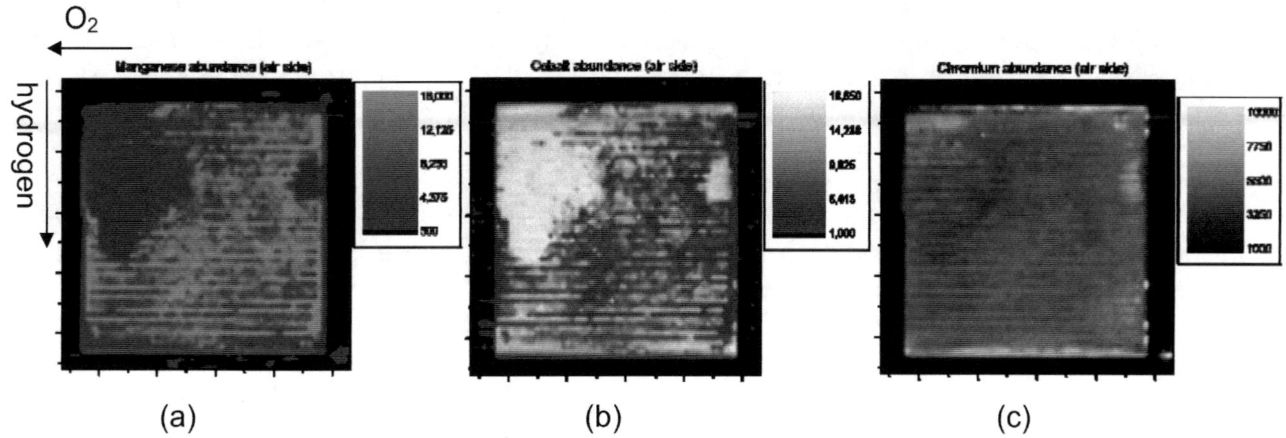


**Fig. 2.** X-ray fluorescence maps of (a) Mn, (b) Co, and (c) Cr abundance in the oxygen electrode of a 10 x 10 cm cell.

shades of color (magenta, yellow and light blue) indicate a greater elemental abundance. The horizontal lines were tracks left by the flow field channels in the bipolar plate.

The Mn and Co maps are virtually the negative image of each other on this scale (1 mm), which shows that over the 1000 h of operation, there was only limited interdiffusion between the two elements. The large patch in the upper-left corner is a residue of the bond layer that sintered fast to the electrode and could not be removed.

Chromium had deposited in all areas of the active electrode, particularly along the edges near the gas seals. In these areas, excess Cr was a result of the bipolar plate edge rail in close proximity to the electrode. Chromium, however, did not deposit on the exposed zirconia edges of the cell. This demonstrates that Cr diffusion is confined to the electrically active region in the electrode. Additionally, the Cr transport appears to be affected somewhat by carrier gas flow, as flow patterns can be seen that are related to the flow of air through the

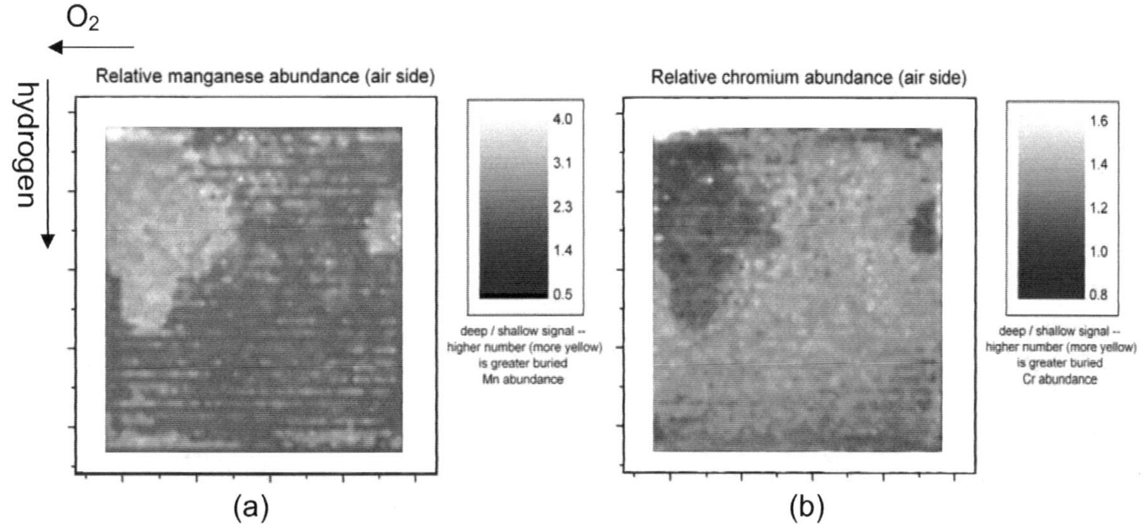


**Fig. 3.** Depth profile of (a) Mn and (b) Cr in the 8x8-$cm^2$ oxygen electrode. The zirconia edges have been masked.

cell. Chromium has been shown to poison the oxygen reduction reaction in solid oxide fuel cells and it is suspected that it causes degradation in steam electrolysis cells as well. This is of particular concern when Cr deposits at the electrode-electrolyte interface where the oxygen reaction occurs. Depth information was needed to determine if, in the cell being examined, Cr had deposited at the electrode-electrolyte interface.

The depth profile illustrated in Fig. 3 was constructed using the Mn or Cr ratio of data from two detectors positioned at high and low take-off angles. The zirconia edges have been masked to cover the background noise of the ratios of small numbers. In this figure, the darker shades of color indicate that Mn or Cr is near the surface and lighter shades indicate that the element has diffused deeper into the electrode. The patch in the upper left corner of Fig. 3(a) shows that the electrode is buried under the bond layer, whereas Cr (Fig. 3(b)) lies near the surface. In the exposed electrode region, Mn is at the surface and, in comparison, Cr has migrated deeper towards the electrolyte interface. The high abundance of Cr near the electrolyte interface may have contributed to stack degradation.

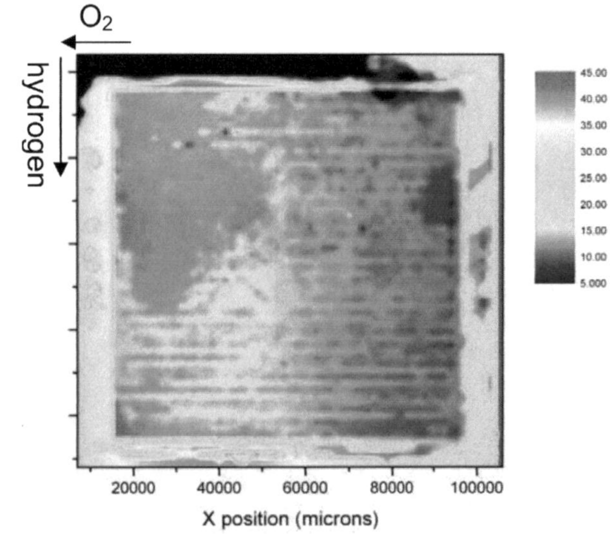


**Fig. 4.** X-ray transmission image of a 10x10 $cm^2$ cell.

An X-ray transmission image, (Fig. 4) was obtained to gain information on defects or cracks in the cell and the uniformity of applied layers. The dark region on the top left edge is the shadow of the clamp holding the cell.

No large cracks or defects were apparent in the cell; however, a thickness or a density gradient can be seen in the electrode area. This gradient could be due to preferential densification of the electrode during operation, or from the fabrication process. Also, there are other areas where X-rays are transmitted more readily than in the rest of the plate, particularly in regions where the seals were made. More transmission images of other cells are being analyzed to gain a better understanding of the significance of these features.

*Raman micro-spectroscopy*

The exposed surface of a cell separated from the stack was examined by Raman micro-spectroscopy to identify any foreign or unexpected phases that may

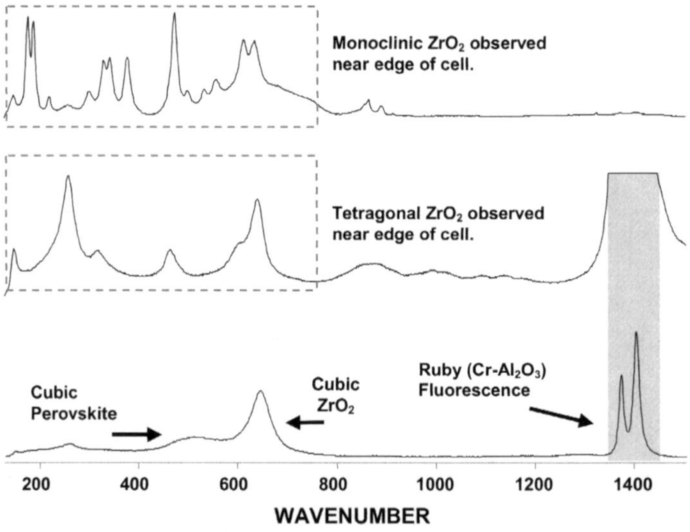


**Fig. 5.** Raman spectra of the oxygen bond layer surface.

have formed during stack operation. Raman spectra of the electrode matrix (Fig. 5) showed the prominent presence of cubic $ZrO_2$. Areas of tetragonal and monoclinic $ZrO_2$ were also identified, found mostly located around the edges of the electrolyte plates. The transition from tetragonal to monoclinic $ZrO_2$ can either be beneficial through a toughening mechanism, or it can be damaging due to cracking, depending on the volume fraction and grain size of the cubic, tetragonal, and monoclinic phases.[4]

Many areas around the edges of the plates contained $Cr^{3+}$-doped $Al_2O_3$ (i.e., ruby), related to the seal material. This excess of Cr near the edges where the seal was applied complements findings from the X-ray maps.

*Scanning electron microscopy*

Using the maps made by the various techniques, 2-cm-long sections were cut out of the cell near the edge of the electrode. Figure 6 is an SEM image of this region. The image shows that the electrode had begun to delaminate from the electrolyte near the edge where the four-point measurements had shown a large increase in resistivity. Cracks can also be seen within the electrode, which would be the cause of increased resistivity.

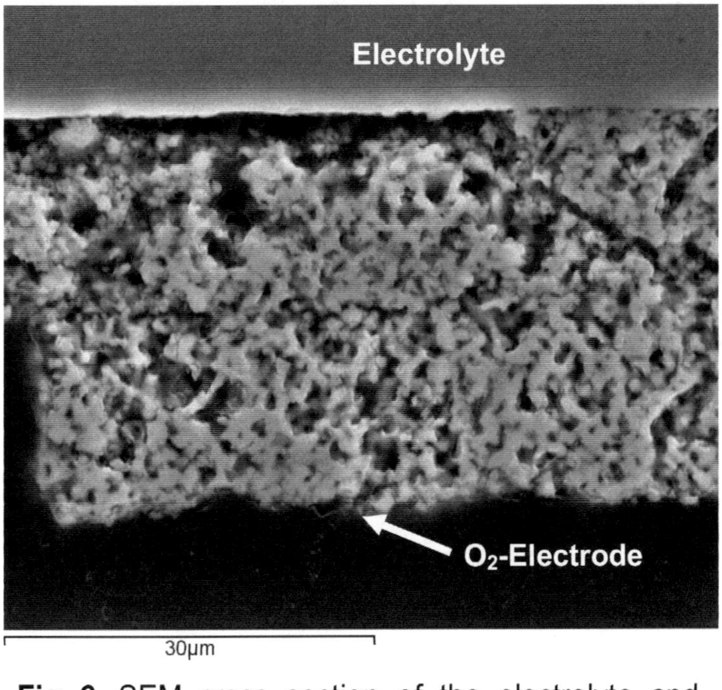


**Fig. 6.** SEM cross section of the electrolyte and oxygen electrode.

Perovskite materials are known to expand upon reduction, which could cause the electrode to delaminate. If hydrogen had leaked into this region, these materials would most likely begin to be reduced. Towards the center of the cell (right), however, adherence to the electrolyte improves and the cracks begin to disappear. Close inspection of this and other cells has revealed that other sections of the $O_2$-electrode have spalled off. Further investigation is needed to determine the cause of this. If this delamination is progressive, it would be a source of cell performance degradation.

*Electrode development*

In addition to degradation analyses of stacks, we have investigated new materials to reduce the cell overpotential. The composition $Pr_2NiO_4$ has been studied as an oxygen electrode for solid oxide fuel cells and other solid state electrochemical devices,[5,6] and it has shown improved performance over manganite electrodes. We investigated these as oxygen evolution electrodes and have obtained promising results.

In initial work, we evaluated $Pr_2NiO_4$ electrodes in single-atmosphere air cells and found that although the composition had good activity as an electrode, better results were obtained

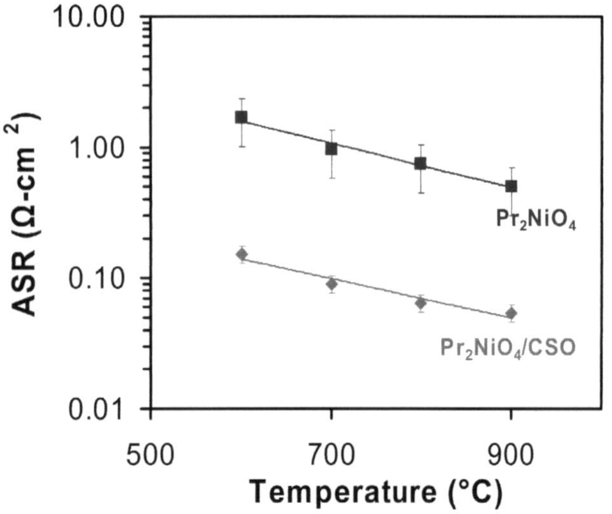


**Fig. 7.** Area-specific resistance of $Pr_2NiO_4$ oxygen evolution electrodes.

when a samarium-doped ceria (CSO) interlayer was deposited on the zirconia electrolyte to prevent interaction and to enhance the oxygen evolution reaction.

In Fig. 7, electrode performance is plotted in terms of the area-specific resistance (ASR) of the electrode as a function of temperature, using impedance spectroscopy and a reference electrode to separate the ohmic resistivity and contributions from other cell components. At the normal operating temperature for steam electrolysis of 800°C, the electrode ASR can be reduced below 0.07 Ω-cm$^2$ when a CSO interlayer is applied between the electrolyte and electrode.

Long term performance studies of these electrodes in air-cells showed that the electrodes deposited with a CSO interlayer had good stability over 300 h. On the other hand, electrodes without the interlayer exhibited some degradation over time.

We also tested $Pr_2NiO_4$ in a full cell setup. A $Pr_2NiO_4$ oxygen electrode was deposited on an InDec™, scandia-stabilized zirconia/Ni cermet electrolyte-supported cell. (At this point we were unable to deposit a CSO interlayer.) This cell was mounted in the test fixture and heated to 800°C in air and $H_2$:$N_2$:$H_2O$ = 20:40:40. Figure 8 illustrates the cell performance in electrolysis mode.

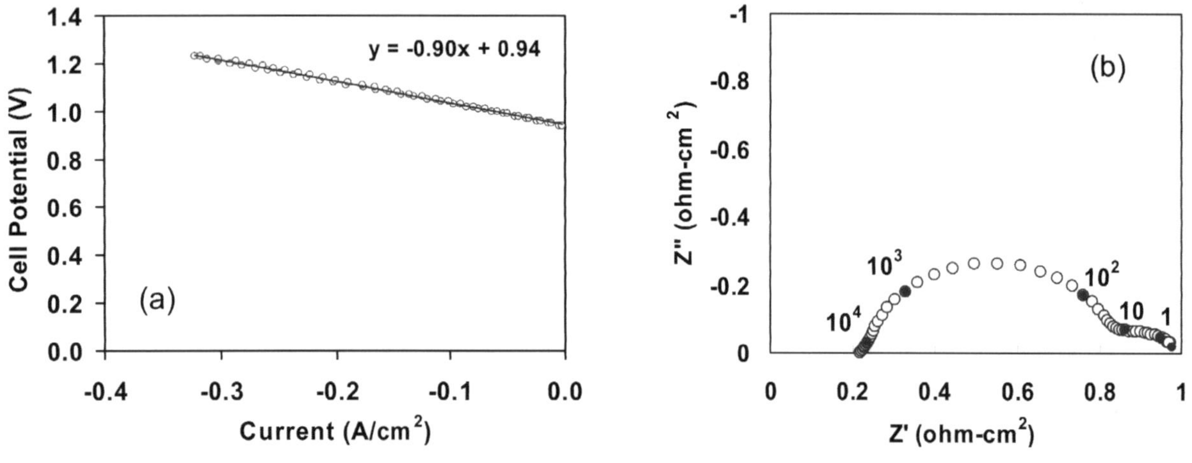


**Fig. 8.** (a) DC polarization and (b) impedance curve of a $Pr_2NiO_4$/Sc-doped $ZrO_2$/Ni-cermet cell at 800°C and in air/$H_2$:$N_2$:$H_2O$ = 20:40:40 % dual atmosphere.

The overall cell resistance can be divided into an ohmic resistance, including contributions from the electrolyte and the two electrodes of 0.2 Ω-cm$^2$, and a polarization resistance containing contributions from both electrodes. The electrode polarization of

0.8 Ω-$cm^2$ compares well with that of the $Pr_2NiO_4$ at 800°C in the air-cell measurements shown in Fig. 7. Further efforts will include testing of cells where CSO is used as an interlayer.

## *CONCLUSIONS*

A high temperature steam electrolysis stack that had operated for 1000 h at ~830°C was disassembled and examined for potential causes of long term degradation. X-ray fluorescence measurements revealed that over the 1000 h, Mn and Co from the oxygen electrode and bond layer showed no macro-scale interdiffusion. On the other hand, Cr had diffused from the ferritic stainless steel bipolar plate toward the electrode/electrolyte interface, preferentially through the active electrode area. Raman micro-spectroscopy identified a monoclinic phase in the exposed regions of the zirconia plate near the edge of the plates, which gives evidence of instability of the cubic phase in the zirconia plate. Also, Cr-doped $Al_2O_3$ was identified in areas near the seal.

Using the mapping approach, we were able to identify regions of interest where performance degradation most likely occurred and then used cross-sectioning and SEM analysis to further assess the causes. SEM images identified delamination and cracking on the edges of the electrodes. We determined that most of the cell degradation occurred along these edges. This information will be used to improve materials and procedures in building future solid oxide steam electrolysis stacks.

Oxygen electrodes based on $Pr_2NiO_4$ were tested in air and in a dual air/steam-hydrogen atmosphere. Area-specific resistances of the oxygen electrode lower than 0.07 Ω-$cm^2$ and cell resistances of 1.0 Ω-$cm^2$ were achieved at 800°C in the electrolysis mode.

## ACKNOWLEDGEMENTS

Thanks to Ceramatec for supplying stack components for this study. This research was funded by U.S. Department of Energy, Office of Nuclear Energy, Science and Technology. Argonne National Laboratory is operated for the U.S. Department of Energy by UChicago Argonne, LLC, under contract DE-AC02-06CH11357.

## REFERENCES

1. W. Doenitz, R. Schmidberger, E. Steinheil et al., "Hydrogen-Production by High-Temperature Electrolysis of Water-Vapor", *Int. J. Hydrogen Energy*, **5**(1), 55 (1980).
2. J. E. O'Brien, C. M. Stoots, J. S. Herring et al., "Performance of planar high-temperature electrolysis stacks for hydrogen production from nuclear energy", *Nuclear Technology*, **158**(2), 118 (2007).
3. J. O'Brien, INL Button Cell Test Results, 11/03 and 11/04, private communication.
4. R. Stevens, *Zirconia and Zirconia Ceramics*, 2nd ed. (Magnesium Elektron Ltd., Twickenham, United Kingdom, 1986).
5. V. V. Kharton, A. A. Yaremchenko, and E. N. Naumovich, "Research on the electrochemistry of oxygen ion conductors in the former Soviet Union. II. Perovskite-related oxides", J. Solid State Electrochemistry, 3(6), 303 (1999).
6. S. Miyoshi, T. Furuno, O. Sangoanruang et al., "Mixed conductivity and oxygen permeability of doped $Pr_2NiO_4$-based oxides", *J. Electrochem. Soc.*, **154**(1), B57 (2007).

# 3D CFD MODEL OF A MULTI-CELL HIGH TEMPERATURE ELECTROLYSIS STACK

*Grant Hawkes, James O'Brien, Carl Stoots, Brian Hawkes*
*Idaho National Laboratory, Idaho Falls, Idaho*

## Abstract

A three-dimensional computational fluid dynamics (CFD) electrochemical model has been created to model high-temperature electrolysis stack performance and steam electrolysis in the Idaho National Laboratory Integrated Lab Scale (ILS) experiment. The model is made of 60 planar cells stacked on top of each other operated as Solid Oxide Electrolysis Cells (SOEC). Details of the model geometry are specific to a stack that was fabricated by Ceramatec, Inc[1]. and tested at the Idaho National Laboratory. Inlet and outlet plenum flow and distribution are considered. Mass, momentum, energy, and species conservation and transport are provided via the core features of the commercial CFD code FLUENT[2]. A solid-oxide fuel cell (SOFC) model adds the electrochemical reactions and loss mechanisms and computation of the electric field throughout the cell. The FLUENT SOFC user-defined subroutine was modified for this work to allow for operation in the SOEC mode. Model results provide detailed profiles of temperature, Nernst potential, operating potential, activation over-potential, anode-side gas composition, cathode-side gas composition, current density and hydrogen production over a range of stack operating conditions. Variations in flow distribution, and species concentration are discussed. End effects of flow and per-cell voltage are also considered.

## Introduction

A research program is under way at the Idaho National Laboratory (INL) to simultaneously address the research and scale-up issues associated with the implementation of planar solid-oxide electrolysis cell technology for hydrogen production from steam. The research program includes an experimental program aimed at performance characterization of electrolysis cells and stacks. Results of some multi-cell tests have been documented in several recent papers [1], [2]. This paper reports the first plenum study of a 60 cell stack when modeling a planar solid oxide electrolysis cells (SOEC) with the FLUENT code and SOFC module [3]. This code was used for detailed SOEC modeling. Fluent Inc. was funded by the US Department of Energy National Energy Technology Laboratory (DOE-NETL) to develop a solid-oxide fuel cell (SOFC) module for coupling to the core mass, momentum, energy, and species conservation and transport features of the FLUENT CFD code. The SOFC module adds the electrochemical reactions and loss mechanisms and computation of the electric field throughout the cell. The FLUENT SOFC user-defined subroutine was modified for this work to allow for operation in the SOEC mode. Model results provide detailed profiles of temperature, Nernst potential, operating potential, anode-side gas composition, cathode-side gas composition, current

---

[1,2] References herein to any specific commercial product, process, or service by trade name, trademark, manufacturer, or otherwise, does not necessarily constitute or imply its endorsement, recommendation, or favoring by the U.S. Government, any agency thereof, or any company affiliated with the Idaho National Laboratory

density and hydrogen production over a range of stack operating conditions. Reference [4] has details of the FLUENT code and numerical model. Results of the numerical model are shown in this paper.

## Numerical Model

The numerical model developed for this paper was based on the geometry of a 60 cell stack fabricated by Ceramatec, Inc. and tested at the INL. A depiction of four 60-cell stacks comprising an ILS module is shown in Figure 1. The stack has a per-cell active area of 64 cm$^2$. It is designed to operate in cross flow, with the steam/hydrogen gas mixture entering the inlet manifold on the right/front in the depiction, and exiting through the outlet manifold located in the center. Air flow enters at the other center manifold (not visible in Fig. 1) and exits at the front/left where the tabs are shown directly into the furnace. The power lead attachment tabs, integral with the upper and lower interconnect plates are also visible in the depiction. Figure 2 shows the piping for the H2/H2O inlet and outlet and air inlet. Figure 3 shows a blown up view of a single cell with its components that are scaled 10x in the z-direction.

The numerical model geometry represents a complete 60 cell stack that is ¼ of the ILS module. Symmetry boundary conditions are implemented. The numerical domain extends from the bottom of the inlet tubes to the outlet flow path of each stream. Inlet flow tubes for the H2/H2O and air side are modeled 5-in. below the inlet of the plenum. This distance allows the flow to develop.

The FLUENT SOFC module treats the electrolyte as a 2-D planar element. Therefore the electrolyte in the model has geometrical thickness of zero. The electrolyte has a thickness of 0.14 mm. This thickness was apportioned to the separator plates so as to keep the total stack height correct. On either side of the electrolyte are the electrodes that are created with 3-D elements. Therefore, the electrolyte/electrode assembly in the model is only as thick as the two electrodes.

Approximately 1.5 million elements are included in this model. Figure 4 shows the top view of

**Figure 1.** Depiction of ILS module with four 60-cell stacks. Green = air inlet, pink = H2/H2O inlet, orange = H2 outlet.

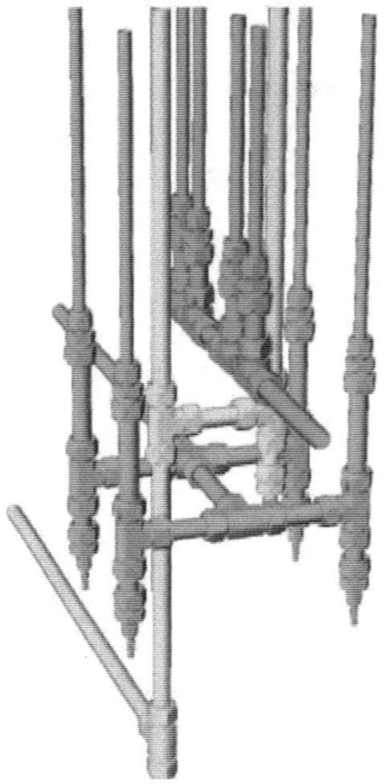

**Figure 2.** Inlet and outlet piping maze.

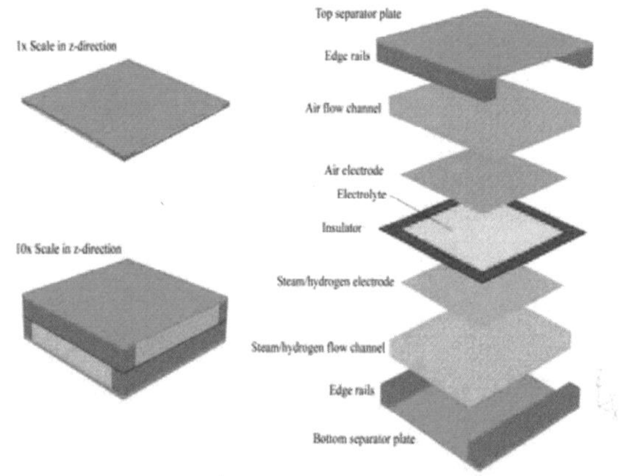


**Figure 3.** Component description for a single cell.

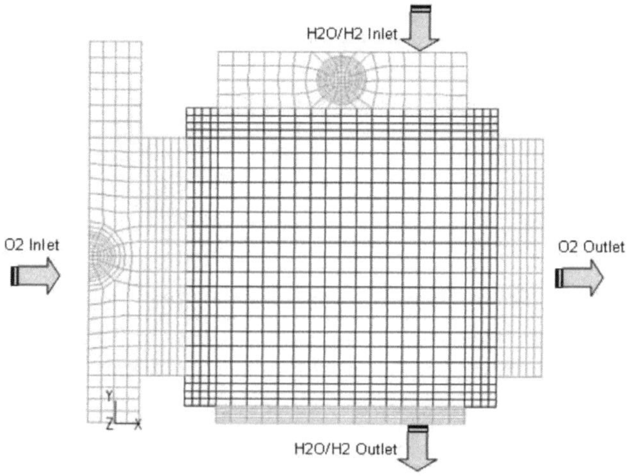


**Figure 4.** Mesh used at each cross section in z-dir.

the grid used. This grid was used at each level throughout the entire height of the model. The numerical grid used in this study included 4 elements each in the flow inlet and outlet regions, and 16 x 16 in the active cell area in the X and Y directions. Each flow channel (current collector) has 6 elements across the flow channel. Single cell numerical models with these numbers of cells give identical results to twice as many, meaning that the model is grid converged in the cell area. More research needs to be done on the mesh in the inlet and outlet plenum areas.

All external surfaces are considered to be adiabatic. Details of the core mass, momentum, energy, and species conservation and transport features of FLUENT are documented in detail in the FLUENT user manual from Fluent Inc. Details of the electrochemical reactions, loss mechanisms, electric field computation, and electrode porous media constitutive relations are documented by the SOFC module in the FLUENT documentation. This reference also documents the treatment of species and energy sources and sinks arising from the electrochemistry at the electrode-electrolyte interfaces.

Mass flow rates, compositions, and other parameters were taken from the base case of Reference [5].

## Results

Results are displayed on Figures 5 through 14. Figures 5 and 6 show the path lines of the flow for the air inlet and H2/H2O inlet at open cell voltage (OCV) respectively. Most of the flow appears to go all the way to the top of the plenum and then recirculate down the outside before entering into each cell. Shown in Figure 7 are temperature contour plots displayed on the H2 side current collector, H2/H2O inlet plenum, and air inlet plenum for open cell voltage (OCV), 60 V, 77.2 V, and 80 V. These values correspond to 1.0, 1.28667, and 1.333 V/cell respectively. Since the inlet temperature is at 1073 K, the thermal neutral voltage is 1.28667 V/cell. The scale for each figure goes from the minimum to the maximum for each particular model. The second figure is dominated by the endothermic reaction and shows temperatures well below the inlet conditions, while the third figure has a mean outlet gas temperature exactly equal to the inlet temperature of 1073 K. The fourth figure shows that temperatures are dominated by the Ohmic heating in the electrolyte and well above the thermal neutral condition.

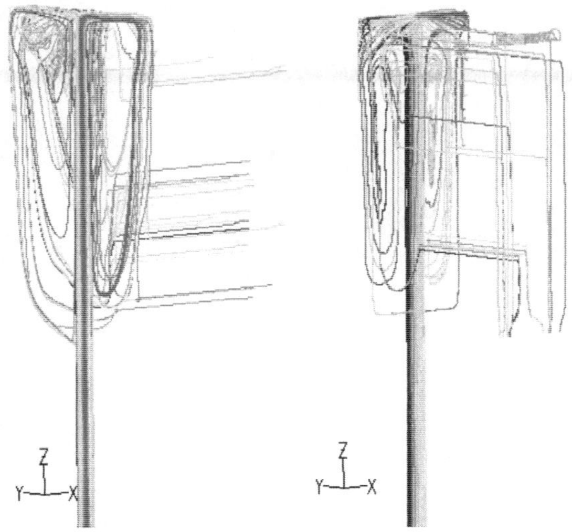


**Figures 5 and 6.** Pathlines for O2 (left) and H2/H2O (right)

Figure 8 shows the Nernst potential on each electrolyte for the same four operating voltages. Nernst potential depends on the gas compositions and temperature. The temperature is dependent on the current density because of the endothermicity of the reaction. Figure 9 shows the current density on each electrolyte in the stack. The highest magnitude of current density and hence H2 production are the most negative. Shown in Figure 10 are the voltages at each separator plate. The colors are all the same because the scales are different for each figure. Figure 11 shows the H2 mole fraction in the H2/H2O inlet plenum, H2 current collector, and the surface on the outlet of the H2/H2O outlet plenum. For the thermal neutral case, the most H2 is produced about ¾ of the way up the stack. Shown in Figure 12 is the O2 mole fraction displayed on the O2 current collector, and air inlet plenum. The amount of O2 produced is higher near the H2/H2O inlet side. This is due to the high concentration of H2O available to be electrolyzed as shown by the Nernst potential.

Figures 13 and 14 show the per cell mass flow rate exiting each cell for the H2/H2O and O2 side respectively. The OCV shows the flow through the stack with no current. The flows all have the highest flow near the bottom cells. Figure 14 shows the O2 production rate is increasing with the increase in voltage and hence current. The flow through the O2 cells during OCV is quite constant.

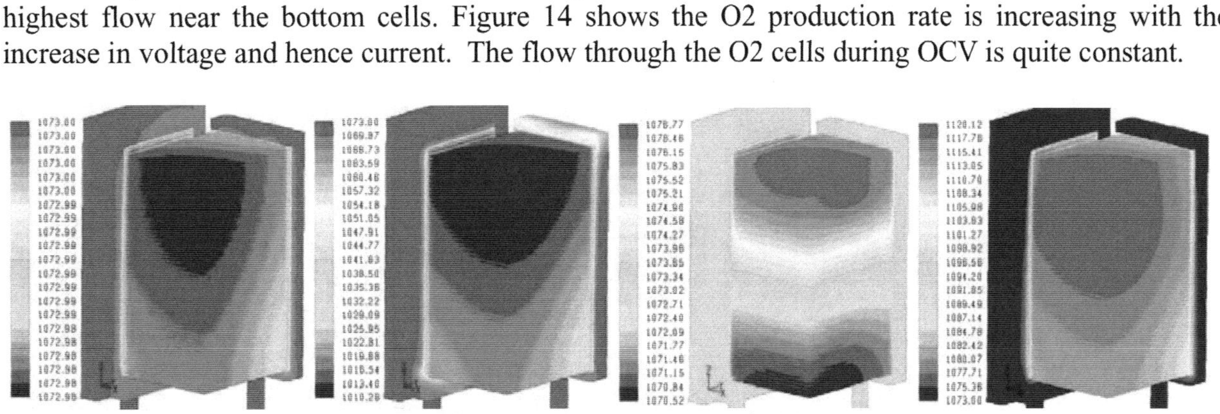

**Figure 7.** Temperature (K) contours for OCV, 60V, 77.2V, and 80V.

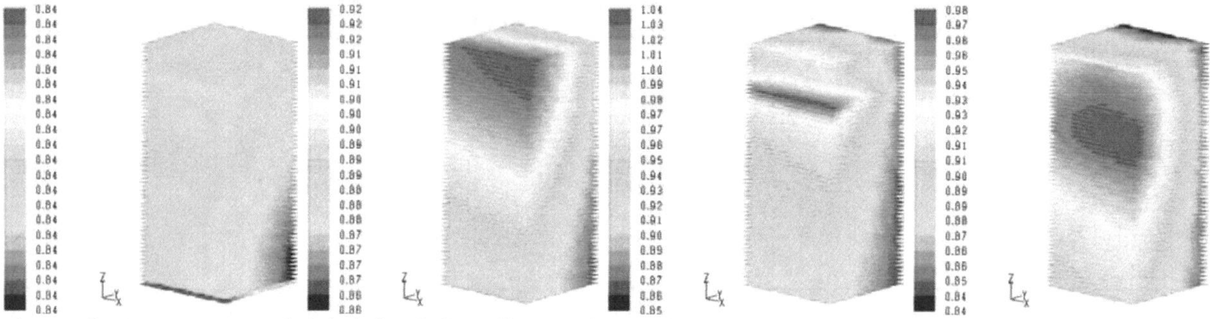

**Figure 8.** Nernst potential (V) for OCV, 60V, 77.2V, and 80V.

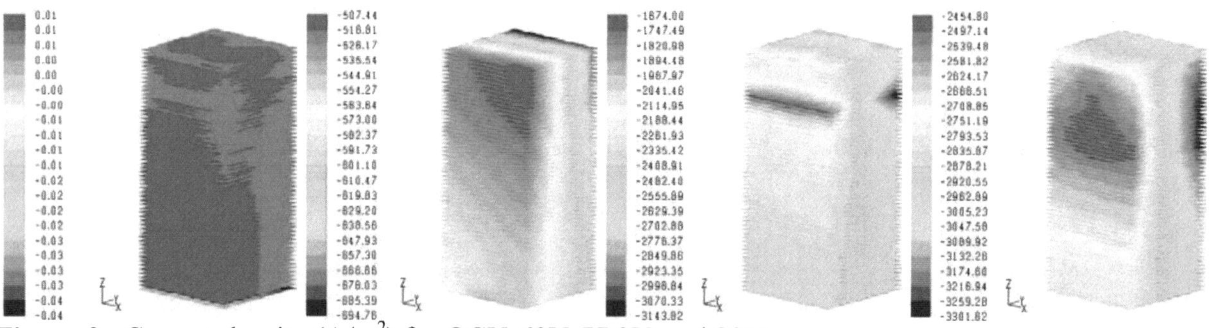

**Figure 9**. Current density (A/m$^2$) for OCV, 60V, 77.2V, and 80V.

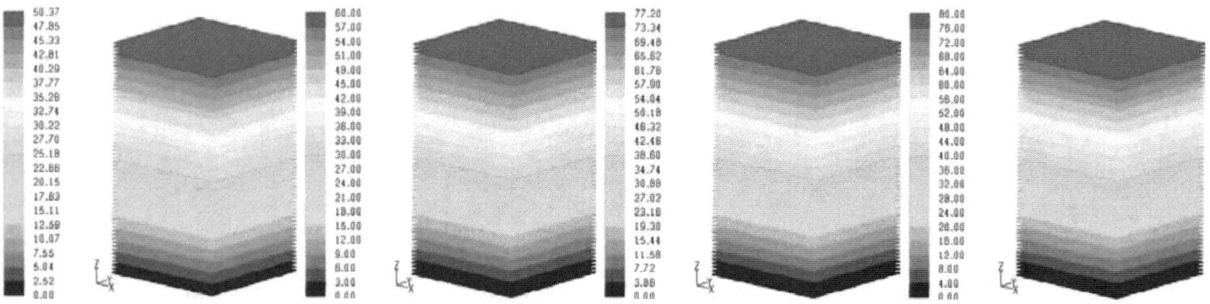

**Figure 10.** Operating voltage (V) for OCV, 60V, 77.2V, and 80V.

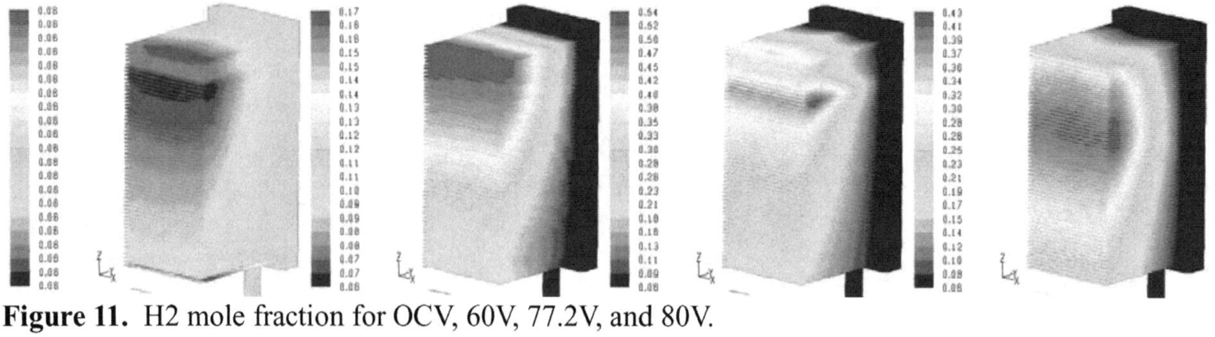

**Figure 11.** H2 mole fraction for OCV, 60V, 77.2V, and 80V.

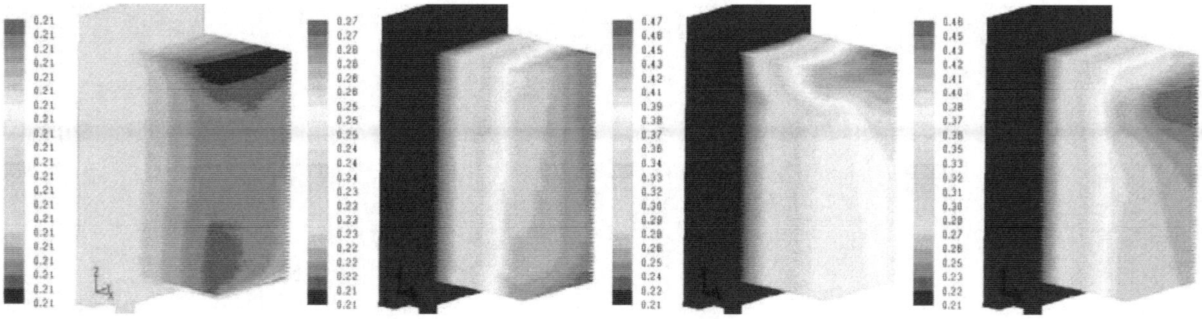

**Figure 12.** O2 mole fraction for OCV, 60V, 77.2V, and 80V.

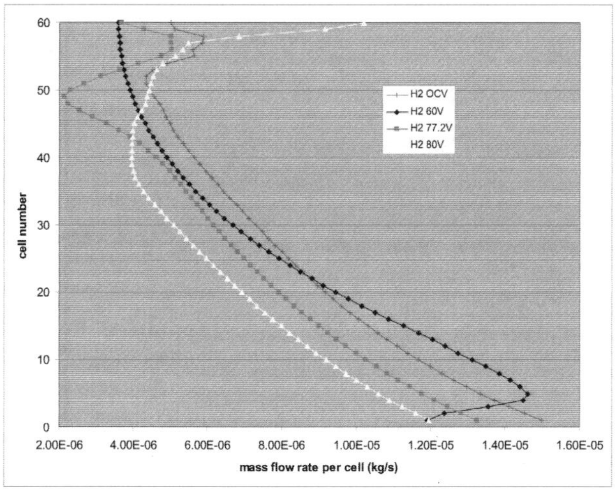


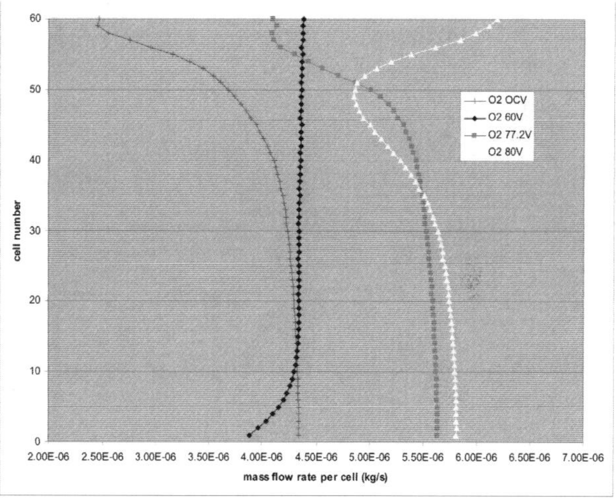


**Figures 13 and 14.** Mass flow rate exiting per cell for H2 and O2 respectively.

## Conclusions

A three-dimensional computational fluid dynamics (CFD) model has been created to model high-temperature steam electrolysis in a planar solid oxide electrolysis cell (SOEC) stack. Effects of the variation of input parameters are shown for this stack and model. The model represents 60-cell stack that represents a ¼ of an ILS module. Details of the model geometry are specific to a stack that was fabricated by Ceramatec, Inc. and tested at the Idaho National Laboratory. Mass, momentum, energy, and species conservation and transport are provided via the core features of the commercial CFD code FLUENT. A solid-oxide fuel cell (SOFC) model adds the electrochemical reactions and loss mechanisms and computation of the electric field throughout the cell. The FLUENT SOFC user-defined subroutine was modified for this work to allow for operation in the SOEC mode. Model results provide detailed profiles of temperature, Nernst potential, operating potential, anode-side gas composition, cathode-side gas composition, current density and hydrogen production over a range of stack operating conditions. Inlet and outlet plenums are included in the model of this stack. Plenum flow characteristics with recirculation were observed. Contour plots of local electrolyte temperature, current density, and Nernst potential indicated the effects of heat transfer, reaction cooling/heating, and change in local gas composition.

## Acknowledgements

The U.S. Department of Energy, Office of Nuclear Energy, Nuclear Hydrogen Initiative Program supported this work. The DOE National Energy Technology Laboratory (NETL) provided the SOFC module to the INL for this research. The Idaho National Laboratory is operated by the Battelle Energy Alliance through DOE Contract DE-AC07-05ID14517.

## References

[1] J. E. O'Brien, C. M. Stoots, J. S. Herring, and J. J. Hartvigsen, "Performance of Planar High-Temperature Electrolysis Stacks for Hydrogen Production from Nuclear Energy," *Nuclear Technology*, Vol. 158, pp. 118 - 131, May, 2007.

[2] J. E. O'Brien, C. M. Stoots, J. S. Herring, and J. J. Hartvigsen, "Hydrogen Production Performance of a 10-Cell Planar Solid-Oxide Electrolysis Stack," *Journal of Fuel Cell Science and Technology*, Vol. 3, pp. 213-219, May, 2006.

[3] Prinkey, M., Shahnam, M., and Rogers, W. A., "SOFC FLUENT Model Theory Guide and User Manual," Release Version 1.0, FLUENT, Inc., 2004.

[4] G. L. Hawkes, J. E. O'Brien, C. M. Stoots, J. S. Herring, "CFD Model of a Planar Solid Oxide Electrolysis Cell for Hydrogen Production from Nuclear Energy," *Nuclear Technology*, Vol. 158, pp. 132 - 144, May, 2007.

[5] G. L. Hawkes, R. W. Jones, "CFD Model of a Planar Solid Oxide Electrolysis Cell: Base Case and Variations," 2007 ASME-JSME Thermal Engineering Conference and Summer Heat Transfer Conference, July 8 – 12, 2007, Vancouver, BC, Canada.

# 457a S-I Cycle Simulation Using Prosimplus

**Olivier Baudouin**[1], Stéphane Dechelotte[1], Philippe Guittard[1], Martin Roeb[2], Nathalie Monnerie[2], Jean-Marc Borgard[3], Giovanni Cerri[4], and Claudio Corgnale[4]. (1) ProSim, Stratege Batiment A, BP 27210, Labege Cedex, F-31672, France, (2) Deutsches Zentrum für Luft und Raumfahrt (DLR – D), (3) DEN/DPC, CEA Saclay, point courier 41, gif sur yvette, 91191, France, (4) Università degli studi – Roma tre (DIMI – I)

Sulfur-Iodine (S-I) cycle is the most promising thermochemical cycle for water splitting to produce hydrogen in order to substitute the fossil fuels in the future. The HYTHEC project (European Commission funded research project SES6-CT-2004-502704) aims at improving and evaluating the S-I cycle and compare it to Westinghouse hybrid cycle. The flowsheeting and simulation work within HYTHEC has been performed using ProSimPlus, the general steady-sate process simulation software edited by ProSim. In particular, flowsheets of the H2SO4 decomposition and HI decomposition sections have been developed and simulated using ProSimPlus within the framework of the HYTHEC project. The benefits of using ProSimPlus for simulation of these processes within a collaborative project will be demonstrated. Emphasis will be put on : – openness of the software (CAPE-OPEN compliance, user added modules coded in VBS,…) allowing each partner to introduce its own know-how or to customize existing unit operation models (use of membranes, experimental vapor-liquid data…); – convergence methods to tackle large flowsheets and complex unit operations: reactive distillation for HI section have been widely investigated and will be presented; – capability to model highly non-ideal systems thanks to Simulis® Thermodynamics, a mixture physical properties and fluid phase equilibria calculation server embedded in ProSimPlus. The ability to introduce new experimental data obtained during the project for HI section to develop a new thermodynamic model will be shown.

# 457b Detailed Modeling Of The Thermodynamics Of The Sulfur-Iodine Thermochemical Cycle

**George M. Bollas**, Department of Chemical Engineering - Process Systems Engineering Laboratory, Massachusetts Institute of Technology, 77 Massachusetts Ave., RM 66-363, Cambridge, MA 02139, Mujid S. Kazimi, Center for Advanced Nuclear Energy Systems, Department of Nuclear Science and Engineering, Massachusetts Institute of Technology, 77 Massachusetts Ave., RM 24-219, Cambridge, MA 02139, and Paul I. Barton, Chemical Engineering, Massachusetts Institute of Technology, 77 Massachusetts Ave., RM 66-464, Cambridge, MA 02139.

Thermochemical water splitting is considered as an energy-efficient, low-cost process for hydrogen production. Among the large-scale, cost-effective and environmentally-attractive hydrogen production processes, the Sulfur-Iodine (SI) thermochemical cycle appears to be one of the most promising. In this cycle, iodine and sulfur dioxide are combined with water, forming hydrogen iodide and sulfuric acid, which are immiscible and can be readily separated. Then, the sulfuric acid is decomposed at about 850° C releasing the oxygen and recycling the sulfur-dioxide. Finally, the hydrogen iodide is decomposed at about 400°C, releasing the hydrogen and recycling the iodine. According to the flowsheet proposed by General Atomics [1] this cycle can be split into the following sections:

• Recycle Acids Generation (Section I):

$[xI_2]_l + [SO_2]_g + [2H_2O]_l \leftrightarrow [2HI_x]_{aq} + [H_2SO_4]_{aq}$

This reaction, named the Bunsen reaction, proceeds exothermically in liquid phase and produces two immiscible aqueous acid phases, with the dense phase containing $HI_x$ and the light phase containing $H_2SO_4$. This phase split is the key to the cycle's success, but the resulting liquid-liquid equlibrium (LLE) is difficult to model.

• Sulfuric Acid Concentration and Decomposition (Section II):

$[H_2SO_4]_{aq} \leftrightarrow [H_2SO_4]_l$

$[H_2SO_4]_l \leftrightarrow [H_2SO_4]_g$

$[H_2SO_4]_g \leftrightarrow [H_2O]_g + [SO_3]_g$

$[SO_3]_g \leftrightarrow [SO_2]_g + [½O_2]_g$

Experience from the sulfuric acid industry can be applied for the simulation of this section. Modeling uncertainty arises from the high temperatures of this section and the complex formation of $H_2SO_4$ and $H_2O$ at high temperatures and/or concentrations.

• HI Concentration and Decomposition (Section III):

$[2HI_x]_{aq} \leftrightarrow [2HI]_g + [(x\text{-}1)I_2]_l + [H_2O]_l$

$[2HI]_g \leftrightarrow [H_2]_g + [I_2]_g$

$[I_2]_g \leftrightarrow [I_2]_l$

The main difficulties that have to be overcome in this section include the extraction of HI from the $HI_x$ mixture because of the presence of an azeotrope in the mixture, the possibility of iodine precipitating as a solid ($T_m$ = 113.6°C) and the strong immiscibility of the $H_2O$-HI-$I_2$ mixture.

It is evident from the above that the greatest challenge in simulating the SI process is modeling of the chemical equilibria in the major unit operations. In particular, special attention has to be paid to predicting the equilibria of the electrolytes' partial dissociation and the stable phase split. A significant bottleneck in the modeling of this process is the empirical nature of the existing models for multi-electrolyte solutions and their restricted validity (low temperatures and pressures). The model applied for the calculation of the chemical equilibria has to be valid for temperatures up to 800°C. Furthermore, the prediction of speciation in the process sections, which is not among the traditional virtues of electrolyte thermodynamic models, is important for the simulation of thermochemical cycles. Finally, hydration and ions-water complex formation add further complexity to the thermodynamic modeling of the system.

Among the existing electrolyte thermodynamic models, two are most likely to succeed in this modeling challenge: the electrolyte NRTL model [2] and the Pitzer model [3]. These two models have the ability to predict reaction and phase (vapor-liquid and liquid-liquid) equilibria, they include the effect of temperature and they can handle multiple-electrolyte solutions. As the ability of these models to extrapolate is of high importance for the SI process, in this work their thermodynamic consistency is extensively examined. Although the thermodynamic consistency of any excess Gibbs free energy model is a priori guaranteed, under its specific assumptions, the effect of these assumptions on the derivation of the excess Gibbs free energy that is used to calculate the activity coefficients can be crucial. A new model has been developed [4] based on the full analytical derivation of the electrolyte NRTL excess Gibbs free energy and without the simplifying assumptions that were used in the original electrolyte NRTL model [2]. The new model reduces to the original electrolyte NRTL for binary mixtures, whereas it retains its thermodynamic consistency and generality for any extension to mixed-solvent, multi-electrolyte solutions. Because of its generality and consistency, the new model possesses better extrapolation abilities and predicts accurately the speciation.

In all the chemical equilibrium models developed in this work, the Gibbs free energy minimization approach is used for finding the point of equilibrium. Because of the nonlinearity of the electrolyte thermodynamic models, the equilibrium constant method (i.e., the calculation of equilibrium using the equality of chemical potentials between products and reactants) may lead to multiple solutions and only those that minimize the total Gibbs free energy are correct. The minimization of Gibbs free energy is formulated as a nonlinear program (NLP), that is, the minimization of the nonlinear Gibbs free energy function that constitutes the electrolyte thermodynamic model subject to mass balance and electroneutrality constraints. The Gibbs free energy minimization approach is also followed for the simulation of the liquid-liquid splits in the process. Phase equilibria of multicomponent mixtures are modeled using a reinterpretation of the Gibbs tangent plane stability criterion via Lagrangian duality [5]. The starting point for this formulation is the natural primal problem of minimizing the Gibbs free energy subject to material balance. The stable phase split is the solution of the corresponding Lagrangian dual problem, providing a necessary and sufficient dual extremum principle.

One of the difficulties with modeling the SI cycle is the absence of parameters for the electrolyte thermodynamic models at the conditions of interest. Recently [6-13], experimental data have become available for some of the process stages, whereas much more are expected in the near future [1]. The fitting of model parameters ($P_i$) to experimental data using the Gibbs energy minimization approach (with respect to the components moles $N_j$) , results in a bilevel optimization problem [14]. That is, the models are fitted to the experimental data, by solving the following bilevel program:

min [discrepancy between predicted and observed system properties]
$P_i$
s.t. min $G$
$N_j$
s.t. elemental balance + electroneutrality

In conclusion, a comprehensive framework has been developed that deals with all aspects of the thermodynamics of the SI cycle. Using this framework, accurate steady-state and dynamic models of this promising process can be developed, which will increase the understanding of the process and provide the ability to set design and operational targets.

**References**

1. Brown, L. C., Lentsch, R. D., Besenbruch, G. E., Schultz, K. R. & Funk, J. E. (2003) Alternative flowsheets for the sulfur-iodine thermochemical cycle. *Technical Report GAA24266.* General Atomics.
2. Chen, C. C. & Evans, L. B. (1986) A Local Composition Model for the Excess Gibbs Energy of Aqueous-Electrolyte Systems. *AIChE Journal,* 32**,** 444-454.
3. Clegg, S. L., Pitzer, K. S. & Brimblecombe, P. (1992) Thermodynamics of Multicomponent, Miscible, Ionic-Solutions. 2. Mixtures Including Unsymmetrical Electrolytes. *Journal of Physical Chemistry,* 96**,** 9470-9479.
4. Bollas, G. M., Chen, C. C. & Barton, P. I. (2007) The significance of mixing rules, hydration and complex formation in the electrolyte NRTL model. *Submitted to: AIChE Annual Meeting.* Salt Lake City, Utah.
5. Mitsos, A. & Barton, P. I. (2007) A Dual Extremum Principle in Thermodynamics. *AIChE Journal,* Submitted for publication.
6. Sakurai, M., Nakajima, H., Amir, R., Onuki, K. & Shimizu, S. (2000) Experimental study on side-reaction occurrence condition in the iodine-sulfur thermochemical hydrogen production process. *International Journal of Hydrogen Energy,* 25**,** 613-619.
7. Sakurai, M., Nakajima, H., Onuki, K. & Shimizu, S. (2000) Investigation of 2 liquid phase separation characteristics on the iodine-sulfur thermochemical hydrogen production process. *International Journal of Hydrogen Energy,* 25**,** 605-611.
8. Kubo, S., Kasahara, S., Okuda, H., Terada, A., Tanaka, N., Inaba, Y., Ohashi, H., Inagaki, Y.,

Onuki, K. & Hino, R. (2004) A pilot test plan of the thermochemical water-splitting iodine-sulfur process. *Nuclear Engineering and Design,* 233**,** 355-362.
9. Kubo, S., Nakajima, H., Kasahara, S., Higashi, S., Masaki, T., Abe, H. & Onuki, K. (2004) A demonstration study on a closed-cycle hydrogen production by the thermochemical water-splitting iodine-sulfur process. *Nuclear Engineering and Design,* 233**,** 347-354.
10. Barbarossa, V., Brutti, S., Diamanti, M., Sau, S. & De Maria, G. (2006) Catalytic thermal decomposition of sulphuric acid in sulphur-iodine cycle for hydrogen production. *International Journal of Hydrogen Energy,* 31**,** 883-890.
11. Belaissaoui, B., Thery, R., Meyer, X. M., Meyer, M., Gerbaud, V. & Joulia, X. (2007) Vapour reactive distillation process for hydrogen production by HI decomposition from HI-I2-H2O solutions. *Chemical Engineering and Processing,* doi:10.1016/j.cep.2007.01.018.
12. Doizi, D., Dauvois, V., Roujou, J. L., Delanne, V., Fauvet, P., Larousse, B., Hercher, O., Carles, P., Moulin, C. & Hartmann, J. M. (2007) Total and partial pressure measurements for the sulphur-iodine thermochemical cycle. *International Journal of Hydrogen Energy,* doi: 10.1016/j.ijhydene.2006.11.039.
13. Giaconia, A., Caputo, G., Ceroli, A., Diamanti, M., Barbarossa, V., Tarquini, P. & Sau, S. (2007) Experimental study of two phase separation in the Bunsen section of the sulfur-iodine thermochemical cycle. *International Journal of Hydrogen Energy,* 32**,** 531-536.
14. Dempe, S. (2003) Annotated bibliography on bilevel programming and mathematical programs with equilibrium constraints. *Optimization,* 52**,** 333-359.

# 457c Modeling The Solubility Of Sulfur Dioxide In Sulfuric Acid Solutions

**Maximilian B. Gorensek**, Computational Science and Statistics Department, Savannah River National Laboratory, Aiken, SC 29808, John P. O'Connell, Chemical Engineering, University of Virginia, 102 Engineers' Way, Charlottesville, VA 22904, and Paul M. Mathias, Fluor Corporation, 3 Polaris Way, Aliso Viejo, CA 92698.

The solubility of sulfur dioxide in sulfuric acid is an important consideration in the design of sulfur-based thermochemical cycles, particularly the Hybrid Sulfur cycle. Data culled from the literature show that for a given temperature, the solubility decreases as the sulfuric acid concentration increases from zero, reaching a minimum value. As the H2SO4 content increases past 50 mol%, the SO2 solubility begins to grow steadily, exceeding its solubility in pure water. Initial attempts to model this behavior using the ELECNRTL properties method in Aspen Plus met with limited success. Results were comparable to those obtained with OLI Systems' Mixed Solvent Electrolyte model, but different enough to cause significant disagreement in the prediction of SO2 absorber behavior using the two properties models. This presentation will document the problem and suggest an approach to fix it.

# 457d Influence Of Some Critical Data On Efficiency Of Extractive Distillation For S_I Cycle

**Jean-Marc Borgard**, DEN/DPC, CEA Saclay, point courier 41, gif sur yvette, 91191, France

Different processes have been proposed for the iodine section of the S_I cycle, essentially the reactive distillation of iodine and extractive distillation using phosphoric acid. The extractive distillation is currently used in the I-Neri demonstration loop (ILS). If the feasibility of the process has been demonstrated by General Atomics in the late 70's, many questions remain on the thermodynamic of the process, mainly related to the effective efficiency of this process, the purity of hydrogen and iodine products, and the recirculation rate of the expensive phosphoric acid. We try in this article to review all relevant available thermodynamical data in the literature and determine how they impact the evaluation of the different steps of this process.

# Process Model-Free Analysis for Thermodynamic Efficiencies of Sulfur-Iodine Processes for Thermochemical Water Decomposition

by

John P. O'Connell and Parinya. Narkprasert
Department of Chemical Engineering
University of Virginia
Charlottesville, VA 22904 USA
and
Maximilian B. Gorensek
Computational and Statistical Science Department
Savannah River National Laboratory
Aiken, SC 29808 USA

Material, energy and entropy balances, which depend only on stream conditions and flows entering and leaving a system, have been used to evaluate different scenarios for thermochemical decomposition of water to manufacture hydrogen with S-I processes. Energy efficiencies have been found for ideal systems with variable stream amounts, as well as for an available flowsheet, to locate the greatest effects on energy requirements and inefficiencies. Aspen, OLI and ProSim property models have been used on the sections of a General Atomics process to reveal the effects of differences in computed energies and entropy generation. While the calculated efficiencies and inefficiencies are generally consistent with reported effects of constraints, differences in stream properties and phase behaviors suggest that optimal process configurations from simulations may have significant uncertainties.

## 1. Introduction

Considerable effort is currently being expended to design efficient processes for manufacturing hydrogen as an energy carrier using chemistries with high temperature heat, such as from a nuclear reactor [1-5]. Because of the immense scale of the facilities to be constructed and operated, reliable estimates must be made of the energy efficiency and process configurations of the competing process concepts. A number of efforts have been done along these lines [6-12]. This paper describes an alternative approach based on thermodynamics that avoids assumptions about the process units, leading to a process model-free analysis. It also allows direct comparisons among various property models used in the process simulations, which heretofore has not been done.

The paper develops completely general relations applicable to hydrogen manufacture and then applies them to Sulfur-Iodine thermochemical water decomposition processes of the type developed by General Atomics [13-17]. The basic flow sheets are taken from the work of Mathias and Brown [18-22]. The property models used in AspenPlus are adopted for the initial calculations to compute energy needed for reversible process and the consequences of introducing irreversibilities into the different GA Sections. Finally some comparisons are made for property models when applied to a particular GA flow sheet.

## 2. Thermodynamic Analysis

Thermodynamics offers many ways, both rigorous and approximate, to evaluate the utilization of energy flows for open systems [23]. One such approach is the use balance equations about a "black-box" system, accounting for material, energy and entropy effects from interactions of the system and its surroundings. Figure 1 illustrates the concept, where inlet and outlet streams at specified temperatures, $T$, pressures, $P$, and stream flows expressed in molar or mass amounts, $\{N\}$, plus work, $W$, and heat, $Q$, flows are related conservation of atoms and energy as well as by entropy conservation for reversible cases and by positive entropy generation, $S_{gen}$, in real systems.

The basic equations for a steady-flow system for a fixed basis, which could be amount of a product, *e.g.*, per mole, or per unit time, are:

Material balance
$$\sum_i N_i - \sum_o N_o = 0 \quad (1)$$

Energy balance
$$\sum_i N_i h(T_i, P_i, \{x\}_i) - \sum_o N_o h(T_o, P_o, \{x\}_o) + \sum_s W_s + \sum_b Q_b + Q_e = 0 \quad (2)$$

Entropy balance
$$\sum_i N_i s(T_i, P_i, \{x\}_i) - \sum_o N_o s(T_o, P_o, \{x\}_o) + \sum_b \frac{Q_b}{T_e} + \frac{Q_e}{T_e} + S_{gen} = 0 \quad (3)$$

Here, $h$ is the molar enthalpy, $s$ is the molar entropy. The summation $_i$ is over input streams and the summation $_o$ is over output streams, so all mole numbers, $\{N_i\}$ and $\{N_o\}$ are positive. The summations for work and heat are associated with utilities; $Q_e$ is the heat which is discharged to the surroundings at $T_e$. Work is usually from electricity for pumps, compressors and turbines. Work may also be obtained from a high energy stream, such as for heat pumps run from helium; however this is accounted for by stream terms, where the stream flow and conditions are used rather than the amount of work. If the work is internally generated from heat that does not cross the system boundary, it is not included in the energy balance (2). Heat enters or leaves the system through heat exchangers with $T_b$ being the absolute temperature at which the heat appears in the surroundings. If the heat is transferred to or from a stream with changing temperature, the log mean temperature may be used. A reversible process with $S_{gen} = 0$ gives the upper limit on efficiency.

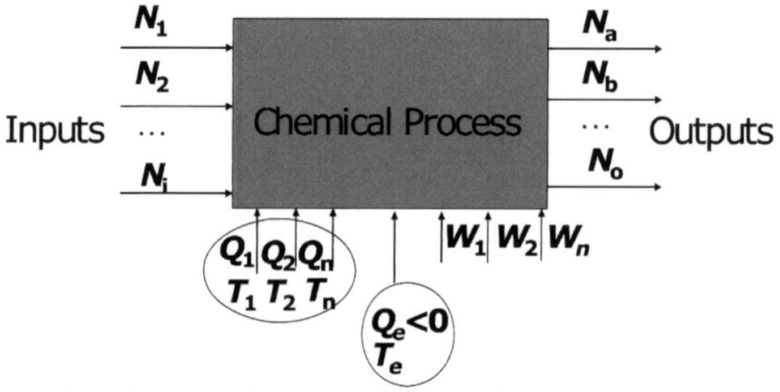


Figure 1. Schematic of a general process for material, energy and entropy balances.

For closed systems , the consequence of $S_{gen} > 0$ from irreversibilities, is that more work must be put in ($W_{real} > W_{rev} > 0$) and more heat removed ($\Sigma Q_{real} < \Sigma Q_{rev} < 0$) for work consuming devices and less work is obtained ($W_{real} < W_{rev} < 0$) and less heat put in ($0 < \Sigma Q_{real} < \Sigma Q_{rev}$) for work producing devices. For open systems, there can be a complex interplay of stream flows and conditions with the work and heat such that elements may be changed in different directions. In the cases of producing output energy or materials of greater useful energy than input, such as work from heated steam or hydrogen from water using a hot gas, a consequence of entropy generation is that, 1) more high energy input either with greater stream enthalpy from higher temperature or greater stream flow, and 2) removal of either more heat or of output streams of higher energy. Thus, if the efficiency of a steam turbine is decreased, more flow, or a higher inlet pressure and/or temperature, will be needed to achieve the same amount of work, and either the outlet temperature is higher or more heat must be removed. For helium of a given pressure and temperature fed to an S-I process producing a given amount of hydrogen, when $S_{gen}$ is increased, the solution to Equations (1) - (3) gives more He flow and more heat rejection.

While there may appear to be three equations, so three quantities might be solved for, the material balance (1) can only be used to make the input and output stream flows consistent. Therefore an energy/entropy analysis will allow only two quantities to be found. This can proceed in different ways. One common problem is to assume reversibility ($S_{gen} = 0$) and, after specifying the material flows, determine the minimum amount of input energy via one heat, one work, or one condition for one flow, plus find the minimum heat to be discharged to the environment. In the S-I process, we have focused on determining the He flow for specified inlet and outlet conditions and the heat rejection to the environment, since these would be the aspects that must be dealt with in a real process. We can also obtain the usual efficiency.

Reversible analysis will give the best possible case. Other analyses can either specify a value of $S_{gen} > 0$ and obtain the changes in process conditions, or use actual or simulated process heat and work effects, made consistent with the energy balance (2), to find $S_{gen}$ from the entropy balance (3). The former case provides insights into the effects of irreversibilities in different parts of a process and suggests areas for potential efficiency improvement, mainly by reducing temperature differences over which heat flows. The latter case allows comparisons of different process configurations with the lowest value of $S_{gen}$ being the most efficient.

This approach is reliable only with accurate values of $h$ and $s$. Since the terms in the stream flows are often large numbers, so the difference in the summations can be relatively small, significant errors might appear in the computed values. This would be most obvious when a calculated $S_{gen}$ is negative. In addition, energy efficiencies are typically calculated as the ratio of output product energy content divided by the total energy input or as the ratio of reversible energy requirement to actual energy requirement. As a result, errors in the $h$ and $s$ values will affect the accuracy of the computed efficiency.

In the next section, the two different uses of the analysis, reversible and real, will be done.

## 3. Analysis of Cases Based on the General Atomics Flowsheet with Aspen Property Modeling

### A. Overall process

The simplest case of analysis is for a reversible process taking in water and producing hydrogen and oxygen at $T_e$ with the energy source being a stream of high pressure, high temperature helium from an gas-cooled nuclear reactor, as shown in Figure 2. Table 1 gives conditions and properties for the input and output components for a basis of 1 kmol of $H_2$. Conditions can differ in various efficiency studies, as can the property values, which depend on the reference state. Therefore, the results here may be somewhat different from others, but the trends will be substantially the same.

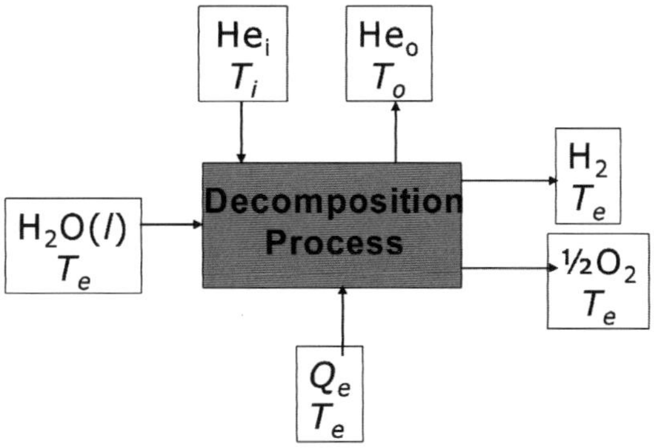


Figure 2. Schematic of thermochemical decomposition of water using nuclear energy provided by He gas.

Table 1. Conditions for overall reversible generation of 1 kmol/hr of hydrogen via thermochemical decomposition of water using the GA system of Figure 2 driven by high temperature Helium. Molar enthalpy, *h*, and entropy, *s*, values from the AspenPlus property model of Mathias [19].

| Component | Flow rate, kmol/hr | *T*, K (Phase) | *P*, MPa | *h*, MJ/kmol | *s*, MJ/kmol -K |
|---|---|---|---|---|---|
| $H_2O$ | 1.0 | 275.00 (*l*) | 0.4 | -287.24 | -0.169 |
| $H_2$ | 1.0 | 386.15 (*g*) | 4.0 | -7.32 | -0.023 |
| $O_2$ | 0.5 | 346.57 (*g*) | 0.4 | -25.99 | -0.007 |

Figure 3 gives the process efficiency when the temperature difference, $T_i$ - $T_o$ = 300 K for a range of $T_i$ values from 900 to 1600 K. As predicted by the Carnot effect, even for a totally reversible process, not all of the He energy can be utilized in the decomposition. But the process becomes more efficient at higher $T_i$ and, as shown below and by Gorensek [7], for smaller $T_i$ - $T_o$.

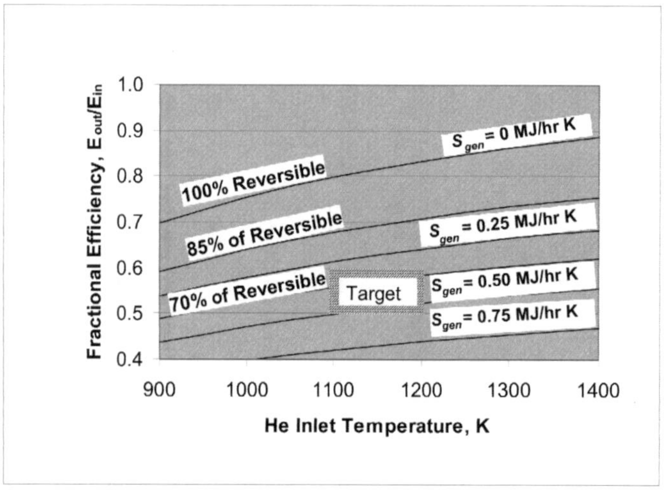


Figure 3. Overall thermal efficiency of reversible and irreversible S-I processes for a Helium gas temperature drop of 350 °C.

Next, one can introduce inefficiencies in this overall process, particularly if a target efficiency and He conditions can be specified. For $T_i$ - $T_o$ = 350 °C, Gorensek [7] suggested an efficiency target range of 50-60% and a $T_i$ range of 1100 - 1200 K. Figure 3 shows how the efficiency decreases for 85% and 70% of reversible, with the latter just meeting the target. Also shown in Figure 3 are results when values of $S_{gen}$ are specified to be 0.25, 0.50, and 0.75 MJ/kmol $H_2$-K; only the first two meet the target. Thus, to meet the target efficiency, the upper bound on irreversibilities for any process is given by $S_{gen}$ 0.50 MJ/kmol $H_2$-K. Note that the same $T_i$ variation is obtained when either a fraction of reversible or a fixed $S_{gen}$ are specified. What is interesting is that the efficiency drop for $S_{gen}$ from 0 to 0.25 MJ/kmol-K is greater than for $S_{gen}$ from 0.25 to 0.50 and from 0.50 to 0.75 MJ/kmol $H_2$-K. This means that initial departures from reversibilities hurt the process the most.

Table 2 gives more details of the calculations shown in Figure 3. It gives the flow rate of He, the rejected heat, $Q_e$, and the efficiency as the ratio of the energy value of the hydrogen product to the input He energy for various $S_{gen}$. The first columns for reversible processes show that the minimum He flow is when $T_i$ and $T_i$ - $T_o$ are largest, while the minimum $Q_e$ is for the largest $T_i$ and smallest $T_i$ - $T_o$. However, the efficiency is not a strong function of He conditions.

Table 2. Helium flows, heat rejection rates, and efficiencies for hydrogen generation system of Table 1.

| Helium | | $S_{gen}$= 0 MJ/kmol $H_2$-K | | | $S_{gen}$= 0.25 MJ/kmol $H_2$-K | | | $S_{gen}$= 0.50 MJ/kmol $H_2$-K | | |
|---|---|---|---|---|---|---|---|---|---|---|
| $T_i$, K | $T_o$, K | He Flow, kmol/hr | Heat Out, $Q_e$, MJ/hr | $E_{out}/E_{in}$ | He Flow, kmol/hr | Heat Out, $Q_e$, MJ/hr | $E_{out}/E_{in}$ | He Flow, kmol/hr | Heat Out, $Q_e$, MJ/hr | $E_{out}/E_{in}$ |
| 1100 | 800 | 58.6 | 74.0 | 0.80 | 76.0 | 183.0 | 0.61 | 93.5 | 292.1 | 0.50 |
| 1100 | 700 | 45.3 | 85.2 | 0.77 | 58.8 | 197.6 | 0.60 | 72.3 | 310.0 | 0.48 |
| 1100 | 600 | 37.6 | 99.7 | 0.07 | 48.8 | 216.4 | 0.57 | 60.1 | 333.2 | 0.47 |
| 1200 | 900 | 56.1 | 58.3 | 0.83 | 72.8 | 162.7 | 0.64 | 89.5 | 267.1 | 0.52 |
| 1200 | 800 | 43.0 | 66.5 | 0.81 | 55.9 | 173.3 | 0.63 | 68.7 | 280.2 | 0.51 |
| 1200 | 700 | 35.4 | 76.6 | 0.79 | 46.0 | 186.5 | 0.61 | 56.5 | 296.3 | 0.50 |

Columns in Table 2 to the right of the reversible cases give results for $S_{gen}$ values of 0.25 and 0.50 MJ/kmol $H_2$ - K. For the latter case, which just achieves the target, the irreversible He flow is 60% greater, while the rejected heat is 4 times greater than for the reversible process. These will have significant consequences for the equipment to used in a process. It also suggests that every effort needs to be made to minimize entropy generation.

**B. Sections of GA Process**

Because the GA S-I process has three distinct sections, these can be analyzed in the same fashion as the overall process. An additional degree of freedom to investigate is the effect of changing the flow rates and compositions of the streams among the sections. Then Equations (1) - (3) can be used to find potential heat flows between sections for the reversible and real cases. Because the process is closed to sulfur and iodine, there are constraints on the external and intersection material and energy flows, which provide a rigorous check on the complex calculations that can arise when changes in intersection flows are made. In particular, the ultimate helium flows and sums of $Q_e$ and $S_{gen}$ for the sections must equal the results for the overall process. However, the section analyses can suggest where are the greatest effects of inefficiencies might arise. Detailed results of this study will be described elsewhere; here the concept is introduced and a few results are given.

Figure 4 shows the schematics of the sections and stream flows for idealized S-I processes. (The analysis used the actual AspenPlus streams; Table 3 lumps them into the streams A - D as in Figure 4.) Table 3 gives stream condition values for four variations of intersection flows where the streams are single-phase and the property values are from the AspenPlus properties model of Mathias [19]. The first case has no recycle of water or iodine compounds, the second has recycle only through Section 2, the third has recoil only through Section 3. The last case has streams from the process studied extensively by Goldstein, et al. [6].

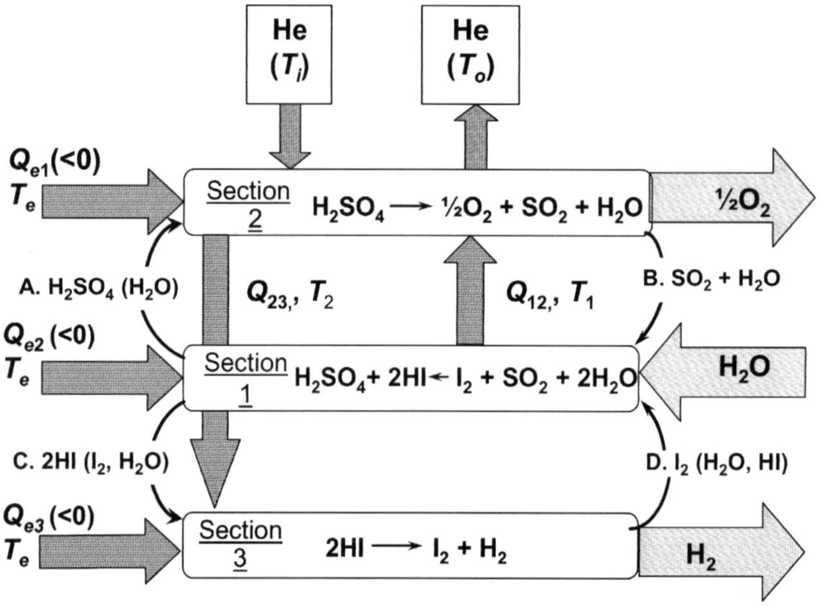


Figure 4. Sections of the idealized S-I Process showing inlet, out and intersection streams and inlet and environmental heat flows.

An important assumption here is that all computed heat flows leaving a section can be utilized in the receiving section, *e.g.*, the Bunsen heat, $Q_{12}$, is completely used in Section 1, which is unlikely. In a real analysis, this issue can most easily be dealt with by fixing the fraction or amount of heat which flows between the sections and then rejecting the balance to the environment. Another assumption is that all specified separations of streams can be achieved in the given sections. Thus, the separation leading to the flows leaving Section 1 from the liquid-liquid split of the $H_2SO_4$ stream to Section 2 and the HI stream to Section 3 is assumed to be perfect. Realistic limitations of separations can be taken into account by changing the stream specifications.

As in Table 2, Table 4 gives He flows, heat rejections and efficiencies for each case of Table 3 in the completely reversible system and one where $S_{gen}$ = 0.5 MJ/kmol $H_2$-K in each section.

Note that some cases have energy flowing into Sections 2 and 3 from the environment ($Q_e > 0$) accompanied by large amounts of heat then being transferred to another section. This would not be normal, but suggests that heat flows between the sections chosen in available designs might not be optimal. However, it is also likely that section temperature profiles, as shown by Gorensek [7], Ewan and Allen [8], and Lewis, et al. [11], would prevent some of the implied heat transfers.

Table 3. Component flows and properties for models of the AspenPlus S-I process of Mathias & Brown [18-22] for different cases of the process of Figure 4.

| Case 1 | Molar Flow (kmol/kmol-$H_2$) | | | | | | Enthalpy, *h* | Entropy, *s* |
|---|---|---|---|---|---|---|---|---|
| Stream | $H_2SO_4$ | $SO_2$ | HI | $I_2$ | $H_2O$ | Total | MJ/kmol | MJ/kmol-K |
| A | 1 | 0 | 0 | 0 | 0 | 1 | 27 | -0.25 |
| B | 0 | 0 | 2 | 0 | 0 | 2 | 29 | 0.09 |
| C | 0 | 1 | 0 | 0 | 1 | 2 | -287 | -0.07 |
| D | 0 | 0 | 0 | 1 | 0 | 1 | 27 | 0.07 |
| Case 2 | Molar Flow (kmol/kmol-$H_2$) | | | | | | Enthalpy, *h* | Entropy, *s* |
| Stream | $H_2SO_4$ | $SO_2$ | HI | $I_2$ | $H_2O$ | Total | MJ/kmol | MJ/kmol-K |
| A | 1 | 0 | 0 | 0 | 4 | 5 | -336 | -0.07 |
| B | 0 | 0 | 2 | 0 | 0 | 2 | 29 | 0.09 |
| C | 0 | 1 | 0 | 0 | 5 | 6 | -281 | -0.12 |
| D | 0 | 0 | 0 | 1 | 0 | 1 | 27 | 0.07 |
| Case 3 | Molar Flow (kmol/kmol-$H_2$) | | | | | | Enthalpy, *h* | Entropy, *s* |
| Stream | $H_2SO_4$ | $SO_2$ | HI | $I_2$ | $H_2O$ | Total | MJ/kmol | MJ/kmol-K |
| A | 1 | 0 | 0 | 0 | 0 | 1 | -727 | -0.25 |
| B | 0 | 0 | 2 | 7 | 10 | 19 | -155 | -0.18 |
| C | 0 | 1 | 0 | 0 | 1 | 2 | -287 | -0.07 |
| D | 0 | 0 | 0 | 8 | 10 | 18 | -142 | -0.05 |
| Case 4 | Molar Flow (kmol/kmol-$H_2$) | | | | | | Enthalpy, *h* | Entropy, *s* |
| Stream | $H_2SO_4$ | $SO_2$ | HI | $I_2$ | $H_2O$ | Total | MJ/kmol | MJ/kmol-K |
| A | 1 | 0 | 0 | 0 | 4 | 5 | -336 | -0.07 |
| B | 0 | 0 | 6 | 23.42 | 30.61 | 60.03 | -149 | -0.17 |
| C | 0 | 1 | 0 | 0 | 5 | 6 | -282 | -0.12 |
| D | 0 | 0 | 4 | 24.42 | 30.61 | 59.03 | -142 | -0.12 |

Table 4. Section heat transfers and rejections to the environment for cases of Table 3 with $T_i$ = 1200K and $T_o$ = 900K. All heat amounts are in MJ/kmol $H_2$. Positive signs for heat transferred between sections, $Q_{i\rightarrow j}$, mean that heat moves from section i to section j. Positive signs for heat rejection, $Q_{ei}$, mean that heat is taken in to section i from the environment. The overall heat rejections and He flows are from Table 2.

| $S_{gen}$ in Section i, $S_{gen,i}$, MJ/kmol $H_2$-K | | | Case | Heat Transferred Between Sections | | Heat Rejection from Section *i*, $Q_{ei}$, MJ/kmol $H_2$ | | | Total Heat Rejection | He Flow |
|---|---|---|---|---|---|---|---|---|---|---|
| $S_{gen,1}$ | $S_{gen,2}$ | $S_{gen,3}$ | | $Q_{1\rightarrow 2}$ | $Q_{2\rightarrow 3}$ | $Q_{e1}$ | $Q_{e2}$ | $Q_{e3}$ | $Q_e$ | kmol/kmol $H_2$ |
| 0 | 0 | 0 | 1 | -480 | 24 | -315 | 309 | -55 | -61 | 56 |
| | | | 2 | -585 | 24 | -264 | 257 | -55 | -61 | 56 |
| | | | 3 | -2075 | -830 | -2330 | 1049 | 1220 | -61 | 56 |
| | | | 4 | -2058 | -755 | -2308 | 938 | 1308 | -62 | 57 |
| 0.5 | 0 | 0 | 1 | -788 | 24 | -932 | 715 | -55 | -271 | 90 |
| | | | 2 | -894 | 24 | -880 | 662 | -55 | -273 | 91 |
| | | | 3 | -2383 | -830 | -2947 | 1455 | 1220 | -271 | 90 |
| | | | 4 | -2366 | -755 | -2924 | 1342 | 1308 | -275 | 92 |
| 0 | 0.5 | 0 | 1 | -480 | 24 | -315 | 99 | -55 | -271 | 90 |
| | | | 2 | -585 | 24 | -264 | 46 | -55 | -273 | 91 |
| | | | 3 | -2075 | -830 | -2330 | 839 | 1220 | -271 | 90 |
| | | | 4 | -2058 | -755 | -2308 | 729 | 1308 | -275 | 92 |
| 0 | 0 | 0.5 | 1 | -480 | 354 | -315 | 429 | -385 | -271 | 90 |
| | | | 2 | -585 | 354 | -24 | 376 | -385 | -273 | 91 |
| | | | 3 | -2075 | -500 | -2330 | 1169 | 890 | -271 | 90 |
| | | | 4 | -2058 | -425 | -2308 | 1056 | 978 | -275 | 92 |

The results of Table 4 suggest the following:

A. A large amount of heat is always rejected from Section 1, increasing with recycle from Section 3, but decreasing with recycle to Section 2. Much of this heat is actually from Section 2 which increases with recycle amounts. Irreversibility in the section increases the amounts.

B. A large amount of heat is always taken up from the environment by Section 2; it also increases significantly with recycle from Section III. Irreversibility in Section 2 decreases the amount, but irreversibilities in Sections 1 and 3 increase this heat flow.

C. The heat rejection of Section 3 is small for no $HI/I_2$ recycle, but changes sign to requiring environmental heat when recycle occurs. Irreversibility in Section 3 puts more heat into the environment directly or by reducing amount taken in for $HI/I_2$ recycle.

D. The relative effects of recycles cannot be inferred from the limited results in Table 4. This will be explored elsewhere. However, the effects of the same level of irreversibility in the different sections can be indicated. For example, the values of $Q_{e2}$, a measure of increased He energy requirements, are more affected by nonzero $S_{gen,1}$ than by nonzero $S_{gen,3}$. Thus, attempts should be made to reduce temperature differences and heat flows associated with Section 1.

Some of the behaviors given in Table 4 are counter to expectations. As discussed below, this may result from inaccurate property values computed with the Aspen model. We are in the process of repeating the above analysis with the OLI model.

## 4. Comparison of Process Model-Free Analyses for Different Property Models

The above analysis can be implemented to determine the effects of different property model results on the energy flows for a proposed flow sheet system which includes computed heat and work transfers between sections and to/from the environment. For the software selected, the process simulation can satisfy the material (1) and energy (2) balance equations, but not necessarily yield positive entropy generation. If the property values from other software are different, the process will not satisfy the energy balance, but could be made consistent with a positive $S_{gen}$. Thus, the purpose of this part of the paper is to determine what differences there are in the software package properties and suggest what the impact might be on process configurations and units.

We utilized the full process developed by Mathias and Brown [18-22] with stream flows between sections reconciled by Summers [24] and by Murphy [25]. The processes are too complex to be shown in this paper, but Table 5 summarizes gives the material, heat and work flows. In addition to our own property calculations with the AspenPlus model and the OLI model [26], we were graciously provided $h$ and $s$ values for the streams from the JAEA modification of the OLI model [27] and from the ProSim model [28]. Table 5 gives the heat and work amounts for the process sections. There may be some misassignments of sources/sinks though the temperatures are consistent, especially for Section 3. However, all property model calculations were evaluated in the same way.

Table 5. Heat and work by Sections of an S-I process from Mathias and Brown [18-22]. Heat source (> 0) or sink (< 0) from environment ($e$), Helium (He), or another Section (i) at temperature $T_j$.

| Section | Work, MJ/kmol $H_2$ | Heat in from $T_1$, MJ/kmol $H_2$ | $T_1$, K | Heat In from $T_2$, MJ/kmol $H_2$ | $T_2$, K | Heat In from $T_3$, MJ/kmol $H_2$ | $T_3$, K |
|---|---|---|---|---|---|---|---|
| 1 | 39 | -17 ($e$) | 300 | -13 ($e$) | 360 | -748 (2) | 393 |
| 2 | -7.8 | -134 ($e$) | 313 | 67 (1) | 373 | 358 (He) | 636-1250 |
| 3 | 9 | 423 ($e$) | 561 | -209 ($e$) | 518 | 56 ($e$) | 545 |

Table 6 gives the energy balance and entropy generation amounts for each property model in each section for the above process. The first four rows are for the directly computed results. It can be seen that for Section I, while the energy is balanced for the Aspen simulation, as it should, the amount of heat leaving that Section calculated by the other programs is too large. For the Aspen simulation, there is a large, negative $S_{gen}$ for the process, indicating that it would be infeasible. We speculate that this arises because of incorrect stream properties leading to excess heat outflow. Examination of the values suggests that the Aspen $h$ and $s$ values are somewhat too negative for streams with large amounts of iodine, making the stream energy differences between the stream inlets and stream outlets become too negative. The other programs do not give the energy in balance, though they do have $S_{gen} > 0$. However, with the energies being different, the $S_{gen}$ cannot be appropriate. In fact, they are far in excess of what the model-free analysis requires.

Table 6. Energy and entropy terms for Sections of an S-I process from Mathias and Brown [18-22]. $S_{gen}$Adjust from reducing $Q_e$ values to yield zero net energy, or to have no leaving energy (*). OLI (1-phase) results by using properties of single phase obtained by Aspen model.

Section 1

| Property Model | Net Energy, MJ/kmol $H_2$ | $S_{gen}$, MJ/kmol $H_2$-K | $S_{gen}$Adjust, MJ/kmol $H_2$-K |
|---|---|---|---|
| AspenPlus [18-22] | 0 | -3.05 | -3.05 |
| SRNL OLI | -344 | 1.05 | 0.21 |
| JAEA OLI [27] | -381 | 1.07 | 0.11 |
| ProSim [28] | -479 | 1.55 | 0.59* |
| SRNL OLI (1-phase) | -398 | 1.05 | 0.09 |
| JAEA OLI[28](1-phase) | -715 | 0.97 | 0.01* |

Section 2

| Property Model | Net Energy, MJ/kmol $H_2$ | $S_{gen}$, MJ/kmol $H_2$-K | $S_{gen}$Adjust, MJ/kmol $H_2$-K |
|---|---|---|---|
| AspenPlus [18-22] | -14 | 0.77 | 0.77* |
| SRNL OLI | -16 | 0.80 | 0.80* |
| JAEA OLI [27] | -11 | 0.79 | 0.79* |
| ProSim [28] | 47 | -0.03 | -0.03* |
| SRNL OLI (1-phase) | -20 | 0.81 | 0.81* |
| JAEA OLI[28](1-phase) | 322 | -0.18 | -0.2* |

Section 3

| Property Model | Net Energy, MJ/kmol $H_2$ | $S_{gen}$, MJ/kmol $H_2$-K | $S_{gen}$Adjust, MJ/kmol $H_2$-K |
|---|---|---|---|
| AspenPlus [18-22] | 0 | 0.43 | 0.43 |
| SRNL OLI | -81 | 0.43 | 0.43* |
| JAEA OLI [27] | -91 | 0.55 | 0.55* |
| ProSim [28] | 60 | -0.20 | -0.20* |
| SRNL OLI (1-phase) | -81 | 0.54 | 0.54* |
| JAEA OLI[28](1-phase) | -91 | 0.55 | 0.55* |

While we did not have access to the flow sheets for these other simulations, we might anticipate that their simulated processes would reject less heat to the environment. Thus, to make a reasonable estimate of $S_{gen}$, we can reduce their $Q_e$ values as far as possible, either to make the energy balance or make $Q_e = 0$ (indicated by *), so the energy is still not in balance. These results are in the last column in Table 6 ($S_{gen}$Adjust). While not all of the energies can be made consistent, most of the mismatches are relatively small, and the $S_{gen}$ values are not far above the appropriate range for a viable process.

It is apparent that discrepancies can arise because the property models do not to provide the same *h* and *s* values. It is possible that none are reliable, especially if they were not tuned to energy data. In addition, it was found that the stream conditions were not always the same from the different property software. All of the streams in the Aspen process were computed to be single phase, as were those from ProSim. However, some streams computed by both the OLI programs were multiphase; interestingly, they were not the same ones. The computations discussed above ignored this effect, and used the multiphase stream properties. An alternative approach would be to calculate the results by assuming that the smaller phase was not present and using *h* and *s* values from the dominant phase only. The last rows of Table 6 show these results for both ways of computing $S_{gen}$. Here the JAEA energies are extremely out of balance while the standard OLI calculations are still relatively consistent. The ultimate consequences of different phase behaviors are uncertain.

The above analysis was also carried out for Sections 2 and 3 as shown in Table 6. In this case, the property models are mostly consistent, and the $S_{gen}$ values are in the range consistent with the process model-free analysis. The ProSim values are somewhat different, giving negative $S_{gen}$ values. One is tempted to conclude that the overall calculations are generally most reliable for the OLI models, but the optimal simulated process configurations probably remain uncertain.

## 5. Conclusions

Thermodynamic analyses of an S-I process for hydrogen manufacture have been carried out with a process model-free approach and with properties computed with different property model software. Efficiencies and ultimate limitations on irreversibilities have been expressed in terms of entropy generation. These suggest where in the process to focus on process improvement as well as highlight the effects of differences in the property models.

## 6. Acknowledgments

This work was completed with funding under DOE Contract DE-FC07-051D14677 to Clemson University. The authors are extremely grateful to Dr. A. Anderko of OLI Systems, Inc. for assistance in, and to Dr. H. Kubo or JAEA for, providing stream properties from their version of the OLI property model, and to Dr. O. Baudouin for providing stream properties from the ProSim property model.

## 7. References

1. G.D. Berry, "Hydrogen Production", in Encyclopedia of Energy, C.J. Cleveland, Editor. 2004, Elsevier: New York, 253-265.

2. R. Agrawal, M. Offutt, M. Ramage, AIChE Journal 51(2005): 1582-1589.

3. S. Chalk, "Challenges and Benefits of a Move Towards Hydrogen Fuel," AIChE Meeting, Orlando, FL, April 2006.

4. A. Le Duigou, J.M. Borgard, B. Larousse, D. Doizi, R.W.H. Allen, B.C.R. Ewan, G.H. Priestman, R. Elder, R. Devonshire, V. Ramos, G. Cerri, C. Salvini, A. Giovannelli, G. De Maria, C. Corgnale, S. Brutti, M. Roeb, A. Noglik, P.-M. Rietbrock, S. Mohr, L. de Oliveira, N. Monnerie, M. Schmitz, C. Sattler, A. O. Martinez, D. de Lorenzo Manzano, J. C. Rojas, S. Dechelotte, O. Baudouin, "HYTHEC: An EC funded search for a long term massive hydrogen production route using solar and nuclear technologies," Int. J. Hydrogen Energy, 32 (2007): 1516-1529.

5. M. Sakurai, H. Nakajima, K. Onuki, K. Ikenoya, S. Shimizu, "Preliminary process analysis for the closed cycle operation of the iodine-sulfur thermochemical hydrogen production process," Int. J. Hydrogen Energy, 24(1999): 603-612.

6. S. Goldstein, J.-M. Borgard, X. Vitart, Int. J. Hydrogen Energy, 30(2005): 619 – 626

7. M.B. Gorensek, "Model-based Performance Comparison of Thermochemical Nuclear Hydrogen Processes," AIChE Meeting, Atlanta, GA, April 2005.

8. B.C.R. Ewan, R.W.K. Allen, "Assessing the Efficiency Limits for Hydrogen Production by Thermochemical Cycles," AIChE Meeting, Cincinnati, OH, November 2005.

9. D.F. McLaughlin, E.J. Lahoda, L.A. Paoletti, W.Kriel, "Thermodynamic Modeling for the Hybrid Sulfur Process in Chemcad," AIChE Meeting, San Francisco CA, November 2006.

10. Rosen, M. A. and Scott, D. S. "Entropy Production and Exergy Destruction: Part II-illustrative technologies." Int. J. Hydrogen Energy 28(2003): 1315-1323.

11. M. Lewis, J. Masin, A.C. Taylor, "Evaluation of Alternate Thermochemical Cycles," AIChE Meeting, San Francisco CA, November 2006.

12. M. A. Rosen, Int. J. Hydrogen Energy, 21(1996): 349-65.

13. G.E. Besenbruch, L.C. Brown, J.H. Norman, D.R. O'Keefe, D.R. "GA Technologies Thermochemical Water-Splitting Process Improvements," San Francisco Operations Office, Department of Energy. November 1984.

14. J.H. Norman, G.E. Besenbruch, D.R. O'Keefe, C.L. Allen, "Thermochemical Water-Splitting Cycle, Bench-Scale Investigations, and Process Engineering," Final Report for the Period February 1977 through December 31, 1981. 1982.

15. L.C. Brown, G.E. Besenbruch, R.D. Lentsch, K.R. Schultz, J.E. Funk, P.S. Pickard, A.C. Marshall, S.K. Showalter, "High Efficiency Generation of Hydrogen Fuels Using Nuclear Power," Final Technical Report for the Period August 1,1999 through September 30, 2002. 2003, General Atomics A24285.

16. R. T. Buckingham, L. Brown, B.E. Russ, G. Besenbruch, W. Sweet, "Process Flowsheet Analysis of Hydrogen Iodide Decomposition in the Sulfur-Iodine Cycle," AIChE Meeting, San Francisco CA, November 2006.

17. G. Besenbruch, L. Brown, J. Funk, S. Showalter, "Thermochemical Hydrogen Production With A High Temperature Advanced Nuclear Reactor." General Atomics Technologies. General Atomics. 29 October 2006. <http://web.gat.com/pubs-ext/MISCONF00/A23510abs.pdf>.

18. P.M. Mathias, "Modeling the Sulfur-Iodine Cycle: Aspen Plus Building Blocks and Simulation Model," AspenTech, 2002.

19. P.M. Mathias, "Report-4 Preliminary Flowsheets for SI Cycle," General Atomics, December 2004.

20. P.M. Mathias, "Modeling the Sulfur Iodine Cycle, Aspen Plus Building Blocks and Simulation Models", General Atomics and Sandia National Laboratories, 2002.

21. P.M. Mathias, L.C. Brown. "Thermodynamics of the Sulfur Iodine Cycle for Thermochemical Hydrogen Production," 68th Annual Meeting of the Japan Society of Chemical Engineers. 2003.

22. L.C. Brown, P.M. Mathias, C-C. Chen, D. Ramrus, "Thermodynamic Model for the HI-I2-H2O System". AIChE Meeting Reno, NV, 2001.

23. J.P. O'Connell, J.M. Haile, Thermodynamics: Fundamentals for Applications. 2005, New York, NY: Cambridge University Press.

24. W.A. Summers, "Centralized Hydrogen Production from Nuclear Power: infrastructure Analysis and Test Case Design Study," 2006, Westinghouse Savannah River Company: Aiken, SC.

25. J.E. Murphy, "Process Analysis and Design of the Sulfur-Iodine Cycle for Thermochemical Water Decomposition to Hydrogen," M.S. Thesis, University of Virginia, 2007.

26. P. Wang, A. Anderko, R.D. Springer, R.D. Young, J. Molecular Liq., 125(2006): 37–44.

27. S. Kubo, JAEA, Personal Communication, August 2007.

28. O. Baudouin, ProSim, Personal Communication, August 2007.

## C. H. Gooding, Economic Analysis of the Hybrid Chlorine Cycle

### Overview

This paper reports the results of the most complete study conducted to date on hydrogen production using the hybrid chlorine cycle. Three alternative flow sheets were developed to produce 200 million kg of hydrogen per year at 35°C and 20 barg. Each was modeled using Aspen simulation software, supplemented where necessary by custom Excel spreadsheets. Major equipment was sized, and feasible materials of construction were identified for all unit operations. Fixed capital investment and energy efficiency were estimated for each of the flow sheets.

The hybrid chlorine cycle consists of two reactions. The reverse Deacon reaction is

$Cl_2(g) + H_2O(g) <=> 2HCl(g) + 0.5O_2(g)$

This reaction is endothermic and reversible with an equilibrium constant of 6.8 at 850°C, but only 0.0001 at 130°C. The second reaction involves the decomposition of hydrogen chloride, which is accomplished electrochemically.

2HCl *(gas or aqueous)* => $Cl_2(g) + H_2(g)$.

Electrolysis dominates the energy requirements of all three flow sheets even with the relatively optimistic assumption of 50% efficiency for conversion of thermal energy to electrical energy. Practical aqueous electrolysis requires operation at 2.0 V (Motupally 1998), which is equivalent to 772 kJ of thermal energy input per mol of hydrogen produced. Gas phase electrolysis can be conducted at 1.5 V (Motupally 1998, Weidner 2007), reducing the equivalent thermal energy input to 579 kJ/mol $H_2$. For aqueous or gas phase electrolysis, the practical energy input is about 102 kJ/mol $H_2$ above the thermodynamic minimum required for the reaction under reversible conditions. This over-potential energy can be recovered as waste heat at the cell operating temperature of 80°C. But under the most favorable circumstances, the net energy consumption for electrolysis is still about twice the lower heat of combustion of the hydrogen produced (242 kJ/mol $H_2$).

The three alternative flow sheets differ in how the reverse Deacon reaction is conducted, how the product gases are separated, and whether the electrolysis step is conducted in the liquid or gas phase. Each process is described and the energy requirements are summarized in the discussion that follows.

### Aqueous quench option

In this flow sheet the reverse Deacon reaction is conducted in a heated, tubular reactor with a slightly chlorine-rich feed (13% excess $Cl_2$). The reaction proceeds to equilibrium at 850°C and 20 barg, converting 51% of the steam that enters into HCl. Higher equilibrium conversion could be achieved at lower pressure, but all of the product hydrogen would have to be compressed before it left the process. Higher pressure is also beneficial in the oxygen recovery that occurs downstream by distillation.

The gas leaving the reactor is quenched rapidly to 380°C in a spray chamber by adiabatic humidification, using a portion of the liquid effluent recycled from the aqueous electrolyzers. The basic concept behind running the reactor this way was demonstrated by Carty (1981), who showed that rapid quenching is required to prevent reversion of products to reactants by the Deacon reaction. Davis (1951) observed low conversions in the Deacon reaction at 400°C with a copper catalyst present, so we concluded that quenching to 400°C or lower should be sufficient to minimize loss of HCl.

The quenched gas is cooled further to 80°C in a partial condenser, producing a liquid stream that contains 45% HCl (weight basis) with minimal dissolved oxygen and chlorine. This is sent to aqueous electrolysis. Aspen indicates that the gas stream leaving the partial condenser in equilibrium with the aqueous liquid contains 43% HCl, 38.5% $Cl_2$, 18.2% $O_2$, and 0.3% $H_2O$ on a molar basis. These results were obtained using the ELECNRTL property package which agrees well with HCl-$H_2O$ data at 80°C near 1 bar, but we have no multi-component data at 20 barg for comparison.

Oxygen is recovered from the other gases by distillation at 20 barg. The RKS equation of state was used for property estimation. Liquid HCl, chlorine, and residual water leave the bottom of the distillation column at 9°C and are vaporized and fed at 80°C to gas-phase electrolysis. A small flow of additional water must be fed to the gas-phase electrolyzers to keep the membranes moist and make up for water vapor that leaves with the separated gases. The hydrogen produced (60.5% of the total leaving the process) is cooled to 35°C and is discharged, saturated with water vapor at 20 barg. Saturated chlorine gas at 80°C is recycled to the reverse Deacon reactor preheat section of the plant.

The liquid stream leaving the partial condenser goes to aqueous electrolyzers. Unlike gas-phase electrolyzers, these units cannot be operated at 100% conversion of HCl because the electrical resistance rises significantly as the ionized HCl content falls. High HCl conversion requires either higher voltage (more energy) or lower current density (more surface area and thus higher capital expenditure). Typically, conventional Uhde chlorine production units are fed 22 weight % HCl and discharge 17 weight % HCl for recycle. The current version of our flow chart is based on 10 weight % HCl leaving aqueous electrolysis. A portion of this stream is recycled to the aqueous quench tower that follows the reverse Deacon reactor. The balance is recycled to the reactor preheat section of the plant, as is the chlorine gas produced by electrolysis. Hydrogen produced (39.5% of the total leaving the process) is cooled to 35°C after it leaves the aqueous electrolyzers and is discharged, saturated with water vapor at 20 barg.

All of the unit operations used in this flow sheet have been demonstrated in other process applications or lab experiments with the chemical species involved, though some have not been demonstrated at the exact conditions indicated. Phase equilibrium models used in the partial condenser and oxygen recovery still may be inaccurate, which would clearly affect cost and energy efficiency estimates, but we are not aware of any physical phenomena or properties that would invalidate the basic concept (e.g., an undetected azeotrope that would prevent a separation).

Table 1 summarizes the current best estimate of energy requirements and efficiency for this flow sheet. Pinch adjustment refers to the quantity of energy in "hot streams" (those that need to be cooled) that cannot be recovered by thermal integration with "cold streams" (those that need to be heated), assuming that the minimum allowable temperature driving force is 20°C. To reduce the pinch adjustment and improve the thermal efficiency of the cycle, the pressure of the combined aqueous stream recycled to the reactor is dropped to 6 barg, allowing vaporization to occur between 165 and 175°C. The vapor is then compressed to 20 barg before it is fed to the reactor. Compression requires electrical input, but the energy tradeoff is favorable. Further optimization of the flow sheet might be possible. Analysis is continuing on other pressure adjustments and on the conditions of the partial condenser since energy efficiency could be improved by electrolyzing a larger fraction of HCl in the gas phase.

Table 1. Energy estimates for the hybrid chlorine cycle with aqueous quench.

| | | (kJ/mol $H_2$ produced) |
|---|---|---|
| Heat and vaporize reactants | | 203 |
| Reverse Deacon reaction at 850°C | | 59 |
| Other separation and recycle requirements | | 64 |
| | total heat in | 326 |
| Cool rDr products and other streams | total heat out | -371 |
| 20°C pinch adjustment | | 190 |
| Electrolysis (thermal equivalent of electricity) | | |
| $2HCl(aq) \Rightarrow H_2(g) + Cl_2(g)$ | [39.5% of total at 2.0V] | 305 |
| $2HCl(g) \Rightarrow H_2(g) + Cl_2(g)$ | [60.5% of total at 1.5V] | 350 |
| Summary | | |
| Net thermal energy input | | 800 |
| Net energy value of $H_2$ (LHV) | | 242 |
| Hybrid chlorine cycle efficiency | | 30 % |

**Membrane Reactor Option**

The efficiency of the hybrid chlorine cycle would be improved if all HCl electrolysis were accomplished in the gas phase. But major challenges must be overcome to allow this strategic advantage. A practical method must be found to separate product gases and excess reactants from the reverse Deacon reaction without creating an aqueous HCl phase. This would be simplified if either HCl or oxygen could be removed from the reactor while the reaction is proceeding. One way to accomplish this would be to permeate oxygen as it is formed through a membrane wall in the reactor. Highly selective, dense ceramic membranes have been developed for this purpose over the last two decades (see Maiya 1997, Baker 2002, Julbe 2005, Baluchandran 2006).

Dense ceramic membranes are well suited for operation at 800 to 900°C; they can permeate virtually pure oxygen; and they are resistant to acid gases. Much of the testing conducted on these membranes has been done in the context of supplying pure oxygen to fast reactions. The oxygen-consuming reaction acts as a nearly perfect sink, providing a large driving force for oxygen transfer. Over a wide range of operation, oxygen flux has been shown to be proportional to $\ln[PO_2h/PO_2l]$, where $PO_2h$ represents the partial pressure of oxygen on the oxygen-rich or high side of the membrane, and $PO_2l$ is the partial pressure of oxygen on the depleted or low side of the membrane. With transfer from air to a fast, oxygen consuming reaction $PO_2h/PO_2l$ can be several orders of magnitude (Baluchandran 2006).

There is no obvious oxygen consuming reaction to use in conjunction with the reverse Deacon reactor in an efficient hydrogen production cycle, but a modest difference in oxygen partial pressure can be achieved by manipulating total gas pressure or by using a sweep gas. Calculations indicate that use of an inert sweep gas will require much less energy to accomplish a given driving force than compression of the reacting gases beyond 20 barg or vacuum pumping on the permeate side of the membrane. So the flow sheet for this option is based on use of a sweep gas. Heated nitrogen is a possible choice, and the resulting mixture could be discharged safely after heat recovery, or it might be feasible to separate the oxygen for sale and the nitrogen for recycle. Another possibility is to use as a sweep the hot gas (presumably helium) from a very high temperature gas-cooled nuclear reactor (VTGR). In this case recovery of the sweep gas and oxygen would most likely be required.

In the current version of this flow sheet the reverse Deacon reactor operates with a 5:1 feed ratio of $Cl_2$ to $H_2O$, each of which enters the reactor at 20 barg and 850°C. The penalty in equilibrium conversion caused by elevated pressure is outweighed by the assistance that higher pressure provides for the flux of oxygen across the membrane. The reactor is divided into two sections. The first section is similar to that used in the aqueous quench option. The reacting gases flow through impermeable tubes, which are heated to provide isothermal operation at 850°C. This section achieves 88% $H_2O$ conversion (due to the excess $Cl_2$ present) before equilibrium slows the reaction progress. The gases then enter the tubular membrane section, where oxygen permeates into the sweep gas. The sweep gas flow is countercurrent to the reacting gases, and the sweep gas must be at 850°C or above to avoid convective/conductive cooling. Even if the sweep gas adds no heat to the reactor, the calculated temperature drop that results from endothermic reaction is less than 2°C for every additional % of $H_2O$ conversion.

Removal of oxygen allows the reaction to continue toward completion. Our calculations of required membrane area have been based on the arbitrary specification that $PO_2h$/Po21 should be > 2 at all points in the membrane section of the reactor. Obviously a larger pressure ratio will reduce the membrane area required while a lower pressure ratio will increase it. With an oxygen partial pressure ratio of 2.0, 0.24 mol of sweep gas is required per 2 mol of HCl produced (equivalent to 1 mol of $H_2$) to achieve 97% conversion of $H_2O$. This leaves a reactor exit gas that is 32 mole % HCl, 67% $Cl_2$, and less than 1% $O_2$ and $H_2O$ combined. This is suitable for processing in gas-phase HCl electrolyzers after cooling to 80°C. Hydrogen is cooled to 35°C before leaving the plant. Chlorine is recycled to the reactor at 80°C. Both gases leave the electrolysis section saturated with water at 20 barg. Energy and efficiency estimates are shown below.

Table 2. Energy estimates for the hybrid chlorine cycle with oxygen permeating membrane reactor (does not include allowance for separation of oxygen and sweep gas).

| | | (kJ/mol $H_2$ produced) |
|---|---|---|
| Heat and vaporize reactants | | 228 |
| Reverse Deacon reaction at 850°C [in heated section] | | 55 |
| Heat sweep gas | | 6 |
| | total heat in | 289 |
| Cool rDr products and sweep gas | total heat out | -184 |
| 20°C pinch adjustment | | 3 |
| Electrolysis (thermal equivalent of electricity) | | |
| $2HCl(g) \Rightarrow H_2(g) + Cl_2(g)$ | | 579 |
| Summary | | |
| Net thermal energy input | | 687 |
| Net energy value of $H_2$ (LHV) | | 242 |
| Hybrid chlorine cycle efficiency | | 35 % |

**HCl Sequestration Option**

This option uses a concept suggested by Amendola (2005). The approach is to induce the reverse Deacon reaction to occur as chlorine and steam flow over a solid sorbent that will sequester HCl as it is produced. The effect is similar to the membrane reactor – one of the products of the reaction is removed as it is formed, allowing the reaction to proceed to completion. The sorbent may also be coated with a catalyst, e.g., cobalt chloride, to speed up the reaction. The combination of HCl sequestration and catalytic action can, in principle, induce the reaction to occur at a reasonable rate at a much lower temperature than 850°C despite the low equilibrium constant. If the feed is stoichiometric and all HCl formed can be sequestered on the sorbent, only oxygen will leave the reactor. When the sorbent is saturated with HCl, valves are used to switch the reactor to the desorb mode, and the temperature is raised to recover the HCl. Meanwhile the reaction can continue in a parallel, regenerated reactor.

We have demonstrated proof of concept of this scheme in bench-scale experiments, but we did not obtain data of sufficient quantity and quality for meaningful reactor scale-up. When chlorine and steam were passed through a column packed with Zeobrite, a mined zeolite, at 105°C and 1 atm, oxygen was detected in the gas leaving the column. But the results were deemed to be unreliable because measured oxygen levels were inconsistent. After reactant flow was stopped, desorption was conducted at 140°C using a small nitrogen sweep to assist HCl removal from the zeolite bed. HCl detection in the exit gas indicated with more confidence that the reverse Deacon reaction had occurred. With a 1:1 molar ratio of chlorine and steam and a space time of 0.2 minutes, the results indicated that roughly 0.2% conversion was achieved.

This option provides HCl gas for electrochemical conversion to chlorine and hydrogen. Byproduct oxygen is made as well. If the concept can be reduced to practice, it should result in lower capital investment than either of the previous options due to lower operating temperatures. It should also yield higher efficiency than the aqueous quench option because all of the HCl is converted to product using gas-phase electrolysis.

Energy estimates for the cycle, summarized in Table 3, indicate an efficiency of 34%. The absolute value of the heat of absorption of HCl into the sorbent is designated by ΔHabs, an unknown value that will depend on the particular sorbent used. This value is subtracted from the heat of reaction, but it is added back to recover the HCl and regenerate the resin. The inherent assumption is that the absorption and desorption cycle will be essentially reversible. This should be a reasonable approximation if the reactor is configured somewhat like a rotary dryer so that the resin can be heated for regeneration by direct contact with the vessel walls. An additional 2 kJ is included to account for raising the HCl temperature from the reaction temperature (105°C) to the desorption temperature (140°C). The sensible heat taken up and released by the sorbent as it cycles between reaction and desorption should cancel. Residual heat that remains in the resin after regeneration will simply assist the endothermic reaction in the next cycle, offsetting part of the heat of reaction that is included in the table. That same quantity of heat input will be needed each time the regeneration cycle begins.

Table 3. Energy estimates for the hybrid chlorine cycle with HCl sequestration.

| | | (kJ/mol $H_2$ produced) |
|---|---|---|
| Heat and vaporize reactants | | 48 |
| Reverse Deacon Reaction at 105°C | | 57.5 - ΔHabs |
| Separation Requirement (heat and recover HCl from sorbent) | | 2.3 + ΔHabs |
| Compress $H_2$ product (0 barg → 20 barg) | | 32 |
| | total heat in | 140 |
| Cool rDr Products | total heat out | - 7 |
| 20°C Pinch Adjustment | | 1 |
| Electrolysis (thermal equivalent of electricity) | | |
| $2HCl(g) \Rightarrow H_2(g) + Cl_2(g)$ | | 579 |
| Summary | | |
| Net energy input | | 713 |
| Net energy value of $H_2$ (LHV) | | 242 |
| Hybrid chlorine cycle efficiency | | 34% |

**Materials of Construction, Capital Estimate, On-going Work**

The demands on the materials of construction used in the hybrid chlorine cycle will be significant, especially for the options that include operation at 850°C. Even at lower temperatures, corrosion resistance to the gaseous and liquid species present is not trivial. Haynes alloy HR-160 and alloy 556 are considered to be suitable for oxidizing, chlorine-bearing environments at 900 and 980°C, respectively (see http://www.haynesintl.com), but they have not been tested for the exact conditions required. Other nickel alloys should be suitable for lower temperature service. Materials of construction for liquid and gas phase electrolysis cells include fluorocarbon polymers, graphite, and nickel alloys.

The current best estimate of capital investment for a hybrid chlorine plant that could produce 200 million kilograms of hydrogen per year is $1 billion (Hibbitts 2007). This estimate applies to both the aqueous quench option and the membrane reactor option. The capital difference between the options is much less than the likely precision of the estimates at this point. We have not attempted an evaluation of capital expenditure for the HCl sequestration option because we have no meaningful way to calculate the reactor size without additional data. Work is continuing to refine all estimates reported in this paper. This effort includes reassessment and verification of calculations done by the student team (Hibbits, 2007), further evaluation of alternative materials of construction, and additional engineering analysis to improve energy efficiency and reduce costs.

**References cited**

Amendola, S., "Thermochemical hydrogen produced from a vanadium decomposition cycle," U. S. Patent Application Publication US 2005/0013771 A1, (January 20, 2005).

Baker, R. W., "Future directions of membrane gas separation technology," *Ind. Eng. Chem. Res., 41*, 1393-1411 (2002).

Balachandran, U., and B. Ma, "Mixed-conducting dense ceramic membranes for air separation and natural gas conversion," *J. Solid State Electrochem, 10,* 617-624 (2006).

Carty, R.., M. Mazumder, J. Schreider, and J. Panborn, "Thermochemical hydrogen Production, Vol. 3, Appendix B., Discussion of experimental and design work," Gas Research Institute, GRI-80/0023, Chicago, IL 60616 (1981).

Davis, C. W., F. A. Ellers, and R. G. Ellis "Process for manufacture of chlorine by oxidation of hydrogen chloride," US Patent 2,547,928 issued April 10, 1951.

Hibbitts, D., K. Forbis, C. Knight, M. Owings, and E. Quin, "Analysis of the hybrid chlorine cycle for producing hydrogen," unpublished final report completed in partial fulfillment of the requirements off ChE 432, capstone design course, Department of Chemical and Biomolecular Engineering, Clemson University, (2007).

Julbe, A., D. Farrusseng, and C. Guizard, "Limitations and potentials of oxygen transport dense and porous ceramic membranes for oxidation reactions," *Catalysis Today, 104*, 102-113 (30 June 2005).

Maiya, P. S., U. Balachandran, J. T. Dusek, R. L. Mieville, M. S. Kleefisch, and C. A. Udovich, "Oxygen transport by oxygen potential gradient in dense ceramic oxide membranes," *Solid State Ionics, 99*, 1-7 (1997).

Motupally, S., D. Mah, F. Freire, and J. Weidner, "Recycling chlorine from hydrogen chloride," *The Electrochemical Society Interface*, *7*(3), 32-36 (1998).

Weidner, John, Department of Chemical Engineering, University of South Carolina, Columbia, SC, personal communication, July 31, 2007.

## Acknowledgments

A portion of this work was completed under contract with the U.S. Department of Energy, via subcontract with Argonne National Laboratory. Michele A. Lewis served as technical supervisor for Argonne.

Clemson seniors (class of 2007) Kelly Forbis, David Hibbits, Chase Knight, Megan Owings, and Eric Quin contributed valuable content through their capstone design study.

Clemson University Professor Emeritus Joe Mullins helped to develop Aspen simulations for several process steps.

Balu Balachandran of Argonne National Laboratory provided reference papers on dense ceramic membranes and patiently answered many questions about their use.

Steven Amendola of Reaction Sciences, Inc. provided additional laboratory data and discussion to clarify and support the claims of U. S. Patent Application Publication US 2005/0013771 A1.

John Weidner of the University of South Carolina provided helpful explanation of his work on gas phase electrolysis of HCl.

# DECOMPOSITION OF HYDROGEN IODIDE IN THE S-I THERMOCHEMICAL CYCLE OVER Ni CATALYST SYSTEMS.

*Annarita Spadoni, Michela Lanchi, Claudio Felici, Paolo Favuzza, Raffaele Liberatore, Pietro Tarquini, ENEA Casaccia Research Centre, Rome, Italy*

*Alina Carmen Tito, Carlo Vittorio Mazzocchia, Polythechnique "Giulio Natta", Milan, Italy*

## Abstract

The Sulphur-Iodine thermochemical cycle for hydrogen production has been investigated by ENEA over the last five years, with a particular focus on chemical aspects. Regarding the hydrogen iodide decomposition, three γ-alumina supported nickel catalysts were produced and characterized, and then tested in terms of catalytic activity and stability by means of a tubular quartz reactor. In particular the relationships between catalytic activity and preparation procedure was investigated. From the experimental data acquired it can be concluded that two of the three catalysts tested demonstrated high catalytic activity, since hydrogen iodide conversion was almost coincident with theoretical equilibrium value. On the other side, for all the catalysts, a gradual but considerable deactivation phenomenon was observed at 500°C, while at a temperature higher than 650°C the catalytic activity was recovered.

## Introduction

The Sulphur-Iodine thermochemical cycle is one of the most promising routes currently under development for sustainable hydrogen production from water. It is based on three main reactions carried out at three different temperature levels. Among these reaction, the hydrogen iodide (HI) decomposition presents a rather low equilibrium conversion even at high operating temperatures (the equilibrium conversion at 750 K is about 0.22) and subsequently a considerable energy expense for the separation and recirculation of the unreacted species is required [1]. It is well known that the homogeneous gas phase decomposition of hydrogen iodide has a very low rate [2,3]. The use of a catalytic system is essential for the improvement of the whole cycle thermal efficiency. Moreover, the reduced temperature resulting from the use of a catalyst attenuates the corrosion problems associated with the presence of hydrogen iodide-iodine ($I_2$ ) mixtures in process equipment.

Heterogeneous catalytic reaction was extensively investigated by NCLI (National Chemical Laboratory for Industry, Japan) and GA (General Atomics, USA) [4]. Different catalytic systems were found effective, in particular the Pt supported catalysts showed very good performances with respect to activation energy values and lifetime [5,6,7]. Some authors proposed less costly catalysts: the cost savings resulting from the use of non precious metals may be an advantage in the design of a plant of large size. O'Keefe reported that nickel was for the most part a poor catalyst, although its particular deposition on an active carbon surface produced a catalyst that was notably good. Nickel alumina supported catalysts were also tested and demonstrated a remarkable catalytic activity. However, operating times were limited to the 100 hours range.

The aim of the present work is the analysis of the catalytic activity of three γ-alumina supported nickel catalysts obtained from different preparation methods, in particular focusing on the relationships between catalytic activity and preparation procedure. Two of them were prepared from γ-alumina by impregnation-calcination method and were obtained from the deposition of two different precursors, such as nickel acetylacetonate and nickel nitrate [8]. The third one is a nickel aluminate catalyst obtained by coprecipitation method, prepared in order to better understand the metal-support interactions in nickel acetylacetonate catalyst (Ni-acac), and, at the same time, to test the effectiveness of this catalyst in HI decomposition .

## Materials and methods

***Catalyst preparation***

Three different γ-alumina supported nickel catalysts with a nickel content from 0.83 to 11.93% were prepared by impregnation-calcination and co-precipitation methods.

For catalysts obtained by impregnation technique, Ni-acac and Ni-nitrate, neutral activated alumina was used (powdered γ-alumina of high purity from Aldrich, BET: 153.37 $m^2/g$; micropore volume:12.80 $mm^3/g$, average particle size: 150 mesh; pH=7.0, density: 0.397 $g/cm^2$; melting point: 2040°C).

Regarding the Ni-nitrate catalyst, it was prepared by conventional wetness impregnation adopting nickel nitrate salt ($Ni(NO_3)_2\_6H_2O$) as precursor. In particular the alumina powder was impregnated with the nickel nitrate aqueous solution, stirred at room temperature for 5 hours, filtered by membrane and finally washed with water. The resulting solid was dried at 60°C for 12 hours in a $N_2$ atmosphere, calcined in air at 550°C for 10 hours (under 5°C/min ramp followed by holding isothermally for 2 hours), and then reduced in presence of a gaseous mixture of $H_2/N_2$ (1:1) at 550 °C (heating rate of 5°C/min and isotherm of 2 hours).

The Ni-acac catalyst was obtained by impregnation of powdered γ-alumina with an acetone solution of nickelacetylacetonate, stirred at 40°C for 5 hours. Then, the resulting solution was dried under vacuum in a rotovapor using a flask with internal breakfluxes, in order to have impregnation even at the first phase of the drying.

The co-precipitated Ni catalyst, designated as Ni-aluminate, was prepared from a mixed aqueous solution of nickel hydroxide $Ni(OH)_2$ and aluminium hydroxide $Al(OH)_3$. The precipitate was obtained by adding aqueous ammonia slowly to the solution to a final pH of 9,5. Then the resulting residue was filtered , washed with water, dried at 100°C in air and calcined in air at 900°C for 5 hours.

***Catalyst characterization***

BET surface area and pore volume ($N_2$ adsorption/desorption) were determined using a Kelvin 042 Sorptometer apparatus. X-ray diffraction (XRD) patterns were collected over fresh catalysts, before testing in the fixed-bed tubular reactor. ICP-OES analysis were also performed in order to determine Ni loading. A SEM analysis over Ni-acac was conducted.

***Catalytic tests***

Catalytic reaction experiments were carried out in a quartz vertical fixed-bed reactor (11.4 mm inside diameter.). The other apparatus parts were mostly made of pyrex. The reactor was positioned vertically in a thermostated oven and fed with 0.1 mL/min of HI aqueous solution at azeotropic concentration (7.5 M, Sigma Aldrich) by means of a microfeeder through a capillary glass inlet. The capillary inlet temperature was kept over 136°C in order to assure the acid solution flash vaporization (B. P. 127 °C). A flow of 50 mL/min of $N_2$ from the top of the reactor was also adopted to drive the vaporized HI throughout the apparatus. The reactor was filled with a catalyst system distributed in five different zones: quartz wool was placed in the top and the bottom layers, granular SiC (Carlo Erba) in the intermediate layers, while 0.25 g catalyst was mixed with 2.25 g of powder SiC (Sigma Aldrich, 230 mesh) in the central layer. The tests were carried out under a total pressure equal to 1.5 bar and at two temperature levels: 500°C and 650°C. All the outlet components except for $H_2$ and $N_2$ were absorbed in a first scrubber containing water. A second scrubber containing titrated thiosulfate ($Na_2S_2O_3$) solution was added to ensure instrumental safety from corrosion due to possible HI and $I_2$ residues in the gas stream. The $H_2$ concentration (% v/v) of the outlet gas was determined by the use of an electrochemical sensor (Orbisphere). The flow rate of the stream coming from the second scrubber ($H_2$+$N_2$) was measured by a massflow meter F, as indicated in figure 1. Every test started with the HI solution injection, after a reactor conditioning step consisting solely of $N_2$ flow for 1 hour at 500°C .

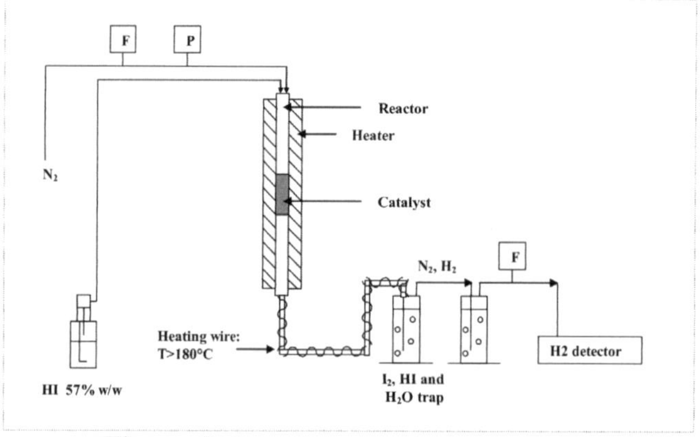


**Figure 1:** **Experimental apparatus layout**

## Results and discussion

***Catalyst characterization***

XRD patterns of fresh Ni-acac catalyst revealed that the phases detected were mainly nickel oxide (NiO) and nickel aluminate ($NiAl_2O_4$). Nickel crystal showed small intensity peak. This experimental evidence suggested that a phase transformation occurred on Ni-acac during calcination step with the result of $NiAl_2O_4$ formation. This compound is hardly reduced to metallic nickel at 500°C. Molina and Poncelet [9] evidenced in the calcination step strong interactions between NiO and alumina support producing the formation of nickel aluminate spinel which required temperatures higher than 500°C to be reduced.

ICP analysis detected a Ni loading of 11.93% (%w. dry basis). SEM analysis (figure 2) showed that there were large nickel particles and that their dispersion wasn't homogeneous. This could be attributed to the impregnation phase in the catalyst preparation. During this step, indeed, metal ions may not be easily diffused into the pores leading to a deposit on the outer surface and to their aggregation in large particles.

XRD analysis over Ni-aluminate fresh catalyst showed that the main phases of nickel crystal were $NiAl_2O_4$ and NiO, but no nickel peaks appeared. A $\gamma$-alumina peak was also reported, indicating that the coprecipitation and the thermal process were correctly executed.

During the preparation of Ni-nitrate several difficulties were encountered in the impregnation step: in particular, nickel ion was removed during washing stage of the solid. Catalyst obtained was found to contain only a few percentage of nickel (0.83 %), despite the expected concentration was around 20%.

**Figure 2.** **SEM micrographs of Ni-acac at two different magnifications**

***Catalytic activity for HI decomposition***

The catalytic performance of nickel catalysts was evaluated operating a fixed-bed reactor in the temperature range of 500-650°C. The conversion of HI was measured after 1 hour from the start, when a steady state was reached. Table 1 reports HI conversions at 500°C and 650°C over the catalysts prepared. It is clearly deducible that both Ni-acac and Ni-aluminate catalysts demonstrated very high HI conversion at 500°C and 650°C. High conversions on γ-alumina supported nickel catalysts were also observed by other researchers. NCLI has shown that there was a rapid approach to theoretical equilibrium values, comparable to Platinum based catalysts for temperatures over 500°C. However, at lower temperatures this catalyst showed a lower activity than platinum based catalyst.

High catalytic activity of both Ni-acac and Ni-aluminate suggested that active sites for hydrogen iodide decomposition may not be ascribed only to nickel crystallites. As above mentioned, XRD characterization showed no nickel peaks existence in the calcined $NiAl_2O_4$ phase of Ni-aluminate catalyst. Hence, it is deducible that both the calcined $NiAl_2O_4$ phase of Ni-aluminate catalyst and the reduced $NiAl_2O_4$ phase in Ni-acac catalyst may be considered as the active site in the HI decomposition reaction.

**Table 1.** **HI conversion at 500°C and 650°C**

| Catalyst type | Conversion at 500°C | Conversion at 650°C |
|---|---|---|
| Ni-acac | 20.7 | 23.3 |
| Ni-nitrate | 12.0 | 21.7 |
| Ni-aluminate | 21.0 | 21.2 |

Moreover, conversion obtained on the Ni-acac catalyst at 650°C were slightly higher than in the case of Ni-aluminate. This result may be ascribed to :

(1) the synergy effect of the reduced $NiAl_2O_4$ and surface nickel particle;

(2) during Ni-aluminate catalyst preparation, $Ni^{2+}$ and $Al^{3+}$ are coprecipitated as hydroxide compounds and $Ni(OH)_2$ are coated by a large amount of $Al(OH)_3$. This configuration could prevent the contact of reactants to the active sites leading to lower conversions.

Regarding the behaviour of Ni-nitrate, the HI conversion was considerably lower than the theoretical equilibrium value at 500°C, but it significantly raised at 650°C. This could be attributed to the lower concentration of active sites, as the nickel loading was very poor compared to other catalysts.

***Deactivation of catalysts***

The catalysts stability experiments were carried out at 500°C. Assuming that the catalysts can be considered deactivated when the conversion is 25% lower than the initial value, the deactivation rates of the three catalysts demonstrated to be in the following order: Ni-aluminate~Ni-acac<Ni-nitrate.

Regarding the Ni-aluminate catalyst, in figure 3 the variation in HI conversion as a function of time is reported and it is evident a gradual decrease of the catalyst activity. The same behavior was experimentally found for the other two catalysts. O'Keefe [4] reported that a copper oxide catalyst was tested for the HI decomposition reaction. As expected, the $Cu_xO$ reacted with the HI to form what appeared to be $Cu_xI_2$, but this latter material was not a particularly active catalyst. In addition, the catalyst was seen to crumble off of the support alumina substrate and there was evidence of $Cu_xI_2$ material being transported downstream with the HI carrier gas flow. Moreover, previous NCLI studies over γ-alumina supported nickel catalysts evidenced solid deposits at the exit of the reactor, indicating the transport of the metal from the catalyst, as well as a significant decrease in catalyst surface area after the HI decomposition reaction. In this study, over the testing period black deposits were noted at the exit of the fixed-bed reactor.

Moreover, it was found that, keeping the deactivated catalysts at a temperature of 650°C for at least 1 hour, the HI conversion seemed to substantially increase. These observations suggested that the deactivation may be ascribed for all catalysts to $NiI_2$ formation during the experiment. Since $NiI_2$ thermal decomposition begins at about 500°C [10] and can be considered quantitative at about 650°C, as observed at ENEA labs, it can be deduced that catalyst thermal treatment produces a considerable $NiI_2$ decomposition and consequently the catalyst activity recovery.

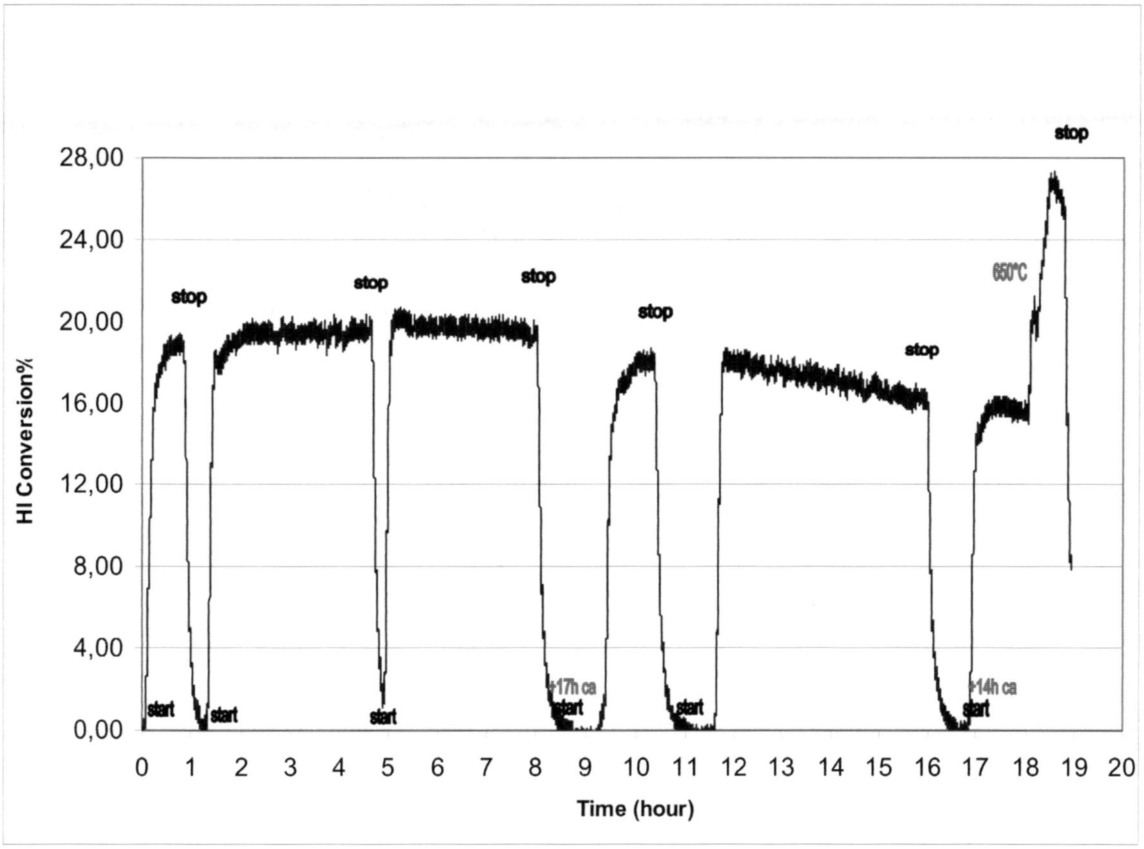


**Figure 3:** **HI conversion vs. time. Operating conditions: T=500°C-650°C; HI=0,1 mL/min. In correspondence of HI feed rate stops the reactor was kept at 500°C in a $N_2$ stream**

## Conclusion

The present experimental investigation has showed that both Ni-acac and Ni-aluminate catalysts have high catalytic activity at 500°C and 650°C, as HI conversion detected quickly approached the theoretical equilibrium value. These results suggest that catalytic activity in HI decomposition may not be ascribed only to nickel crystallites, and in particular it is deducible that the calcined $NiAl_2O_4$ phase of the Ni-aluminate catalyst and the reduced $NiAl_2O_4$ phase in Ni-acac catalyst may be considered as the active site in the HI decomposition reaction. In addition, the slightly higher activity of the Ni-acac compared to Ni-aluminate at 650°C could be attributed mostly to the synergy effect of the reduced $NiAl_2O_4$ and surface nickel particle. Catalysts stability experiments at 500°C showed that the deactivation rates of the three catalysts were in the following order: Ni-aluminate~ Ni-acac<$NiNO_3$. The experimentally demonstrated deactivation behavior was probably due to $NiI_2$ formation. In fact, since $NiI_2$ thermal decomposition occurs near 650°C, the catalysts deactivation has been found to be reversible over this temperature.

## References


1. R. H. Elder, G. H. Priestman, B. C. Ewan and R. W. K. Allen (2005), "The separation of $HI_x$ in the sulphur-iodine thermochemical cycle for sustainable hydrogen production", *Trans. IChemE, Part B, Process safety and Environmental Protection*, 83 (B4), pp. 343-350.
2. G. B. Kistiakowsky (1928), "Homogeneous gas reaction at high concentrations. 1. Decomposition of hydrogen iodide", *The Journal of the American Chemical Society*, vol. 50, pp. 2315-2332.
3. M. Bodenstein (1899), "Gasreaktionen in der chemischen Kinetik", Z. Phys. Chem., vol. 29, pp. 295.
4. D. R. O'Keefe, J.H. Norman, D. G. Williamson (1980), "Catalysis Research in Thermochemical Water-Splitting Processes", *Cat. Rev .Sci. Eng*, vol. 22, n°3, pp. 325-369.
5. C. N. Hinshelwood and R.E. Burk (1925), "The relation of homogeneous to catalysed reactions. The catalytic decomposition of hydrogen iodide on the surface of platinum", *J. Chem. Soc. Trans.*, vol. 127, pp. 2896-2900.
6. Y. Shindo, N. Ito, K. Haraya, T. Hakuta, H. Yoshitome (1984), "Kinetics of the catalytic decomposition of hydrogen iodide in the thermochemical hydrogen production", *International Journal of Hydrogen* Energy, vol. 9 n°8, pp. 695-700.
7. Y. Oosawa, T. Kumagai, S. Mizuta, W. Kondo, Y. Takemori, K. Fujii (1981), *Bull. chem. Soc. Japan,* vol. 54 n°3, pp. 742-748.
8. A. C. Tito, C. Mazzocchia, A. Spadoni, P. Favuzza, M. Lanchi, P. Tarquini (2007), "Synthesis of the catalyst $Ni/Al_2O_3$ for the production of $H_2$ from HI decomposition in the thermochemical water splitting", in *"2nd World Congress of young scientists on hydrogen energy systems"* , June, Turin, Italy.
9. R. Molina, G. Poncelet (1998), "α-Alumina supported nickel catalysts prepared from nickel acetylacetonate: a TPR study", *J.Catal.,* vol. 173, pp.257-267.
10. S. Sato, S. Shimizu, H. Hakajima, K. Onuki and Y. Ikezoe (1986), " Studies on the Nickel-Iodine-Sulfur process for hydrogen production-III", *Int. Journal of Hydrogen Energy,* vol. *11,* n°9, pp. 571-575.

# 532c Method to Rank Thermochemical and Hybrid Cycles According to Energy Efficiency

**Miguel J. Bagajewicz**[1], DuyQuang Nguyen[2], Thung Cao[2], Robbie Crossier[2], Terrel Fish[2], and Matthew Behring[2]. (1) Chemical, Biological and Materials Engineering, University of Oklahoma, 100 E. Boyd St., T-335, Norman, OK 73019, (2) CBME, University of Oklahoma, 100 E. Boyd St., T-335, Norman, OK 73019

A methodology has been developed for the evaluation of thermochemical and hybrid water-splitting cycles based on energy efficiency. Efficiency is referenced to the standard enthalpy of formation of water at 298K and includes minimum hot utility requirements, electrical work to power electrochemical reactions as well as power and heat required for separation processes. In addition, equilibrium conversions were considered. The degrees of freedom of each cycle (temperatures, pressures and excess reactants) were considered and these values were varied to optimized efficiency. Different separation schemes were also considered: recycling reactants to the same reactor or considering other strategic options. Results are shown for three different configurations considered for each of ten chosen cycles. These results show that separation decisions can greatly improve cycle efficiency.

## Testing of Particle Bed Electrodes for the Production of Hydrogen


### Abstract

In this study, the characteristics of particle bed systems suitable for use in hydrogen generating electrolysis cells are investigated. These beds furnish higher electrode surface areas per unit volume and permit electrode adjustments without disassembly. Such adjustments include particle removal, addition of particles and modification of the particle size distribution.

When fluidized by the upward flowing liquid electrolyte, the resulting fluidized particle bed electrodes can reduce reaction polarization, thereby reducing voltage and consequent power requirements. In some cases this effect can also avoid undesirable side reactions.

Particle bed materials attractive for these electrodes include copper, boron carbide, carbon and copper-coated polymer or glass. Particle sizes and shapes are related to fluidization characteristics and to pressure loss for liquids flowing through the beds.

Electrolytes currently under investigation include sodium and potassium hydroxide, sulfuric acid and sodium chloride. Study of some weak electrolytes is also considered.

In conjunction with the electrolysis investigation, some testing has been started on the use of scrap aluminum and sodium hydroxide reactants to generate hydrogen. The scrap aluminum is available from General Pneumatics manufacturing machining operations. This material has historically been sold as scrap. Initial qualitative tests show vigorous reaction rates. An added advantage of using this material is that the cutting oil residue, which represents a contaminant in aluminum recovery operations, does not significantly impede the hydrogen generation reaction. This effect is thought to be the result of a saponification reaction which removes the oil by converting it to soap.

An alternative use for the aluminum under consideration is as a storage material for hydrogen in the form of aluminum hydride or metal aluminum hydrides such as sodium aluminum hydride also known as sodium alanate. Aluminium hydride may be a useful material for storing hydrogen in hydrogen-fueled vehicles. $AlH_3$ contains up to 10% hydrogen by weight and can store up to 148g/L, twice the density of liquid hydrogen. Work on hydride storage has been getting increased attention in current US Department of Energy programs. DOE has reportedly set a goal of developing alternative fuels that have a "hydrogen mass density" of 6 percent by the year 2010 and 9 percent by 2015. A Sandia National laboratory web site describes a metal hydride center of excellence (MHCoE) that involves eight universities, three companies, and six federal laboratories located throughout the country. Each MHCoE partner is reportedly responsible for a well-defined aspect of the center's research, which is coordinated by Sandia National Laboratories. Sandia provides direction and manages the MHCoE for the Department of Energy.

A test loop has been set up in a separate room in the General Pneumatics main building consisting of a DC power supply, an electrolyte head tank, replaceable test cell units, a peristaltic pump, a flow meter and a separator vessel. All fluid components are interconnected with tygon tubing. All hydrogen-containing components are housed in a specially fabricated laboratory hood. A very small experimental photovoltaic unit has been procured and is to be modified for testing in our test apparatus.

Four test cell design configurations are being evaluated. These are referred to as cylindrical, rectangular, hexagonal and annular. These designations are based on the geometry of the cathode half-cells. Figure 4 shows diagrams of the annular and cylindrical test cells. Figure 3 shows diagrams of the hexagonal and rectangular configurations.

.

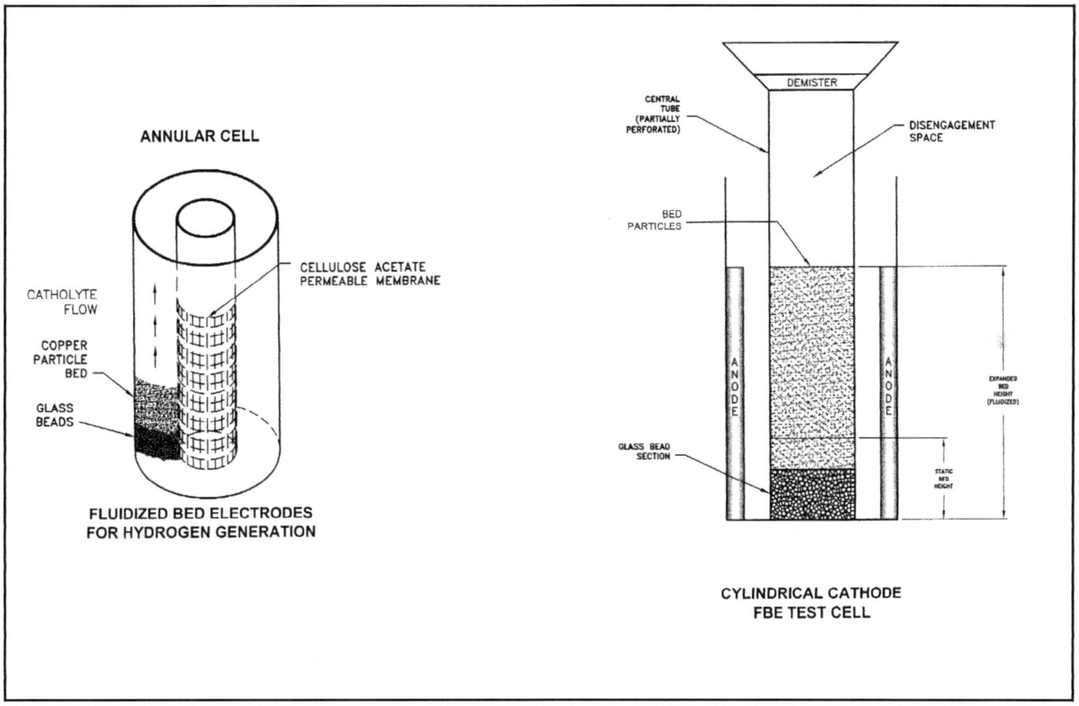


**FIGURE 4: Annular and Cylindrical Cathode FBE Test Cell**

Initial tests indicate that the annular cell is most convenient for this phase of the investigation. It consists of an outer transparent cylindrical tube, a current feeder wrapped around its inner wall, a cathode comprised of an annular bed of copper particles (or copper coated polymer beads), a carbon anode array located in the center tube of the test cell and a membrane separator supported on the center tube. A bed of glass beads is located at the bottom of the cell to provide even flow distribution when the electrolyte is flowing and to support the particulate bed when flow is stopped. Sparger pipes designs are planned for larger size cells

The large cathode surface area of the particle bed electrode is designed to yield increased hydrogen production rates compared to solid plate electrodes. Comparative test runs to verify the increase in production rates are now being planned. Also, the agitation achieved under fluidization is expected to result in improved disengagement of the evolved hydrogen gas bubbles and decreased cathode polarization for the fluidized bed electrode.

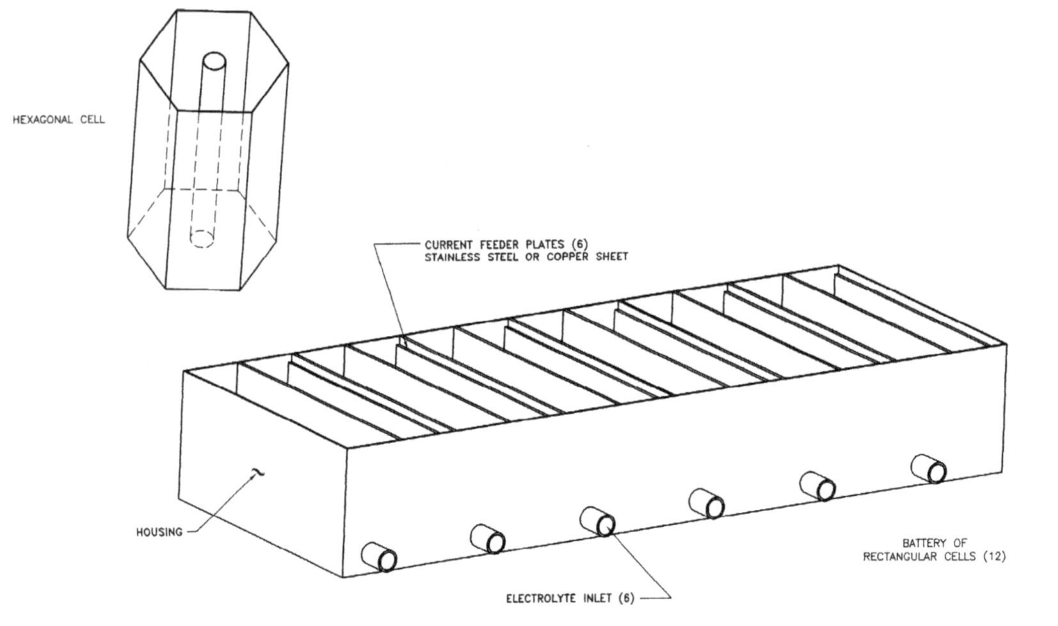


**Figure 3: Alternative Hexagonal and Rectangular Cell Configurations.**

The US Department of Energy has set a targeted untaxed price of \$2.00 – \$3.00 per gasoline gallon equivalent on an energy basis for hydrogen production and delivery (Ref.

2). The department has noted that one of the barriers to cost reduction is the lack of manufacturing capacity for producing the significant quantities of small-scale systems that will be needed for distributed production of hydrogen systems.

Technical Discussion

The required emf for the reversible generation of hydrogen at the cathode of an electrolytic cell as a function of concentration is given by the relation:

$$E = E_O \frac{RT}{2F} \Sigma V_j \,\ell n\, C_j$$

Where,

| | |
|---|---|
| E | = emf |
| $E_0$ | = standard electrode potential |
| R | = universal gas constant |
| T | = absolute temperature |
| $C_j$ | = equilibrium concentration of species j |
| F | = Faraday = 96,500 coulombs |
| $V_j$ | = stoichiometric factor of species j |

This is used to estimate the effects of electrolyte concentrations for the cell testing.
For the estimating of emf requirements as a function of hydrogen pressure we note that:

$$Ep = E_O + \frac{RT}{2F} \ell n (P / P_{atm})$$

Where,

| | |
|---|---|
| P | = pressure in atmospheres |
| F | = Faraday = 96,500 coulombs |
| R | = Gas Constant |
| Ep, Eo | = Electromotive force at pressure P and at standard conditions, respectively |

A plot of voltage versus pressure for the cell- H2: HCl (0.1N), HgCl; Hg shown in Ref. 3. demonstrates the dramatic effect on developed hydrogen pressure that can be obtained from relatively small increases in impressed cell voltage. In this case, a cell voltage increase from 0.4 volts to 0.46 volts increases the pressure from 1 ($10^0$) to over 100 ($10^2$) atmospheres. This characteristic could be useful for fuel cell hydrogen feed units. The ability to generate sizeable hydrogen gas pressure without the use of a compressor is described in reference 6.

Flow Rate - Minimum fluidization velocity will be tested for various cell configurations and for a number of particle materials as guided by calculated velocities. The initial calculations are being made using the Ergun equation for pressure drop in a packed bed in the form:

$$\frac{\Delta p}{L} = \frac{150\mu(1-\varepsilon)^2 u_0}{\varepsilon^3 d_p^2} + \frac{1.75(1-\varepsilon)\rho u_0^2}{\varepsilon^3 d_p}$$

with $\Delta P =$ the pressure drop,
$L =$ the height of the bed,
$\mu =$ the fluid viscosity,
$\varepsilon =$ the void space of the bed,
$u_0 =$ the fluid superficial velocity,
$d_p =$ the particle diameter
and $\rho =$ the density of the fluid.

By setting this equation equal to the gravitational force of the particle bed, we can determine the minimum fluidization velocity (umf). In this case, the above equation has been rearranged for catalyst particle beds in the form:

$$g(\rho_s - \rho) = \frac{1.75\rho u^2}{d_p \varepsilon^3} + \frac{150\mu(1-\varepsilon)u}{d_p^2 \varepsilon^3}$$

with $\Delta P =$ the pressure drop,
$L =$ the height of the bed,
$\mu =$ the fluid viscosity,
$\varepsilon =$ the void space of the bed,
$u =$ the fluid velocity,
$d_p =$ the particle diameter,
$\rho =$ the density of the fluid,
$\rho_s =$ the catalyst density,
$g =$ the acceleration due to gravity.

where u now stands for umf, the minimum fluidization velocity, and $\rho_s$ is the particle density.

Phase I of the program includes technical analysis, design and physical configuration studies, manufacturing and production analysis, and small scale proof of concept study/demonstration to evaluate feasibility of the proposed technology. Phase II involves one or more prototype systems to be fabricated for operational evaluations of the fluidization parameters, safety validations and sustained stability/capacity. This is to be followed by Phase III that covers marketing and production for commercial and governmental applications.

Electrolysis is one of the approaches under investigation in the current government programs. In the National Academy of Science report, NAS recommendation ES-7 states: "DOE should increase emphasis on electrolyzer development" (Ref. 1). Small, distributed hydrogen generating units have the advantages of lower developmental costs, savings in distribution expenses, more tractable safety issues.

Electrolysis has the added advantage of accommodating renewable energy source devices such as small wind turbines and photovoltaic units. This advantage is well described in Reference 6. The low voltage requirements of the electrolysis units make it possible to use less expensive electricity generators for evaluation testing.

In addition to hydrogen production, fluidized bed electrodes can be useful for other electrochemical processes. The information obtained from this investigation is expected to be applicable to other technical areas such as environmental cleanup. An advantage in these applications is that the pollutant concentrations are frequently low which tends to result in efficient energy use.

*Safety Analyses*

Evaluations of the potential hazards and safety measures associated with the substances, processes and operation of the hydrogen generation units will be performed periodically at various stages of the project. This discussion summarizes the initial safety analysis that was done for the pre-testing phase.

The initial safety analysis addresses the hazards of fire, explosion and toxic releases. Fire and explosion hazards from hydrogen and oxygen are evaluated using the Dow fire and explosion index given in Ref 4. In the case of hydrogen the lower flammability limit in air is 4% and the upper limit (ufl) is 75% (Ref 5). It is planned to continue to collect hydrogen at concentrations exceeding 95%, which will maintain the hydrogen well above the upper limit.

Quantities of hydrogen allowed to collect have been limited so as to limit the size of any potential explosion. This limit has tentatively been set at 350 standard cubic centimeters.

Potential leakage of hydrogen into the surrounding air must be considered and steps taken to assure that such leakage does not occur. Reviews of potential leak paths will be performed for each configuration of the test apparatus. Local enclosed venting in the vicinity of hydrogen-containing sections of the test apparatus is being provided with the use of a ducted, hooded enclosure.

Sparking in this low voltage apparatus has a low probability. Precautions to avoid cell voltage reversal will be incorporated into the design and the operating procedures. Nevertheless, humidification of the test area will be maintained as a precaution against static electricity. Possible effects from potential detonations will be limited based on the hydrogen quantity limits discussed above. In addition, a shield to protect against any potential projectiles has been installed at the front of the apparatus protective enclosure.

References Cited:


1. NATIONAL RESEARCH COUNCIL AND NATIONAL ACADEMY OF ENGINEERING, "THE HYDROGEN ECONOMY: Opportunities, Costs, Barriers, and R&D Needs," THE NATIONAL ACADEMIES PRESS Washington, D.C, (2004), page 6.

2. U.S. Department of Energy, "Small Business Innovation Research Program and Small Business Technology Transfer Program, FY 2006 Funding Opportunity Notice, OFFICE OF BASIC ENERGY SCIENCES, PROGRAM AREA OVERVIEW, Topic 20, page 1.

3. MacInnes, Duncan A., "The Principles of Electrochemistry, Dover Publications, (1961) pg.117.

4. Crowl, Daniel A., and Louvar, Joseph F., "Chemical Process Safety", Section 10-2, "Hazard Surveys", (1990), page 311.

5. Air Products and Chemicals, Inc, Hydrogen Gas Material Safety Data Sheet No. 300000000074, (2004), page 3.

6.

# 532e Alternative Thermochemical Cycles Being Considered By Industry For Deployment

**T. Bond Calloway** and William A. Summers. Savannah River National Laboratory, Aiken, SC 29808

The paper considers thermochemical cycles previously tested by industry for vitrifing nuclear waste. Many of these cycles are now being considered for production of hydrogen and carbon monoxide for eventual production of ethanol.

# Analysis of Kinetics of $CaBr_2$ Hydrolysis in a Direct Steam Sparging Contactor

C.B. Panchal,
Robert W. Lyczkowski, Steven A. Lottes, Jianhong Yang, and Richard D. Doctor

Argonne National Laboratory
9700 S. Cass Avenue
Argonne, Illinois 60439

September 17, 2007

# Analysis of Kinetics of $CaBr_2$ Hydrolysis in a Direct Steam Sparging Contactor

C.B. Panchal, Robert W. Lyczkowski, Steven A. Lottes, Jianhong Yang, and Richard D. Doctor*
*Argonne National Laboratory, Argonne, IL — USA; *correspondence: rdoctor@anl.gov*

## ABSTRACT

The objective of the $CaBr_2$ hydrolysis study at Argonne National Laboratory is to experimentally demonstrate that a sparging contactor can be used to develop an economical process for hydrogen production. The first step in Ca-Br cycle consists of reacting one mole of steam with $CaBr_2$ to form two moles of HBr. This is an endothermic reaction with an optimal range of reaction temperature between 1000 and 1050 K (730 — 780ºC). Analysis of the experimental hydrolysis reaction indicates that a second order hydrolysis reaction in $CaBr_2$ may be possible because of the formation of a complex involving $CaBr_2$ and CaO. Analyzing the reaction using second order kinetics finds $k_f\,(a/V_b)$ falling between 2.03-4.06 (l/min cm) where the overall reaction system is represented by the equation below:

$$3\ CaBr_2 + H_2O \rightarrow (CaBr_2)_2{\cdot}CaO + 2\ HBr$$

This reaction is highly endothermic and at a design temperature of 1050 K (777°C), the heat of reaction is 181.5 kJ/mol (43.38 kcal/mol), and the Gibbs free energy change is positive at 99.6 kJ/mol (23.81 kcal/mol). This reaction system behavior is critical for understanding and evaluating the scale-up of the Ca-Br cycle.

## 1. Descriptive Summary of $CaBr_2$ Hydrolysis Experiments

The proof-of-concept $CaBr_2$ hydrolysis experiment briefly described in programmatic communication to the US DOE sponsors of the Nuclear Hydrogen Initiative, November 20, 2006 [1] where the system operates in a batch mode. A schematic of the hydrolysis test reactor is shown in Figure 1. The bench-scale reactor consists of a one-end-closed mullite tube (OD 76 mm, ID 70 mm, and 457 mm long) the other end of which is sealed with a stainless steel flange assembly. $CaBr_2$ melt is contained in a replaceable quartz or sintered alumina crucible (70 mm diameter by 200 mm long), which is placed inside the mullite reactor tube. The reactor tube is placed in a 2.5 kW Lindberg high-temperature furnace with electric resistance heat elements. Steam is continuously sparged into the $CaBr_2$ melt through a 6.3 mm OD diameter tube (lance) made of SiC.

The product gases containing unreacted steam and HBr leave the reactor through a port in the stainless steel flange (or Pyrex glass flange) assembly. Then the product gases pass through a heat-traced line, a glass trap, and a glass condenser chilled to ambient with cooling water. The system is designed so that all gases introduced to it pass through the condensate/neutralization tank where HBr is captured with (10 N) sodium hydroxide solution in sufficient volume (1.25 l) and reaction time to convert all of it into sodium bromide. In order to assure that even trace levels of HBr will not be released from the receiver to the vent regular checks are conducted during the operation of the experiment and none have ever shown any HBr "slip."

As steam bubbles rise into the $CaBr_2$ melt, reaction occurs at the interface. The product gas HBr flows with steam and CaO goes into the melt solution. Samples of molten salts (calcium bromide-calcium oxide melts; ~2 mL) were collected on the exterior of an alumina probe thrust to the bottom of the molten salt bath. Melt samples are taken at the beginning of the test run before steam is turned on and at the end of test run after steam is shut off, and after purging with argon for a 5 minutes. The melt samples were also collected for analysis.

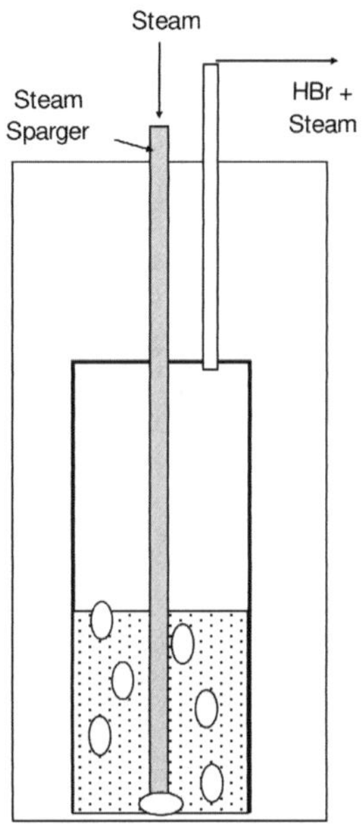


Figure 1. Schematic of $CaBr_2$ hydrolysis test apparatus.

During the contact time, $t_c$, of steam bubble with the $CaBr_2$ melt, the steam is consumed forming HBr. The contact time, $t_c$, is relatively small ($\approx$ 1 sec) as compared to the reaction time length (hours) sampled in 10 to 20 minute intervals. Therefore, a pseudo-steady-state can be assumed for each bubble transit and the conversion of steam can be calculated assuming constant melt composition. Subsequently, the rate of $CaBr_2$ conversion can be determined by integrating the rate of steam conversion or HBr formation over the reaction time.

## 2. Reaction Model

Figure 2 shows the HBr formation rate as a function of reaction time. The HBr formation rate plot implies that the melt composition is changing as the reaction progresses. This could be explained on the basis of the reverse reaction occurring for the forward reaction given by:

$$CaBr_2 + H_2O \rightarrow CaO + 2HBr \qquad (1)$$

However, the experiments were conducted with pure $CaBr_2$ and steam conversion (HBr formation) is low (<5%), therefore a reverse reaction cannot explain the observed behavior.

Furthermore, the phase diagram, shown in Figure 3, indicates that a $2CaBr_2{\cdot}CaO$ complex is formed according to the reaction:

$$2CaBr_2 + CaO \rightarrow (CaBr_2)_2{\cdot}CaO \quad (2)$$

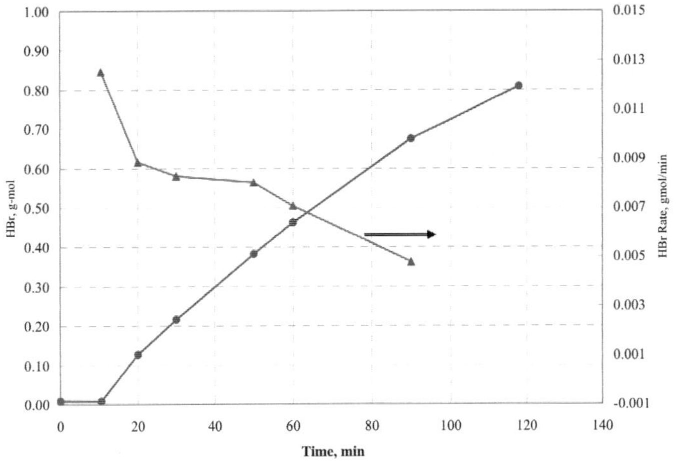


Figure 2. HBr formation rate as a function of reaction time (Test CBH-4).

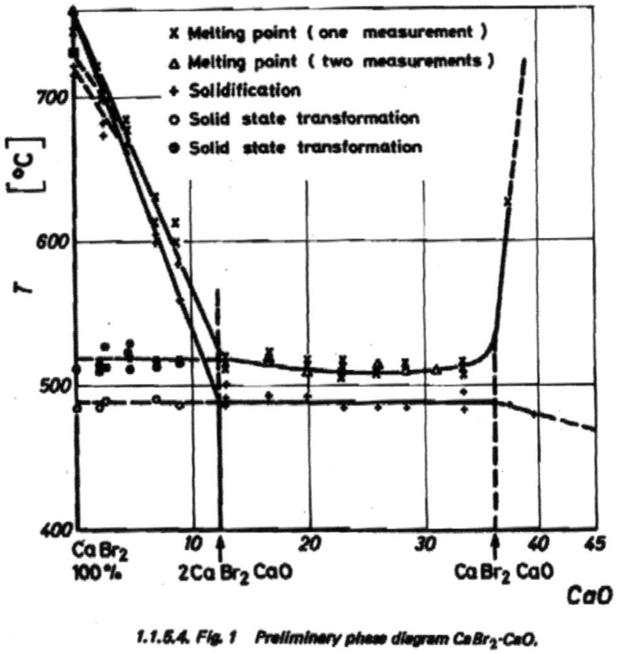


Figure 3. $CaBr_2$-CaO Eutectic phase diagram [2].

It should be noted that there is an error in the original publication source [2] for Figure 3 and the second entry in the abscissa should read $CaBr_2$ 2CaO. The complex $(CaBr_2)_2{\cdot}CaO$ may not be reactive, effectively tying up a portion of the $CaBr_2$, making it unavailable for reaction with steam. Assuming this hypothesis, adding Equations (1) and (2) produces:

$$3\,CaBr_2 + H_2O \rightarrow (CaBr_2)_2{\cdot}CaO + 2\,HBr \quad (3)$$

The combined reaction given by Equation (3) implies that for each mole of CaO produced three moles of $CaBr_2$ are consumed and hence the conversion of $CaBr_2$ might be limited. The conversion of $CaBr_2$ was limited to 20 to 25 weight percent in some the longer run hydrolysis experiments, where the rate of HBr formation was significantly lower towards the end of the reaction time. Based on these observations the reaction given by Equation (3) was used in analyzing the experimental kinetic data.

Applying the reaction given by Equation (3) to a single bubble yields the following rate expression for HBr formation:

$$V_b \frac{d[HBr]}{dt_c} = k_f \bar{a}\left[P_{H_2O}\right]\left[CaBr_2\right]^n \quad (4)$$

The reaction is assumed to be first order in steam partial pressure and $n^{th}$ order (to be determined from the kinetic data) with $CaBr_2$ concentration. Similarly the rate equation for consumption of $H_2O$ is given by:

$$V_b \frac{d[H_2O]}{dt_c} = -\frac{k_f}{2}\bar{a}\left[P_{H_2O}\right]\left[CaBr_2\right]^n \quad (5)$$

Which can be simplified to the following form for integration over the contact time, $t_c$:

$$\frac{dP_{H_2O}}{dt_c} = -k' P_{H_2O} \quad (6)$$

Where:

$$k' = \frac{k_f (\bar{a}/V_b)}{2\rho_m}\left[CaBr_2\right]^n \quad (7)$$

It should be noted that the grouping $k_f \bar{a}$ is an effective reaction rate as commonly used in gas-liquid reactors [3].

Integration of Equation (6) yields the rate of steam conversion for a given contact time $t_c$.

$$P_{H_2O} = P_{H_2O}(t_c = 0)e^{-k't_c} \quad (8)$$

The steam conversion, $x_c$ is given by:

$$x_c = 1 - P_{H2O}/P_{H2O}(t_c{=}0) \quad (9)$$

The rate of steam consumption is equal to the steam flow rate times the steam conversion, and the rate of HBr formation is 2 times the rate of steam consumption.

The formation of HBr at a given reaction time $t_R$ depends upon the melt composition. A 2nd order reaction of $CaBr_2$ was found to match the experimentally measured reaction rates far better than a first order reaction. Figures 4, 5, 6, and 7 show the predicted and experimental rates of HBr formation for four hydrolysis tests. The derived reaction constants for these 4 tests are shown in Table 1.

| Tests | Reaction Temperature, C | Reaction constant, 1/(min cm) $k_f\,(a/V_b)$ |
|---|---|---|
| CBH-2 | 778.5 | 2.03 |
| CBH-4 | 772.9 | 3.43 |
| CBH-5 | 777.3 | 4.06 |
| CBH-10 | 747.3 | 1.28 |

The value of kinetic constant $k_f$, which is a function of reaction temperature, should be the same for Tests CBH-2, 4 and 5. Therefore, the variation in the combined reaction rate constant ($k_f \cdot (a/V_b)$ is due to the contact area per unit volume for the different tests, which represents different effective steam bubble sizes. The results show that by increasing the melt height, the distribution of steam improved. The steam distribution further improved with the sparger tube inserted into alumina beads at the bottom of the reactor crucible. The steam distribution for tests CBH-4 and CBH-10 was comparable; therefore, the reaction constant represents the Arrhenius rate constant. The activation energy derived from these two test runs is 341 kJ/mol, which is greater than the heat of reaction, 181 kJ/mol.

In summary, the analysis of the experimental hydrolysis reaction data provides further insight into possible reaction mechanisms of melt-phase reaction. However, the number of data points is limited so that a definitive conclusion as to the $CaBr_2$ hydrolysis reaction mechanism cannot be reached. Additional tests are required to validate the hydrolysis tests and to determine the conversion of the complex $(CaBr_2)_x{\cdot}CaO$ formed in the melt phase back to $CaBr_2$ using HBr.

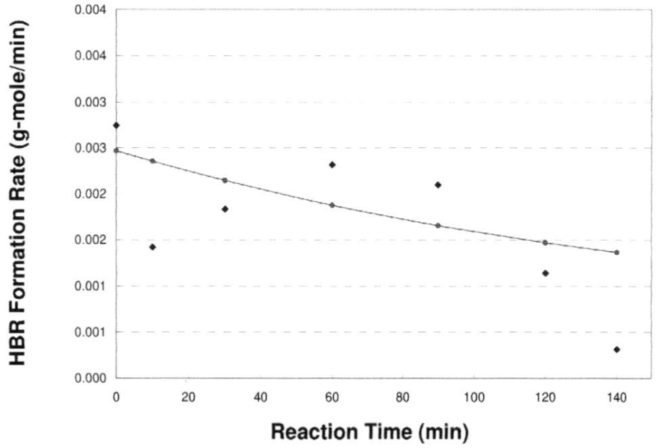


Figure 4. Comparison of predicted and experimental rate of HBr formation (CBH-2)

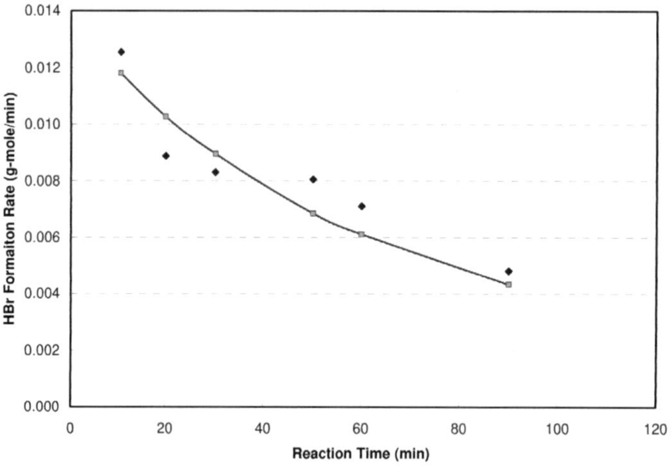


Figure 5. Comparison of predicted and experimental rate of HBr formation (CBH-4)

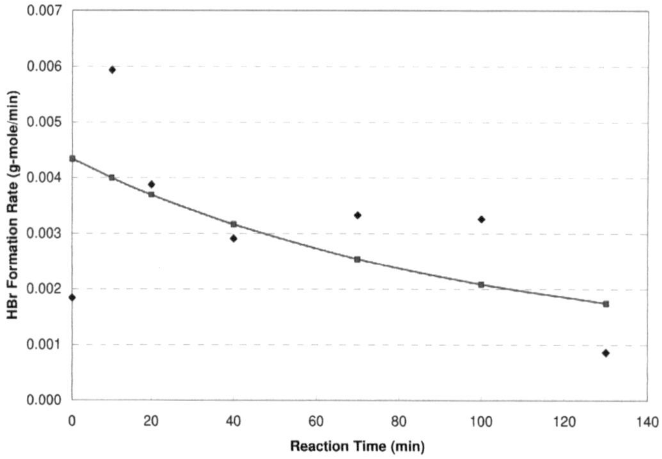


Figure 6. Comparison of predicted and experimental rate of HBr formation (CBH-5)

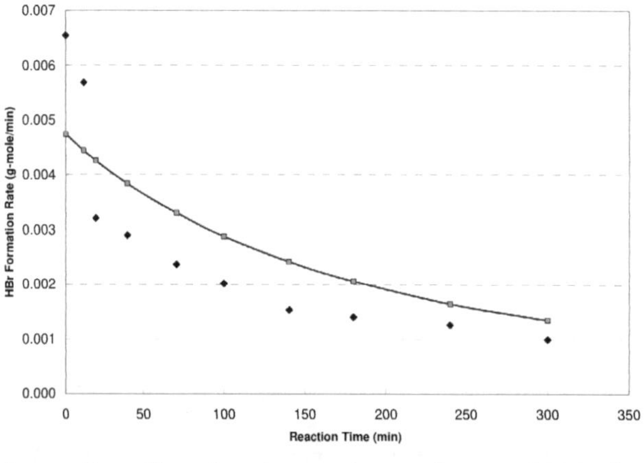


Figure 7. Comparison of predicted and experimental rate of HBr formation (CBH-10)

## 3. Conclusions and Recommendations

A number of conclusions are set forth subject to further experimental validation:

- Experimental data is consistent with a second order hydrolysis reaction in $CaBr_2$ involving the formation of a complex involving $CaBr_2$ and CaO.
- Formation of a complex, $(CaBr_2)_2.CaO$, which does not participate fully in the hydrolysis reaction, seems to explain high dependence of the rate of HBr formation on melt composition.
- The activation energy, based on the analysis of two test runs, 341 kJ/mol, is high compared to the heat of reaction, 181 kJ/mol. An additional test at an intermediate temperature is required to validate this value of activation energy.
- If the basic conclusion of the existence of the complex $(CaBr_2)_x.CaO$ formation is correct, then it is essential to determine the reactivity of this complex in the regeneration reaction. A hydrolysis test followed by regeneration is strongly recommended. Premixed solution of $CaBr_2$ and CaO should not be used for the regeneration tests.

The possibility that a complex is being formed has not thus far appeared in any analysis of $CaBr_2$ hydrolysis and this reaction system insight is critical for understanding and evaluating the scale-up of the Ca-Br cycle.

## Nomenclature

| | |
|---|---|
| [ ] | Concentration of species, g mole/cc |
| $\bar{a}$ | Interfacial area of bubble, $cm^2$ |
| $k_f$ | reaction constant, g-mole/(min $cm^2$ bar) |
| k' | reaction constant defined in Eq. 5, $min^{-1}$ |
| $P$ | Partial pressure, bar |
| $t_c$ | Contact time of gas bubble in the melt, min |
| $V_b$ | Gas bubble volume, cc |
| $\rho_m$ | Molar density, g mole/cc |

## REFERENCES


[1] Doctor, Richard D., Jiahong Yang, C.B. Panchal, Steven A. Lottes, and Robert W. Lyczkowski, *Nuclear Hydrogen Initiative: Proof of Concept of the CaBr₂/CaO Hydrolysis Sparging Contactor for HBr Production from Water-Splitting.* Argonne National Laboratory for the U.S. DOE Nuclear Hydrogen Initiative FY2007: Level-2 Milestone Report N-AN07TC0401 (November 20, 2006).

[2] Avogadro, A et al. *Hydrogen Production from Water Using Nuclear Heat*, Progress Report No. 3, Joint Nuclear Research Centre, Ispra Establishment, Italy, EUR 5059e ,1974.

[3] Danckwerts, P. V., Gas-Liquid Reactions, McGraw-Hill, Inc., New York (1970).

## 566a Tantalum Applications For Use In Scale Sulfur-Iodine Experiments

**Thomas Drake**, Benjamin E. Russ, Lloyd Brown, and Gottfried Besenbruch. General Atomics, 3550 General Atomics Court, San Diego, CA 92121

The development of the S-I cycle Integrated Lab Scale (ILS) loop has been performed as an international effort involving the US Labs and Industry and the Commissariat à l'Énergie Atomique (CEA) of France. The S-I cycle consists of three reaction sections which are currently being developed by three different organizations. Sandia National Laboratory (SNL) is developing the high temperature sulfuric acid decomposition section, General Atomics (GA) is developing the HI decomposition (hydrogen generation) section, and the CEA is developing the primary Bunsen reaction section.

The construction of the ILS loop has posed several challenges stemming from the corrosive nature of the chemicals and the size limits for equipment involved with producing a laboratory scale demonstration which operates at prototypical pressures and temperatures. As such, tantalum has been identified as a material suitable for use in the ILS system. This paper will discuss machining a coating techniques and applications of tantalum for applications in small size sulfur-iodine experiments.

# 566b General And Stress Corrosion Behavior Of Construction Materials For Hi Gaseous Decomposition

**Bunsen Y. Wong**[1], L. C. Brown[1], Gottfried Besenbruch[1], Ajit Roy[2], Joydeep Pal[3], and R.S. Koripelli[4]. (1) General Atomics, 3550 General Atomics Ct., San Diego, CA 92121, (2) Mechanical Engineering, University of Nevada, Las Vegas, 4505, Maryland Parkway, Las Vegas, NV 89154, (3) Mechanical Engineering, University of Nevada, Las Vegas, 4505, Maryland Parkway, Las Vegas, NV 89154, (4) University of Nevada, Las Vegas, 4505 Maryland Parkway, Las Vegas, NV 89154

In the sulfur-iodine thermochemical cycle, hydrogen iodide (HI) gas is catalytically decomposed into hydrogen and iodine gases at around 450°C. Development work has been on going to identify and qualify construction materials that can be used for the fabrication of heat exchanger and other process components. A number of candidates have been screened via immersion in the decomposition environment in excess of 100 hours. Hastelloys C-22, C-276, B2 and B3 have been found to have the lowest general corrosion rate among the common engineering materials tested. To qualify these materials for use, their stress corrosion behavior was investigated. Their susceptibility to stress corrosion cracking was studied through testing of C-ring and U-bend specimens. No crack has been observed in the specimens after 1000 hours of testing. To probe their vulnerability to crack growth, wedged loaded pre-cracked double cantilever beam specimens have been immersed in the gaseous environment and this will provide a stress intensity factor of the various Hastelloys. This talk will present a summary of the material screening work in addition to stress corrosion test results from the Hastelloy specimens tested.
This work was supported by the UNLV under RF-05-HTHX-001.

# Resistance To Corrosion Of Silicon-Based Ceramic Materials In Sulfuric Acid Containing Environments For Hydrogen Production

*Charles Lewinsohn, Merrill Wilson, Hyrum Anderson, Ceramatec, Inc., Salt Lake City, UT*
*Allen Johnson, Department of Chemistry, UNLV, Las Vegas, NV*
*Thomas M. Lillo, Idaho National Laboratory, Idaho Falls, ID*

## Introduction

The Sulfur-lodide (SI) process generates hydrogen through the thermo-chemical decomposition of water. This high temperature processes requires materials of construction that are resistant to corrosion by sulfuric acid. Creep and oxidation limit the use of superalloys to 800-900°C. Ceramic materials are not only capable of operating at elevated temperatures that are compatible with high-temperature reactors but they also have excellent creep corrosion resistance. Therefore, in cooperation with the DOE, Idaho National Laboratory, and the University of Nevada, Las Vegas (UNLV), Ceramatec, Inc. is developing and testing ceramic based micro-channel decomposer components. In order to assess the viability of ceramic materials, extended high temperature exposure tests have been made in gaseous and liquid environments to characterize the corrosion behavior of silicon-based ceramic materials. Results from weight change measurements, strength measurements, surface analysis, and electron microscopy will be presented. Few studies of the corrosion of silicon-based ceramics after exposure to environments of high temperature sulfuric acid have been performed. The work described here is an investigation of the behavior of materials in this environment to allow selection of materials in hydrogen generation systems that use the SI process.

In 1981, Irwin and Ammon of Westinghouse Electric Corporation conducted a materials screening program to identify potential structural materials for an acid vaporizer. Candidate materials, which included both metals and ceramics, were exposed to sulfuric acid, $H_2SO_4$, at temperatures ranging from 361°C to 452°C for 250-hour increments up to 1000 hours to assess corrosion compatibility with the process stream. Irwin and Ammon found that silicon and materials containing significant amounts of silicon, such as silicon carbide and silicon nitride, have the greatest resistance to attack by boiling sulfuric acid.[1] Fernanda Coen-Porisini of the Commission of the European Communities JRC Ispra Establishment performed corrosion studies in sulfuric acid at 800°C in 1979.[2] These corrosion tests were mostly conducted on metal samples, but two of the tests also used a few ceramic samples. Both of these tests showed that alumina, mullite, and zirconia, which were the few ceramics tested, were unchanged or only displayed a deposited coating on the surface while all metals tested displayed considerable to severe corrosion.

---

[1] Irwin, H.A., Ammon, R. L. *Status of Materials Evaluation for Sulfuric Acid Vaporization and Decomposition Applications.* Adv. Energy Syst. Div., Westinghouse Electric Corp., Pittsburg, PA, USA. Advances in Hydrogen Energy (1981), 2(Hydrogen Energy Prog., Vol. 4), 1977-99.

[2] Coen-Porisini, Fernanda. *Corrosion Tests on Possible Containment Materials for H2SO4 Decomposition.* Jt. Res. Counc., ERATOM, Ispra, Italy. Advances in Hydrogen Energy (1979), 1(Hydrogen Energy Syst., Vol. 4), 2091-112.

Materials testing more relevant to thermochemical production of hydrogen was conducted by T. N. Tiegs of Oak Ridge National Laboratory in 1981.[3] Tiegs identified a number of materials from which he chose the leading candidates for testing. The candidate materials were reported in decreasing order of preference as: SiC, MgO, MAS, $Al_2O_3$, $Si_3N_4$, sialon and BeO. SiC, Sialon, MgO, $ZrO_2(MgO)$ and $ZrO_2(Y_2O_3)$ were selected for screening. These materials were then tested in an environment designed to simulate the decomposition of sulfuric acid. Tiegs identified silicon carbide as the best material at 1000° and 1225°C in the simulated sulfuric acid decomposition environment. Tiegs recommended further testing of the SiC materials at conditions more representative of the sulfuric acid decomposition environment, that is, at temperatures of 800 to 900°C and pressures up to 3 MPa.

Ishiyama et al. reported the percent weight change and corrosion rate of samples resulting from 100 hours of exposure to high-pressure boiling sulfuric acid. SiC was found to be the most corrosion resistant followed by Si-SiC and then by $Si_3N_4$. Also in Ishiyama's overall rating of the materials after 1000 hours of exposure the three materials mentioned above were listed as the least affected by the long exposure.

Based on the findings of these studies, silicon carbide and silicon nitride seem to be the most appropriate candidates for heat exchanger/decomposer materials and warrant further corrosion testing. This paper will describe the results from a series of corrosion tests to investigate the influence of the concentration of sulfuric acid and temperature on the corrosion of these candidate ceramic materials.

## Experimental Methods

### *Experimental Setup: Ambient pressure, vapour exposure*

The corrosive environments were selected to mimic the decomposition environment of sulfuric acid and the conditions expected in the SI process. The corrosion test apparatus consists of a long quartz tube partially housed inside a split tube furnace. The long quartz tube itself holds three large quartz cups and three small quartz cups (Figure 1). Starting at the top is a large quartz cup filled with quartz chips which acts as an evaporator and gas pre-heater. Below the evaporator cup sit the three small cups that hold the samples. Below the three sample cups are two large condenser cups, the top of which is filled with zirconia media and the bottom with SiC media. The long quartz tube is capped by a Teflon manifold with a pliable Teflon gasket so that the sulfuric acid vapor and decomposition products stay in the tube. This manifold is fitted with a gas (air/oxygen) feed and a liquid (sulfuric acid) feed. The condensate is collected and disposed of in a waste barrel.

ASTM Standard C 1161-02C bend bars were scribed, weighed and randomly positioned within three small sample cups for sulfuric acid exposure. The simulated conditions were 60% $H_2SO_4$, 30% $H_20$ and 10% air at 900°C. These sample cups were then loaded into the quartz tube and the furnace was heated up to 900°C with flowing argon gas. Once at temperature, the simulated sulfuric acid environment was attained by switching over to air or oxygen from argon and by dripping in the acid solution. At predetermined intervals (100, 200, 500 and 1000 hours), samples were removed, weighed and fractured according to ASTM Standard C 1161-02C procedures. Table I describes the materials that were tested.

---

[3] Tiegs, T. N. (1981, July). *Materials Testing for Solar Thermal Chemical Process Heat.* Metals and Ceramics Division, Oak Ridge National Laboratory. Oak Ridge, Tennessee. ONRL/TM-7833, 1-59.

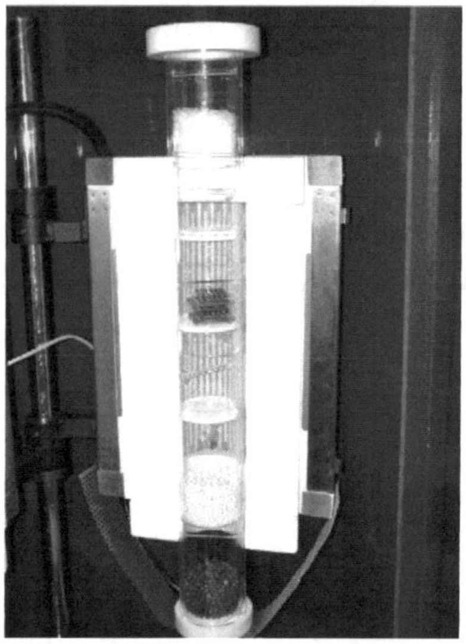

**Figure 1.** Sulfuric acid exposure rig.

**Table 1** – Materials used in this study

| Material | Fabrication Process | Vendor | Identifier |
|---|---|---|---|
| Silicon Nitride | Hot Pressed | Ceradyne | SN-HP |
| Silicon Nitride | Gas Pressure Sintered | Ceradyne | SN-GP |
| Silicon Carbide | Pressureless Sintered | Morgan | SiC-PS |
| Silicon Carbide | Tape Laminated | Ceramatec | SiC-LS |

***Experimental Setup: High-Temperature, pressurised, liquid exposure***

A pressure containment vessel made of alloy C-276 was used to conduct experiments in elevated temperature, pressurized, liquid sulfuric acid. Precautions were taken to minimize contact between the sulfuric acid and acid vapors with the pressure containment vessel. The experiment setup is shown in Figure 2. The typical volume of sulfuric acid in the experiment was 60 ml while the surface area of the samples was on the order of 160 $mm^2$. The vessel was purged with UHP argon three times by pressurizing the vessel to approximately 500 psi and then releasing the pressure to approximately 50 psi. At the target temperature of 375°C very little oxygen is expected to be present due to the large amount of activated carbon in the system. Heating to temperature took 1-2 hours and then was stable ($\pm$ 5°C of set point) throughout the experiment.

The samples were cleaned by a brief ultrasonic agitation in ethanol to remove residual grease and/or loose particles. The sample dimensions were measured to $\pm$0.01 mm and weighed to $\pm$0.00008 grams. After exposure experiments, the sample was rinsed, dried and weighed to yield a corrosion rate, according to ASTM G 31. The sample was then loaded into the vessel with fresh acid and the experiment was continued.

***Experimental Setup: High-Temperature, boiling liquid exposure***

Corrosion is often fastest at the triple point of vapour, solid, and liquid. Apparatus to investigate corrosion under these conditions is shown in Figure 3. The apparatus consists of a vessel for boiling sulfuric acid and a reflux condenser. Quartz sample holders hold samples in the liquid, at the interface between the liquid and vapour, and in the vapour.

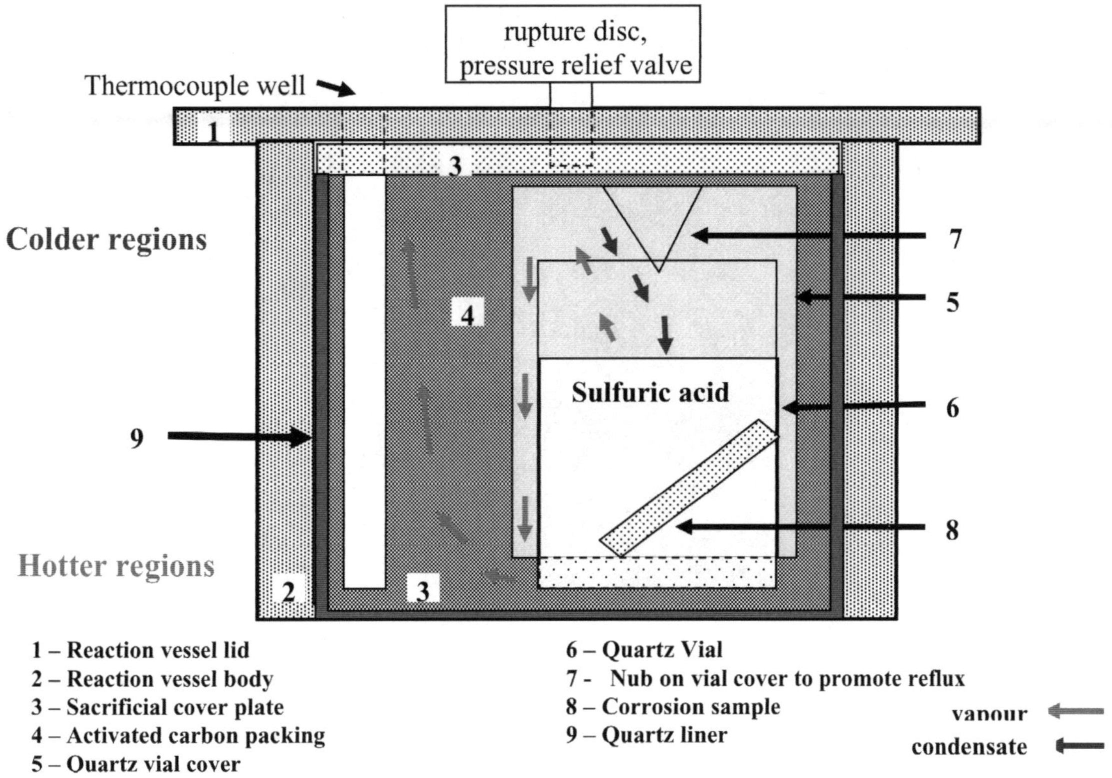


**Figure 2.** Schematic of reaction vessel internal components

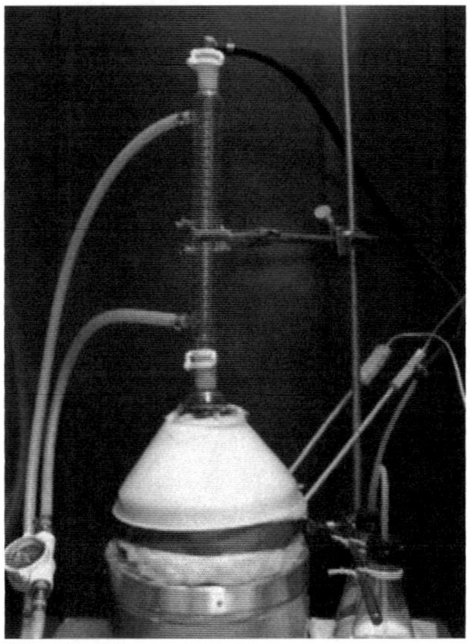

**Figure 3** - Boiling acid exposure apparatus.

## Results and Discussion

### *Exposure to sulfuric acid vapour*

All of the materials experienced an increase in weight after exposure to sulfuric acid vapour at 900 °C (Figure 4). Large standard deviations were obtained due to the small magnitude of the weight changes. Spalling of corrosion products occurred whilst handling specimens of hot pressed silicon nitride after 1000 h, leading to an apparent low weight.

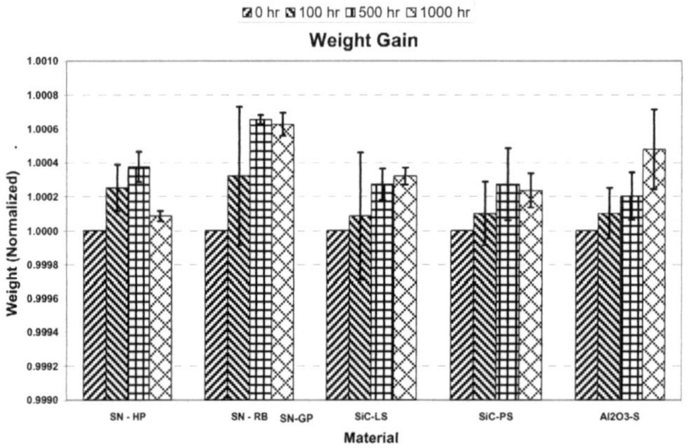


**Figure 4** - Weight Gains for 1000 hr Exposure Samples

The flexural strength of SN-HP, SiC-LS and SiC-PS increased over the first 100 hours of exposure and then remained constant (Figure 5). The flexural strength of SN-GP showed a similar trend, but didn't increase until after 100 to 500 hours exposure (Figure 5). This slight increase in strength is believed to be a result of the blunting of surface defects by silica formed by oxidation during exposure.

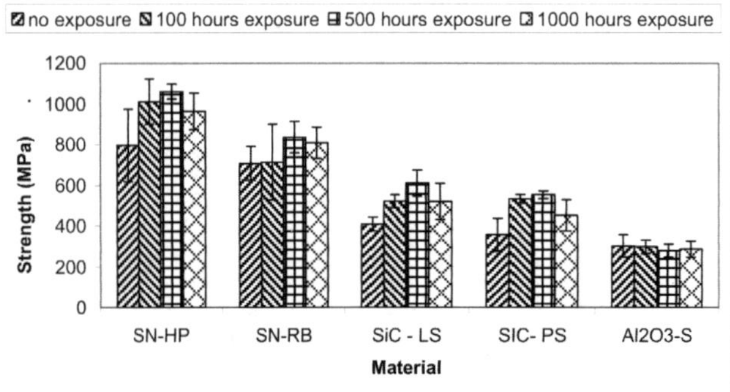


**Figure 5** - Strength Change for 1000 hr Exposure Samples

SEM images indicate the presence of a reaction product after 1000 hour exposures (Figure 6). Energy dispersive X-ray (EDS) and X-ray photo-electron spectroscopy (XPS) analyses, were conducted on an unexposed bar and a bar after each of the different exposure times. XPS analysis, conducted at the University of Nevada, Las Vegas, confirmed that the reaction product was $SiO_2$ (Figure 7). Semi-quantitative EDS analysis, Figure 8, shows that the amount of oxygen on the exposed surfaces increases with exposure time. These results imply that the presence of sulfuric acid did not prevent the formation of silica on the surfaces of the silicon based materials that were tested and that this silica reaction product healed surface flaws leading to higher flexural strength values after corrosion testing.

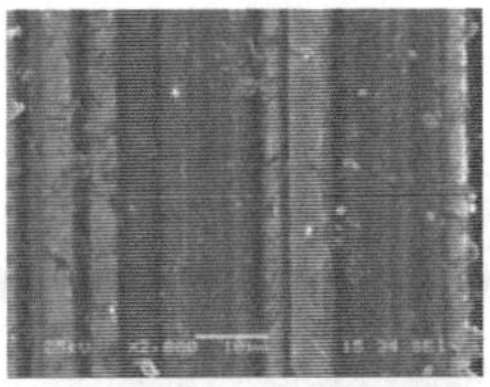

a) Unexposed SN-HP

b) SN-HP exposed 1000 h

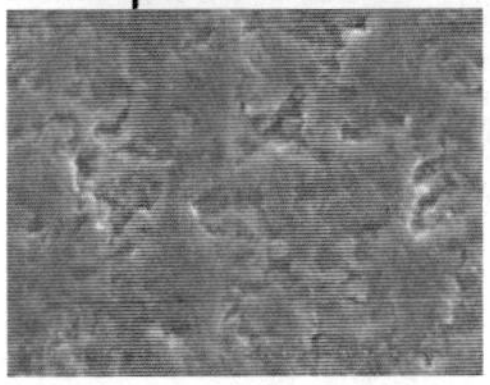

c) Unexposed SiC-LS

d) SiC-LS exposed 1000 h

c) Unexposed SiC-PS

d) SiC-PS exposed 1000 h

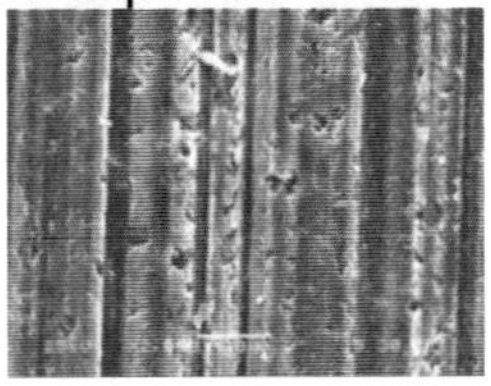

c) Unexposed SN-GP

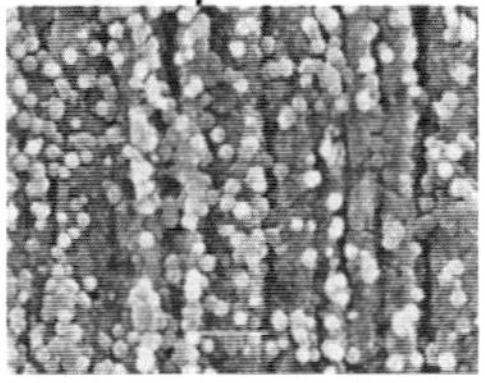

d) SN-GP exposed 1000 h

**Figure 6** - SEM Micrographs of unexposed and exposed samples (2000x)

XPS-UNLV

| | | | | | | | |
|---|---|---|---|---|---|---|---|
| Project | HXproject2 | | | Acquired | 4/9/2006 9:02:52 PM by ESCA | | |
| Experiment | MSiC1000hr10nmSp | | | | | | |
| Function | Survey | Region | 1 | eV/Step | 0.999001 | Cycle No. | 1 |
| Scans/Time | 16 | Res | 4 | Spot | 4 | Etch Time | |

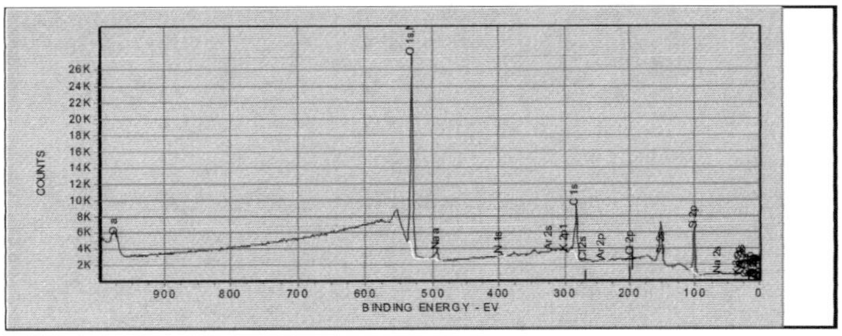


COMPOSITION TABLE

| XPS Line | Adj'ed Be | CrossSec | Norm Area | Atom % |
|---|---|---|---|---|
| Ar 2p | 240.995 | 3.04 | 10.361 | .337 |
| O 1s | 529.927 | 2.93 | 1256.407 | 40.896 |
| Si 2p | 100.786 | 0.817 | 823.564 | 26.807 |
| Na a | 494.226 | 3.4 | 65.037 | 2.117 |
| C 1s | 282.325 | 1 | 856.123 | 27.867 |
| N 1s | 395.641 | 1.8 | 19.924 | .649 |
| Cl 2p | 196.851 | 2.285 | 40.748 | 1.326 |

**Figure 7** - XPS analysis of SiC specimen after 1000 h of testing.

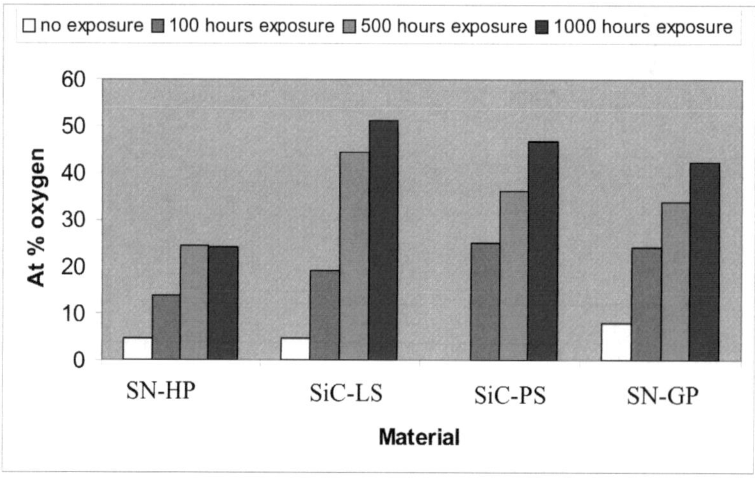


**Figure 8** - Variation in Presence of Oxygen on Exposed Surfaces of Silicon Based Materials

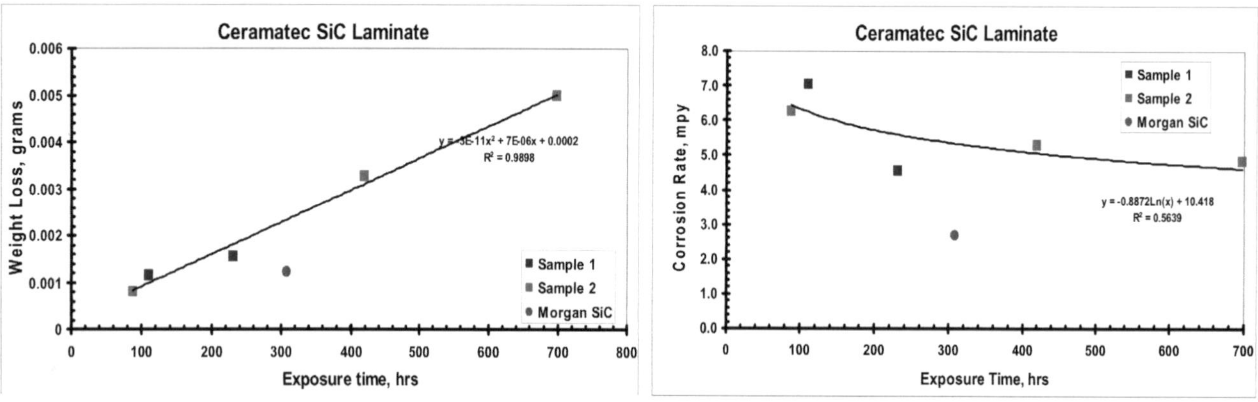


**Figure 9** - Corrosion data for the SiC-LS material. Curve fits utilize data from two samples.

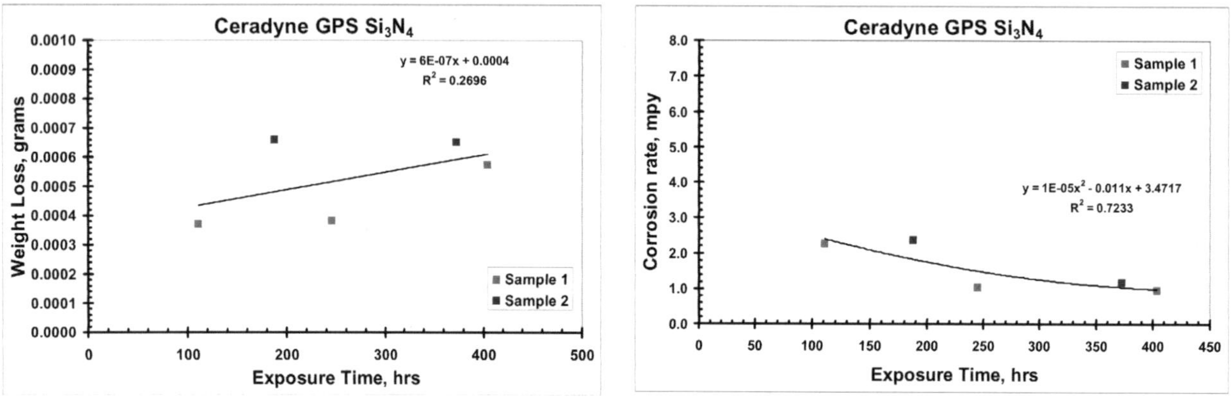


**Figure 10** - Corrosion data for the SN-GP material. Curve fits utilize data from two samples.

### *Exposure to pressurised, boiling sulfuric acid*

Two samples each of SiC-LS and SN-GP were tested in the pressurised corrosion apparatus at The Idaho National Laboratory. The specimens were exposed to sulfuric acid at about 400°C and 14 bar (200 psi) for approximately 250 h. The corrosion rate of SiC-LS was approximately 125 μm/year; SN-GP 62 μm/year. Most metals would exhibit an order of magnitude higher corrosion rate. The weight loss and corrosion data were fitted to linear and parabolic equations to determine whether the corrosion behavior was passivating (Figure 9 and Figure 10). The data could be fit equally well by either the linear or parabolic equations, but microscopy did indicate that whereas the laminated silicon carbide showed pitting after a long exposure the silicon nitride did not (Figure 11). These results suggest that the corrosion rates of the candidate materials are higher at either (1) higher pressures of sulfuric acid; (2) in hot, liquid sulfuric acid; or (3) at the triple points between solid, liquid, and vapors.

(a) SiC-LS (micron bar = 200 μm)

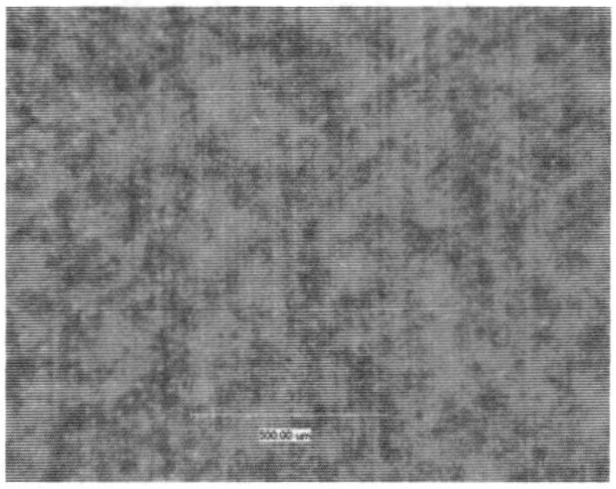

(b) SN-GP (micron bar = 500 μm)

**Figure 11** - Micrographs of (a) silicon carbide after 698 h and, (b) silicon nitride after 307 hours.

### *Exposure to (Unpressurised) Boiling Sulfuric Acid*

To date, one experiment has been completed in the boiling test apparatus. Five bars of SN-GP were exposed to boiling (1 atm) sulfuric acid (75 mol% $H_2SO_4$, 25 mol% $H_2O$) for 100 hours. Weight loss and reduced strength of the silicon nitride were measured after exposure. The recession rate was approximately 120 μm/year. This is about twice the recession measured in the pressurised, liquid sulfuric acid at INL. Additional experiments are required to confirm these results and to understand the actual behavior of the candidate materials.

## Summary

Silicon based ceramic materials, such as silicon carbide and silicon nitride, are promising candidates for use in high-temperature components for the SI process. Four different materials were examined and found to oxidize in high-temperature sulfuric acid vapour. The oxidation product, silica ($SiO_2$), appeared to strengthen the material by blunting cracks and was stable in the vaporous sulfuric acid environment. Material recession and evidence of pitting was seen after exposures at elevated temperatures in liquid sulfuric acid, under pressure, but corrosion rates were still an order of magnitude lower than for metallic materials. Silicon nitride specimens were weakened by exposure to boiling, liquid sulfuric acid, which suggests that corrosion may be accelerated in liquid sulfuric acid or at the triple point between the substrate, liquid, and vapour. Additional experiments and characterization will be performed to understand this behavior.

## Acknowledgements

This work was done under UNLV's High Temperature Heat Exchanger Program (HTHX) funded by the DOE Nuclear Hydrogen Initiative (NHI) (#RF-05-HTHX-006).

# 566d Dynamic Simulation Of The Heat Transfer Loop In A Nuclear Hydrogen Production Plant

**Patricio D. Ramirez-Munoz**[1], Mujid S. Kazimi[2], and Paul I. Barton[1]. (1) Chemical Engineering, Massachusetts Institute of Technology, 77 Massachusetts Ave., RM 66-363, Cambridge, MA 02139, (2) Center for Advanced Nuclear Energy Systems, Department of Nuclear Science and Engineering, Massachusetts Institute of Technology, 77 Massachusetts Ave., RM 24-219, Cambridge, MA 02139

The study of nuclear hydrogen production operation and safety is an important next step in assessing the desirability of bringing this technology to reality. The overall system for nuclear hydrogen production consists of the nuclear plant, the power conversion unit (PCU), the heat transfer loop and the hydrogen plant. This system has been studied using steady-state models and has been shown to have the potential to achieve high efficiencies for hydrogen production [9]. Steady-state models have allowed optimization of the plant configuration and operating conditions [1, 2, 5, 7, 8]. The dynamic interaction of all the components in the system should be investigated next, to produce a more robust and stable system, avoiding accidents and improving operability. Questions regarding maximum stresses exerted on materials during transients and how to modulate the heat delivered to the chemical plant in a safe manner must be addressed.

Nuclear power and hydrogen production systems have distinct dynamic characteristics and their interaction can produce results hard to predict intuitively. Therefore, it is necessary to use mathematical models describing the transient behavior of these systems. Dynamic models have assisted engineers for several decades in the assessment of safety of nuclear power plants [4, 10]. On the other hand, hydrogen production alternatives utilizing heat and/or electricity from nuclear power plants have been mostly studied within the steady-state regime to make design decisions. Some efforts have been made to understand the dynamics of this system [6, 11]. The hydrogen production plant adds extra complexity to the system because it includes chemical reactions with nonlinear dynamics. The simulation environment used should be able to capture these physical and chemical phenomena. The long term goal of this research is to develop a suite of mathematical models that enable dynamic simulation, sensitivity analysis and optimization studies of the interaction of a nuclear reactor and a hydrogen generation plant.

This paper focuses on dynamic simulation of the heat transfer loop. This loop removes the heat from the nuclear reactor coolant and uses it to generate high temperature steam to feed the hydrogen plant. High-temperature electrolysis is considered as the electrochemical process to produce hydrogen. The main components of the heat transfer loop are two heat exchangers and a compressor. The heat exchanger removing the heat from the nuclear reactor coolant is a compact heat exchanger and the heat exchanger heating the steam is of the tube-in-shell kind [1]. The compressor regulates the gas flowrate and it is responsible for smooth operational changes. Helium is considered as the heat transfer fluid. The use of helium in the intermediate loop simplifies the heat exchanger design, and avoids chemical consequences of leakage of different coolants into the reactor loop.

Previous attempts to simulate similar systems involving a nuclear reactor and PCUs have used numerical methods that introduce errors and produce unphysical numerical artifacts. The most common error is to solve the partial differential equations for gas dynamics using the method of lines in an implicit integrator without limiting the time step according to the CFL condition. A simulation of gas dynamics performed in this way cannot capture the fast dynamics of a gas flowing in a pipe. The second error is to use numerical methods without high-resolution corrections; the consequence of this is artificial oscillations in the simulations. Here, this issue is addressed by developing accurate methods for the simulation of the helium gas dynamics in a pipe. The gas dynamics simulator is based on CLAWPACK [3], which is a transient simulator for hyperbolic equations based on an explicit integrator with high-resolution methods. The CLAWPACK boundary conditions have to be reformulated to be able to model closed loop systems involving compressors. Simulation results for the nuclear reactor coupled to the heat transfer interface during start-up and shut-down will be presented.

The dynamic profiles of the different variables involved are fundamental in understanding possible issues related to safety and operability of the plant. The thermal and pressure profiles dictate the necessary characteristics of the materials to be used. The analysis of the power profiles helps to determine and size the auxiliary heat sources/sinks, and the best operation of the plant. Also these simulations help to identify the most important process parameters that can be modified to improve the operation of the plant.

1. C. B. Davis, R. B. Barner, S. R. Sherman, D. F. Wilson. Thermal-Hydraulic Analyses of Heat Transfer Fluid Requirements and Characteristics for Coupling a Hydrogen Product Plant to a High-Temperature Nuclear Reactor. Technical Report INL/EXT-05-00453, Idaho National Laboratory, 2005.

2. Y. H. Jeong, K. Hohnholt, M. S. Kazimi and B. Yildiz. Optimization of the hybrid sulfur cycle for hydrogen generation. Technical Report MIT-NES-TR-004, Center for Advanced Nuclear Energy Systems, Massachusetts Institute of Technology, 2005.

3. R. J. Leveque. Finite Volume Methods for Hyperbolic Problems. Cambridge University Press, Cambridge, U. K., 2002.

4. Michitsugu Mori. Simulation of a BWR-5 plant: Verification and validation of RETRAN-03. Nuclear Technology, 121:245-259, 1998.

5. C. H. Oh, R. Barner, C. B. Davis and S. R. Sherman. Evaluation of working fluids in an indirect combined cycle in a very high temperature gas-cooled reactor. Nuclear Technology, 156:1-10, 2006.

6. S. Oh, N. Brown and S. T. Revankar. Transient Model for the Chemical Chamber in Sulfur Iodine Thermo-Chemical Process, Technical Report PU/NE-06-010, School of Nuclear Engineering, Purdue University.

7. S. B. Rodríguez Jr., R. O. Gauntt, S. T. Revankar and K. Vierow. Development of design and

simulation model and safety study of large-scale hydrogen production using nuclear power. 2005 AIChE Annual Meeting and Fall Showcase, Conference Proceedings, 2005, 8018-8023.

8. B. Yildiz, K. Hohnholdt, and M. S. Kazimi. Hydrogen production using high temperature steam electrolysis supported by advanced gas reactors with supercritical CO2 cycles. Technical Report MIT-NES-TR-002, Center for Advanced Nuclear Energy Systems, Massachusetts Institute of Technology, 2004.

9. B. Yildiz and M. S. Kazimi. Efficiency of hydrogen production systems using alternative nuclear energy technologies. International Journal of Hydrogen Energy, 31:77-92, 2006.

10. K. Vierow, Y. Liao, J. Johnson, M. Kenton, and R. Gauntt. Severe accident analysis of a PWR station blackout with the MELCOR, MAAP4 and SCDAP/RELAP5 codes. Nuclear Engineering and Design, 234:129-145, 2004.

11. R. B. Vilim. Dynamic Modeling Efforts for System Interface Studies for Nuclear Hydrogen Production. Argonne National Laboratory, ANL-07/16, 2007.

# 566e Investigation Of $SO_3$ Electrolysis As An Alternative Step In The Sulfur-Iodine Thermochemical Process For Hydrogen Production

**J. David Carter**, Jennifer R. Mawdsley, and Magali Ferrandon. Chemical Engineering Division, Argonne National Laboratory, Bldg. 205, 9700 S. Cass Ave., Argonne, IL 60439

The standard sulfur-iodine thermochemical cycle for the production of hydrogen includes a step for the thermal decomposition of $SO_3$, that requires temperatures of ≥850°C. The other reactions in the thermochemical cycle can be performed at ≤500°C, and it is highly desirable to be able to carry out the entire process with maximum temperatures less than about 500ºC.

We have investigated the feasibility of using an oxide ion-conducting electrolysis cell to electrochemically decompose $SO_3$ to $SO_2$ at the desired lower temperatures,

$$SO_3 + \text{electricity} \rightarrow SO_2 + \tfrac{1}{2}O_2 \qquad 500\text{-}600^{\circ}\text{C}, E^{\circ}=0.11 \text{ V}$$

We have built a single-atmosphere $SO_3$ electrode test system to evaluate new electrodes for the electrolytic reduction of $SO_3$. Initial tests were performed with Pt electrodes, which confirmed the reduction of sulfuric acid in steam through cyclic voltammetry. Tests with sample solid oxide fuel cell-type electrode materials gave inconclusive results, with voltammograms showing poor cathodic currents and no indication of the formation of $SO_2$ or $O_2$ with the application of voltage.

We have obtained some very promising results with a $GaVO_4$/Au electrode, however. We selected this material based on calculated stability diagrams and knowing that $V_2O_5$ is a good oxidation catalyst for sulfur. After a cyclic voltammetry experiment, which reduced the electrode to –1.3 V, we noticed an interesting change in the voltammogram, in which the cathodic current began to increase at an increasing rate as voltage was applied. More significantly, we also observed that the concentrations of both $SO_2$ and $O_2$ began to increase rapidly beyond what could be explained by just the faradaic oxygen flux through the cell. In this paper, we will discuss the materials selection criteria we have developed, the results of our most recent experiments on the $SO_3$ electrode and the development of the electrochemical cell.

The submitted manuscript has been created by UChicago Argonne, LLC, Operator of Argonne National Laboratory ("Argonne"). Argonne, a U. S. Department of Energy Office of Science laboratory, is operated under Contract No. DE-AC02-06CH11357. The U. S. Government retains for itself, and others acting on its behalf, a paid-up nonexclusive, irrevocable worldwide license in said article to reproduce, prepare derivative works, distribute copies to the public, and perform publicly and display publicly, by or on behalf of the Government.

# 566f Sulfuric Acid Decomposition Under Pressurized Conditions In A Pt-Lined Tubular Reactor

Chang Soo Kim[1], Ki Yong Lee[2], Kwang Ho Song[2], Gyeong-Taek Gong[1], Kye Sang Yoo[1], Byoung Sung Ahn[1], Kwang-Deog Jung[1], and **Honggon Kim**[1]. (1) Energy and Environmental Technology Division, Korea Institute of Science and Technology, Hawolgokdong 39-1, Seongbukgu, Seoul, 139-791, South Korea, (2) Department of Chemical and Biological Engineering, Korea University, Anamdong, Seongbukgu, Seoul, 136-701, South Korea

Sulfur-Iodine (SI) cycle is one of the most promising thermochemical water splitting cycles to produce hydrogen using high-temperature heat transferred from a VHTR (very high temperature gas-cooled reactor). It was originally proposed by General Atomics Co. in 1970s, which was composed of the following three steps: (1) Bunsen reaction: x I2 + SO2 + 2 H2O = 2 HIx + H2SO4, (2) HIx decomposition: 2 HI = H2 + I2, (3) H2SO4 decomposition: H2SO4 = H2O + SO2 + 1/2 O2. Decomposition of sulfuric acid to SO2, O2 and H2O is the step to require the highest temperature heat from a VHTR in the whole SI cycle. This reaction consists of two series reactions: 1) decomposition of sulfuric acid to SO3 and H2O at around or above 350°C with or without catalysts, and 2) catalytic decomposition of SO3 to SO2 and O2 at 750°C or higher temperature. He, a cooling gas of the VHTR pressurized over 40 bars, transfers the high-temperature heat to the reactor. In order to lessen the pressure difference at the reactor wall, the reactions need to be conducted under a pressure as high as the wall can endure even though the equilibrium conversions to SO3 and H2O, and SO2 and O2 unfavorably reduce according to the Le Chatelier's principle. Since the acidic reactants and products either in liquid phase or in gas phase are highly corrosive, only a few materials may be applied to this severe reaction condition of high temperature and pressure. Silicon carbide (SiC) is known a good candidate which is resistant to sulfuric acid and SO3 at high temperature, but it possesses ceramic-like properties so that several problems such as flexible connection of parts and high pressure sealing are still remaining. Pt is another material resistant to the acidic reactant and products at high temperature. Pt is also reported catalytically active for decomposition reactions of both sulfuric acid and SO3 even though its long-term stability at high temperature over 850°C is still questionable. In addition, Pt cannot be solely used for the high-pressure reactor because of its softness. In this study, a Pt-lined Hastelloy-C tubular reactor was prepared and the sulfuric acid decomposition was tested with and without additional catalysts at pressurized condition over 10 bars. The Pt-lined tubular reactor of 19 mm ID and 2440 mm length was placed in a heater with six independently working heating zones. A 10 m coiled tube of Inconel 600 in another heater is placed in front of the Pt-lined tube in order to evaporate and heat the reactant up to 600°C. The decomposition reactions were conducted at 800°C to 900°C and at 1 bar to over 10 bars with 95 wt% H2SO4. The surface of Pt lining showed excellent catalytic activity over a summed-up period of 150 h even though its overall surface area is superficially not as large as the supported Pt catalysts commercially available. The extent of dissociation reaction to SO2 and O2 was evaluated referring to the variation of temperature, pressure and the reactant residence time. Additional charge of 20g commercial Pt/alumina (2 wt% Pt) catalyst pellets did not remarkably enhance the yield of SO2 and O2, indicating that the heat transfer area was the reaction determining factor rather than the amount of catalytic species in this case. This preliminary result has proposed possibility of adopting Pt-lined metallic tubes as reactor parts for high-temperature high-pressure decomposition reaction of sulfuric acid.

# 584a Solar Production of Hydrogen Using a Cadmium Based Thermochemical Cycle

**Lloyd C. Brown** and Bunsen Y. Wong. General Atomics, 3550 General Atomics Ct., San Diego, CA 92121

The Molten Cadmium thermochemical cycle consists of two chemical reactions:

$Cd(l) + H_2O(g) => CdO(s) + H_2(g)$ [350°C]

$CdO(s) => Cd(g) + ½O_2(g)$ [1450°C]

The first experimental hurddle has been overcome. Hydrogen has been produced at reasonable rates in a steam-hydrogen mixture at close to the thermodynamically predicted stoichiometry. The products of the high temperature reaction must be quenched to prevent back reaction. Work has begun on determining the required quench rate. Flowsheets have been developed, based on alternative quench strategies, to aid further process development.

# 584b The Rapid Dissociation Of Manganese Oxide To Produce Solar Hydrogen

**Todd M. Francis**, Department of Chemical and Biological Engineering, University of Colorado, 1111 Engineering Drive, Boulder, CO 80309-0424 and Alan W. Weimer, Chemical and Biologicial Engineering, University of Colorado, 1111 Engineering Drive, Boulder, CO 80309-0424.

A three-step metal oxide cycle to produce renewable hydrogen has been proposed. This cycle employs manganese oxide and solar energy to produce hydrogen:

$\frac{1}{2}Mn_2O_3 \rightarrow MnO + \frac{1}{4}O_2$ (1)

$MnO + NaOH \rightarrow \frac{1}{2}H_2 + NaMnO_2$ (2)

$NaMnO_2 + \frac{1}{2}H_2O \rightarrow \frac{1}{2}Mn_2O_3 + NaOH$ (3)

--------------------------------------------------

$H_2O \rightarrow H_2 + \frac{1}{2}O_2$ (4)

Most work on the manganese oxide cycle has focused on the hydrogen generating step (Equation 2) and the product recovery step (Equation 3) with little work done on the reduction step (Equation 1). It is essential to understand the reduction step because the entire feasibility of the manganese oxide cycle is based on this reaction having a high overall conversion. Specifically, the viability of the MnO and O2 phase separation must be investigated to determine how much recombination might occur during the quench step. Because of the importance of the reduction step, a kinetic study in an Aerosol Flow Reactor (AFR) has been done, as it can handle a high throughput of powder, while allowing reaction conditions to be tightly controlled. In addition, to the kinetic study in the AFR, a mechanistic study has been performed in a Thermogravimetric Analyzer. This presentation will provide results from these studies.

# 584c Thermochemical Performance of Ferrite Materials for Carbon Dioxide Splitting Processes

**Nathan P. Siegel**, Solar Technologies, Sandia National Laboratories, PO Box 5800, Albuquerque, NM 87185

A class of iron oxide-based materials (ferrites), usually partially substituted with a second metal such as cobalt or nickel, and fabricated into composite monolithic structures with a yttria-stabilize zirconia (YSZ) carrier, has been applied to decompose carbon dioxide in a cyclic two-step thermochemical process using a concentrated solar energy thermal input. The resulting carbon monoxide fuel may be used to produce hydrogen via the water gas shift reaction. The realization of a practical carbon monoxide and/or hydrogen producing reactor based on these materials is dependent on their thermochemical performance, specifically on reaction extent and rate for both the high temperature thermal reduction step and the lower temperature carbon dioxide oxidation step. In this paper we present experimental results for reaction extent and rate for cobalt ferrite ($Co_xFe_{3-x}O_4$, x=0.5-1) and discuss the expected performance of a solar carbon dioxide splitting device based on this material.

# 584d Iron-Cobalt Spinel Oxides For Thermochemical Hydrogen Production

**Jonathan R. Scheffe**, Department of Chemical and Biological Engineering, University of Colorado, 1111 Engineering Dr., ECCH 111, Boulder, CO 80309-0424, Hans Funke, Chemical and Biological Engineering, University of Colorado, 1111 Engineering Drive, Campus Box 424, Boulder, CO 80309-0424, and Alan W. Weimer, University of Colorado, 1111 Engineering Drive, Boulder, CO 80309-0424.

Spinel ferrites, mixed metal oxides formed by substitution of iron in Fe3O4 by other metals, are capable of splitting water at relatively low temperatures.

We have investigated the role of specific surface area of CoxFe3-xO4, where x ranges from 0.25 to 1.0, on the amount of oxygen evolved per mole of ferrite, and its subsequent reactivity with steam to generate hydrogen gas. Our results indicate that, by reducing particle size and thus increasing specific surface area, the amount of hydrogen generated per mole of ferrite can be significantly increased. XPS and EDS were utilized as a means to analyze the reduced samples, and confirm that oxygen atoms are evolved primarily from the surface, rather than the bulk lattice.

**Hydrogen Production via Hydrolysis of Zinc Nanoparticles**

*Tareq Abu Hamed, Jane H. Davidson and Julia F. Haltiwanger*
*University of Minnesota, Minneapolis, MN, USA*

## Abstract

Simultaneous synthesis and hydrolysis of Zn nanoparticles is investigated as the second step in a ZnO/Zn redox solar water splitting process. Nanoparticles were synthesized and hydrolyzed in a condensation nucleation process using a mixture of argon and steam to quench zinc evaporated in a hot wall tubular reactor. Hydrogen concentration in the effluent gas stream was measured via gas chromatography. Aerosol size distribution and concentration were measured in-situ using a scanning differential mobility sizer. The composition of particles collected on the reactor surface and on a filter at the exit of the reactor was characterized with X-ray diffraction. Experiments carried out with a $H_2O$:Zn molar ratio of 36, an evaporation temperature of 1023 K, and a quench temperature of about 600 K yielded hydrogen conversions of 51 and 56%.

## Introduction

A process that shows great promise as a means to store solar energy and produce hydrogen without the use of fossil fuel is thermochemical water splitting by means of metal hydrolysis. In a two step metal redox cycle water is the feed stock and concentrated solar is the energy source. Zinc oxide has been shown to be a favorable material for this process due to self-disassociation at temperatures achievable in concentrating solar reactors [1]. The process is composed of two steps. In the solar step, zinc oxide is reduced to zinc and oxygen:

$$\mathrm{ZnO} \rightarrow \mathrm{Zn} + \frac{1}{2}\mathrm{O_2} \quad (1)$$

The endothermic reaction ($\Delta H_{298K}$= 351 kJ/mol) proceeds at reasonable rates above 1900 K. At 2335 K, the Gibbs free energy is zero and the reaction can be carried out using concentrated solar thermal energy without electricity input or carbothermal reduction. The dissociation chemistry and reactor technology are summarized in several papers, including [1-7]. Effective separation of the gaseous products is required to avoid reoxidation of Zn. In a non solar step, zinc is hydrolyzed to produce hydrogen and ZnO in the exothermic reaction:

$$\mathrm{Zn} + \mathrm{H_2O} \rightarrow \mathrm{ZnO} + \mathrm{H_2} \quad (2)$$

Reaction (2) is thermodynamically favorable at or below 1490 K, but there are kinetic constraints. The product ZnO is recovered and recycled to the first step. The net reaction is the splitting of water.

The present study focuses on the hydrogen production step (2). The desire is to produce $H_2$ at high conversion and to recover ZnO in a continuous process for reprocessing in step (1). The studies most relevant to the present work are those that have measured hydrolysis of Zn powder in the absence of oxygen [8,9] and prior efforts to synthesize and hydrolyze zinc nanoparticles [9-14]. Vishnevetsky and Epstein [8] measured hydrolysis of zinc powder produced by the solar carbothermal reduction of ZnO and that of commercial zinc powder (<10 μm). The powder produced in their solar reactor comprised 2 to 7 μm conglomerates of 200 to 500 nm spherical particles. In a batch process at temperatures above 673 K, a fast surface reaction was followed by a slower diffusion limited reaction. Hydrogen conversion for the solar Zn was approximately three times that of the commercial powder at 833 K. The difference was attributed to the difference in powder size.

Ernst [9] carried out a thermogravimetric analysis of hydrolysis of commercial zinc nanoparticles (164 ± 10 nm) from 603 to 633 K. As in [8], a fast surface reaction was followed by a slower diffusion limited reaction. The heterogeneous reaction rate was first order with respect to Zn and half order with respect to $H_2O$. The reaction followed an Arrhenius law with a pre-exponential constant of $2 \times 10^{-5}$ mol/cm$^2$s and an activation energy of 43 ± 7 kJ/mol. The rate limiting step was attributed to diffusion of Zn ions within the ZnO lattice at the surface. The rate followed a parabolic law. The measured rate constant was consistent with that for the reaction of Zn with oxygen. The hydrolysis rate was enhanced for submicron Zn particles. The impact of particle size on the rate of reaction is in agreement with the results of Park et al. [15] who found the activation energy for oxidation of aluminum nanoparticles increases with decreasing particle size.

Co-synthesis of hydrogen and zinc oxide via simultaneous synthesis and hydrolysis of Zn nanoparticles as an effective means to accomplish reaction (2) was first attempted by Weiss et al. [10]. Additional experiments in hot wall flow reactors have been conducted at the Swiss Federal Institute of Technology ETH-Zurich [9,11-13] and at the University of Minnesota [14]. Initial experiments were conducted above the saturation temperature of Zn and thus particle yields were very low [10,11,14]. High hydrogen conversions (70 to 96%) were attributed to heterogeneous reactions at the reactor wall. This process is undesirable as the product is difficult to recover. The desire is to entrain zinc oxide nanoparticles in a carrier gas flow and thus permit continuous removal of products. The group at ETH developed a modified reactor that was divided into three temperature zones in an attempt to separate the Zn evaporation, condensation and hydrolysis steps [9,12]. Above the Zn saturation temperature, hydrolysis again occurred on the wall of the reactor producing hydrogen conversion of approximately 90%, but negligible particle yield. At temperatures below Zn saturation, hydrogen conversion decreased with decreasing reactor temperature. Total particle mass collected on a filter at the exit of the reactor remained very low and less than 20% of the hydrogen conversion could be attributed to hydrolysis of gas borne Zn particles. The recent reactor design at ETH was designed to substantially increase the quench of evaporated zinc [13]. Measured hydrogen conversion was 56 to 76%. About 10% of the evaporated zinc was recovered as ZnO at the filter. The ZnO mass fraction on the filter was attributed in part to scouring of ZnO particles from the walls in the quench region. Improved yield of nanoscale aerosol is essential for continuous recovery and regeneration of the product ZnO.

The paper reports initial experiments in a reactor in which Zn-evaporation and steam-quenching and hydrolysis are isolated into two zones. The objective of the new design was to achieve a rapid quench and thus improve nanoparticle synthesis.

## Apparatus and Experimental Method

Figure 1 is a sketch of the reactor and instrumentation to measure hydrogen concentration and particle size and mass at the reactor exit. The reactor consists of three parts: a 1.5 cm i.d. inner quartz tube, a stainless steel quencher ring, and a 2 cm i.d outer

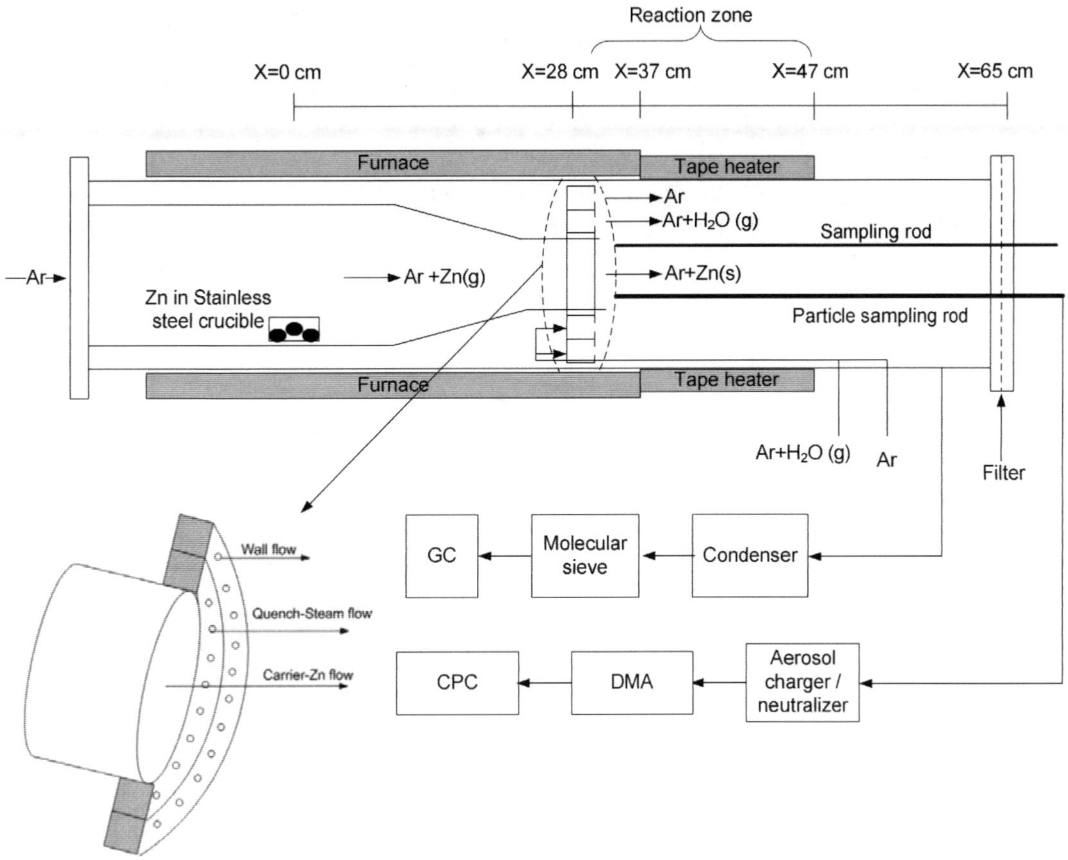


Figure 1. Experimental Apparatus

quartz tube. The 2 cm i.d. quartz tube was placed in an 88 cm long concentric cylindrical electric furnace (Thermolyne 97500). Zinc pellets (99.99% purity) of known weight were loaded in a stainless steel crucible and placed within the 1.5 cm i.d. quartz tube, which was placed within the larger tube. The crucible was positioned at X = 0 cm as shown in Figure 1. Ultra pure Ar was fed to the inner quartz tube to carry the evaporated Zn downstream to the quencher ring, positioned at X = 28 cm. The inner quartz tube narrows to 0.5 cm i.d. prior to the quencher ring to direct the flow through the center of the quencher. The quencher consists of two concentric stainless steel rings welded to each other. The inner ring supplies Ar and steam at 473 K to quench the evaporated zinc below its melt temperature (693 K). The outer ring supplies additional Ar at 298 K to reduce depositions on the wall. Steam was generated by an electrical heater fed with deionized water. The evaporated Zn, the quench/steam flow, and the wall Ar flow enter the outer quartz tube as individual jets in concentric rings, as depicted in Figure 1. The flow rate of the three flows are controlled and measured independently. Outside the furnace at 37 ≤ X ≤ 47 cm, the quartz tube is wrapped with electrical tape to reduce thermophoretic effects and condensation at the wall. At X = 47 cm, the quartz tube is connected to a 2.2 cm i.d. stainless steel tube. A glass fiber filter at X = 65 cm captures particles for XRD analysis. Considering the total volumetric flow rate, the transit time from the quench ring to the filter is 0.9 s.

The outlet gas stream was treated to remove moisture by a condenser and molecular sieve. The hydrogen concentration was analyzed at 2 minute intervals using an inline gas chromatograph (GC, Agilent 6890). Each sample is acquired over 5 s. The composition of material deposited on the quartz reactor surface, on a sampling rod inserted into the reactor, and on the filter at the end of the reactor was analyzed by X-ray diffraction (Bruker-AXS D5005 Diffractometer with 2.2 kW sealed Cu Source). Analysis of the aerosol was attempted in-situ (at X = 33 cm). Particle number concentration and size distribution were analyzed using a nanoscale Scanning Mobility Particle Sizer (SMPS) with 1/70 dilution with clean air. The SMPS system consists of a Po-210 radioactive aerosol charger/neutralizer, a nano Differential Mobility Analyzer (DMA – TSI 3085) with recirculating air sheath flow and an ultrafine Condensation Particle Counter (CPC – TSI 3025). Size distribution was measured in the range 5 to 125 nm where size is the diameter of an equivalent sphere with the same mobility (or drag).

In the present study, at steady state, the furnace temperature was 1023 K. The carrier Ar flow rate was 0.75 l/min. The $H_2O$:Zn molar ratio was 36. The quench and wall flow rates were 1.42 and 5.7 l/min, respectively. These flow rates were selected during a preliminary set of experiments in which the objective was to maximize the yields of $H_2$ and ZnO at the filter.

Prior to each experiment, the mass of the filter, the crucible and the Zn were recorded. The stainless steel crucible was placed in the reactor. The reactor was purged with Ar. Once the furnace temperature reached 773 K, the valves for the steam and the argon carrier gas, wall flow and quench gasses were opened. In the present study, we report the results of two experiments, each conducted for 30 minutes. The centerline temperature distribution in the reactor was measured using a K-type thermocouple in separate experiments.

## Results

Table 1 lists the operating conditions for the reported experiments. The axial temperature distribution within the reactor is plotted during steady state operation in Figure 2. The Ar/steam quench reduced the operating temperature across the quench ring from 946 to 747 K. Temperatures within the reactor zone were below the saturation temperature of 808 K. In-situ measurements with the nano-SMPS indicate very high particle concentrations (greater than 700,000 particles/$cm^3$) at X = 33 cm. Unfortunately, with the dilution ratio used here the particle concentration saturated the CPC. Data obtained while the furnace was still heating (693 to 743 K) indicate that the particle size distribution peaks at 28 nm. From the inlet of the reaction zone, temperature decreased from 747 to 603 K at X ≈ 46 cm. Ernst [9] observed faster rates of hydrolysis at temperatures above 603 K. Beyond the heated reactor zone, the gas temperature continues to drop and hydrolysis is expected to be very slow.

**Table 1.** Summary of operating conditions and results.

| Run | Evaporation Temperature (K) | $Q_{Ar}$ (l/min) | $Q_{steam}$ (ml/min) | $\Delta t$ (min) | Evaporated Zn (g) |
|---|---|---|---|---|---|
| 1 | 1023 | 7.9 | 0.2 | 30 | 0.242 |
| 2 | 1023 | 7.9 | 0.2 | 30 | 0.327 |

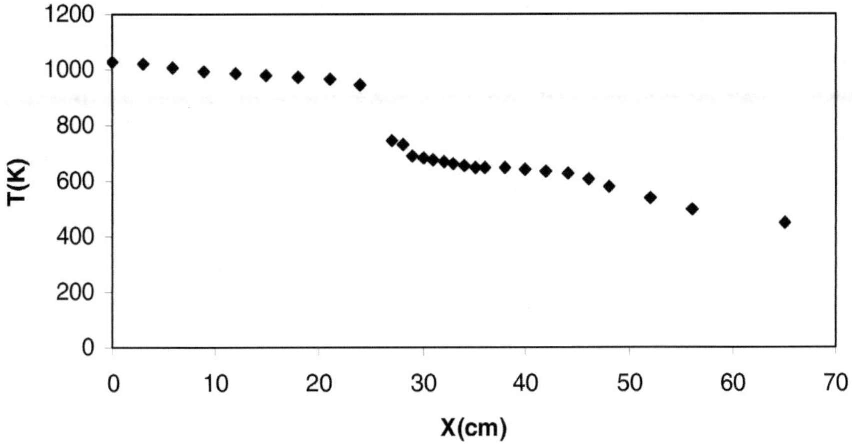


Figure 2. The reaction zone centerline temperature distribution.

**Table2.** Measured temperature and ZnO content at different positions on the sampling rod.

| X (cm) | Temperature (K) | ZnO (%) | Color | Particle size (nm) |
|---|---|---|---|---|
| $28 \leq X \leq 31$ | $728 \leq T \leq 678$ | 79 | White to grey | 47-80 |
| $31 \leq X \leq 38$ | $678 \leq T \leq 648$ | 62 | Grey to black | 39-42 |
| $38 \leq X \leq 53$ | $648 \leq T \leq 535$ | 31 | Grey to black | 27-31 |
| $58 \leq X \leq 64$ | $535 \leq T \leq 450$ | 4 | Black | 67 |
| 65 | 450 | 0 | Black | 73 |

In all experiments, deposits were observed on the wall of the quartz reactor for $28 \leq X \leq 65$ cm and along the particle sampling rod. Unlike our prior experiments which were conducted above the saturation temperature of Zn, the material on the reactor surfaces was easily removed with a brush. The deposits decreased with increasing X, corresponding to decreasing temperature. The material deposited on the sampling rod and reactor walls was white to light gray near the quencher, indicating the presence of ZnO, and black near the end of the reactor, indicative of Zn. Table 2 summarizes XRD of the materials scraped from the sampling rod at different positions and temperatures. The decreasing ZnO content with increasing downstream position (increasing X) is consistent with the visual observations of the color of the deposits and our expectations based on temperature. The deposits on the filter were pure zinc. The decrease of the ZnO content with increasing downstream position is related to the decrease of the reaction zone temperature. Particles collected on the filter represent 9 to 14% of the evaporated zinc. Particles on the filter were pure zinc.

Transient $H_2$ production (mol/min) is plotted in Figure 3. The steam is introduced at t = 0 when the furnace temperature is 773 K. The furnace reaches 1023 K at t = 4 min. Based on prior work, the hydrogen production follows the evaporation rate. Hydrogen is detected at t = 6 s and increases from about $5 \times 10^{-5}$ to $17 \times 10^{-5}$ mol/min within 12 min. It remains steady until the experiment is terminated at t = 30 min. The area under the curve yields the total hydrogen production. Hydrogen conversion is the volume of measured hydrogen divided by the theoretical maximum amount of $H_2$ under complete hydrolysis of the Zn evaporated during the experiment. The hydrogen conversions for runs 1 and 2 are 56 and 51%, respectively. These values are lower than those obtained in prior experiments at temperatures above Zn saturation [10,11,14]. In these prior studies, hydrogen production was attributed to heterogeneous hydrolysis of zinc vapor at the hot walls of the reactor. In the present study, hydrogen production is most likely due to a combination of hydrolysis of Zn aerosol and Zn particles deposited on the walls of the reactor and the sampling rod. The relative ease with which particles could be removed from the wall precludes chemical vapor deposition. Hydrolysis occurred primarily in the region near the quench ring where temperatures were above ~650 K. This result is consistent with the low reaction rates of commercial Zn nanoparticles observed by Ernst [9] below 573 K and the results of Piatkowski [13] in a reactor with high quench rates. The lack of ZnO particles on the filter is not fully understood.

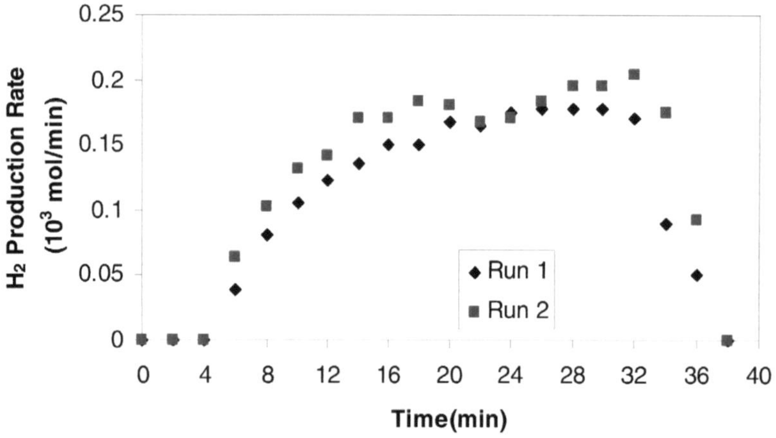


Figure 3. Hydrogen production rate (mol/min).

## Conclusion

A laboratory reactor was designed to provide suitable reaction conditions for simultaneous synthesis and hydrolysis of Zn nanoparticles. Preliminary experiments yielded hydrogen conversions of 51 to 56 %. Hydrogen production is attributed to hydrolysis of Zn nanoparticles in the carrier gas stream and deposited on the reactor walls. Zinc oxide content of the deposited material decreased with distance downstream of the quench flow. This reduction is attributed to lowered reaction rates with decreasing temperature. Particles scraped from the walls of the reactor were predominately ZnO for temperatures above 648 K. The ZnO content was only 4 % for $648 \leq T \leq 535$ K.


## Acknowledgement

The project is funded by the University of Minnesota Initiative for Renewable Energy and the Environment.

## References


1. Steinfeld, A., (2002), "Solar Hydrogen Production via A Two-Step Water-Splitting Thermochemical Cycle Based on Zn/ZnO Redox Reactions," *International Journal of Hydrogen Energy*, Vol.27, pp. 611-619.
2. Steinfeld, A., (2005), "Solar Thermochemical Production of Hydrogen—a Review," Solar Energy, 78, pp. 603-615.
3. Palumbo, R., Lédé, J., Boutin, O., Elorza Ricart, E., Steinfeld, A., Moeller, S., Weidenkaff, A., Fletcher, EA., and Bielicki, J., (1998), "The Production of Zn From ZnO in A Single Step High Temperature Solar Decomposition Process," Chemical Engineering Science, 53, pp. 2503-2518.
4. Haueter, P., Moeller, S., Palumbo, R., and Steinfeld, A., (1999), "The Production of Zinc by Thermal Dissociation of Zinc Oxide – Solar Chemical Reactor Design," Solar Energy, 67, pp.161-167.
5. Moeller, S., and Palumbo, R., (2001), "The Development of A Solar Chemical Reactor for The Direct Thermal Dissociation of Zinc Oxide," Journal of Solar Energy Engineering, 123, pp. 83-90.
6. Schunk, L. O., Haeberling, P., Wepf, S., Wuillemin. D., Meier, A., and Steinfeld, A., (2007), "A Rotary Receiver-Reactor for the Solar Thermal Dissociation of Zinc Oxide," Proceeding of the ASME Energy Sustainability Conference, Paper No. 36078, Long Beach, CA, June 27-30.
7. Perkins, C., Lichty, P., and Weimer A. W., (2007), "Determination of Aerosol Kinetics of Thermal ZnO Dissociation by Thermogravimetry," Chemical Engineering Science, 62, pp. 5952-5962.
8. Vishnevetsky, I. and Epstein, M., (in press 2007), "Production of Hydrogen from Solar Zinc in Steam Atmosphere," International Journal of Hydrogen Energy.
9. Ernst, F.O., (2007), "Cosynthesis of $H_2$ and Nanocrystalline ZnO Particles by Zn Aerosol Formation and In-situ Hydrolysis, Ph.D. Dissertation No. 17272, ETH-Zurich, Zurich, Switzerland.
10. Weiss, R. J., Ly, H. C., Wegner, K., Pratsinis, S. E., and Steinfeld, A., (2005), "$H_2$ Production by Zn Hydrolysis in A Hot-Wall Aerosol Reactor," American Institute of Chemical Engineers Journal, 51, pp. 1966 -1970.
11. Wegner, A., K., Ly, H.C., Weiss, R.J., Pratsinis, S.E. and Steinfeld, A., (2006), "In Situ Formation and Hydrolysis of Zn Nanoparticles for $H_2$ Production by The 2-Step ZnO/Zn Water-Splitting Thermochemical Cycle," International Journal of Hydrogen Energy, 31, pp. 55–61.
12. Ernst, F. O., Tricoli, A., Pratsinis, S. E., and Steinfeld, A., (2006), "Co-Synthesis of $H_2$ and ZnO by In-Situ Zn Aerosol Formation and Hydrolysis," American Institute of Chemical Engineers Journal, 52(9), pp. 3297-3303.
13. Piatkowski, N., (2007), "Zn-nanoparticle in-situ Hydrolysis for Hydrogen Production in a High Quench Rate Reactor," Master Thesis, ETH-Zurich, Zurich, Switzerland.
14. Abu Hamed, T., Davidson, J.H., Stolzenburg, M., (2007), Hydrogen Production via Hydrolysis of Zn in A Hot Wall Flow Reactor. Proceeding of the ASME Energy Sustainability Conference, Paper No. 36176, Long Beach, CA, June 27-30.
15. Park, K., Lee, D., Rai, A., Mukherjee, D., and Zachariah, M. R., (2005), "Size Resolved Kinetic Measurements of Aluminum Nanoparticle Oxidation with Single Particle Mass Spectometry," Journal of Physical Chemistry B, 109(15), pp. 7290-7299.

# 612a Inorganic Membranes to Improve the Efficiency of the Production of Hydrogen Using Nuclear Energy

**Brian L. Bischoff**, Dane F. Wilson, Lawrence E. Powell, and K. Dale Adcock. Oak Ridge National Laboratory, PO Box 2008, Oak Ridge, TN 37831-6044

There are several processes being investigated for using the heat from the next generation nuclear reactors to produce hydrogen. The efficient separation of gases at high temperatures can improve the efficiency of at least two of the leading candidates, thermochemical water splitting and high temperature electrolysis. Through a series of reactions, the Sulfur-Iodine (SI) thermochemical process facilitates the splitting of water into hydrogen and oxygen. In the SI process all the chemicals are recycled except for water. The SI process is efficient, scalable to large sizes, and consumes no expensive chemical reagents; however, it has one major disadvantage. The thermal decomposition of sulfuric acid requires high temperatures (800 to 900 °C). However, there is a potential to lower the peak temperature by 200+ °C if the decomposition products of sulfuric acid, O2, H2O, and SO2, can be separated from SO3 using an inorganic membrane. Silicon carbide supported membranes have been fabricated and we will present preliminary separation data showing the potential to separate the product gases from SO3. We will also present results of the stability of these membranes under simulated operational conditions.

High temperature electrolysis employs a solid oxide electrolyzer to split water into hydrogen and oxygen. As the oxygen is produced, it transports through the electrolyte layer of the electrolyzer leaving behind on the feed side a mixture of only the product hydrogen and the residual steam. The efficiency of this process can be improved by recovering the unreacted steam and feeding it back to the inlet to the electrolyzer without costly cooling and reheat steps. Inorganic membranes have the potential to separate the hydrogen from the steam at the operating temperature of 800°C. Long term stability and separation test data will be presented.

# 612b Numerical Study of Sulfur Trioxide Decomposition in Bayonet Type Heat Exchanger and Chemical Decomposer

**Vijaisri Nagarajan**[1], Valery Ponyavin[1], Yitung Chen[1], Milton E. Vernon[2], Paul Pickard[2], and Anthony E. Hechanova[3]. (1) Mechanical Engineering, University of Nevada, Las Vegas (UNLV), 4505 Maryland Parkway, Las Vegas, NV 89154, (2) Advanced Nuclear Concepts, Sandia National Laboratories, POB 5800, Albuquerque, NM 87185, (3) Nuclear Engineering, University of Nevada at Las Vegas, Las Vegas, NV 89119

The present study is concerned with the use of bayonet type heat exchanger as silicon carbide integrated decomposer (SID) which produces sulfuric acid decomposition product - sulfur dioxide. Thermochemical cycles produce hydrogen through a series of chemical reactions resulting in the production of hydrogen and oxygen from water at much lower temperatures than direct thermal decomposition. Energy is supplied as heat in the temperature range necessary to drive the endothermic reactions, generally 750°C to 1000°C or higher. All process chemicals in the system are fully recycled. In the cycle heat energy enters to the thermochemical process through several high temperature chemical reactions. Some amount of the heat is rejected via exothermic low temperature reaction. The whole cycle includes the three following reactions:

I2 (l) + SO2 (g) + 2H2O = 2HI (aq) + H2SO4 (aq) (1)

2HI (g) = H2 (g) + I2 (g) (2)

H2SO4 (g) = H2O (g) + SO2 (g) + ½O2 (g) (3)

One of the most critical processes in the cycle is the decomposition of sulfuric acid decomposition process via very high temperature and highly corrosive environment. Many researches on bayonet heat exchangers have been done in the past. However there are no reports about application of the bayonet heat exchanger as a decomposer for the catalytic sulfuric acid decomposition. The present study considers a bayonet type high temperature heat exchanger and decomposer made of silicon carbide and quartz for the catalytic decomposition of sulfuric acid in packed bed using porous media approach.

Silicon carbide integrated decomposer also called as bayonet heat exchanger was designed by Sandia National Lab (SNL). The integrated acid decomposer combines the function of boiler, superheater, decomposer and recuperater in a single silicon carbide unit. In this scheme teflon is used for the low temperature region and silicon carbide is used for high temperature regions. The advantage of the silicon carbide decomposer over the others is the presence of the recuperator which heats the incoming acid gases. This would minimize the total input energy required to the system. 35 to 40 mol % concentrated sulfuric acid is coming from a sulfuric acid concentrator and then pumped into the inlet of bayonet heat exchanger.

The bayonet heat exchanger can handle both high temperature and low temperature regions in a single unit. The inlet is made of teflon and maintained at temperatures below 200ºC. The water vapor and the sulfuric acid enter the boiler which would heat the sulfuric acid to 450ºC to produce a sulfuric acid vapor. The superheater would heat the sulfuric acid vapor from 450ºC to 700ºC and the decomposer would heat the vapors to the maximum operating temperature plus provide the heat necessary to dissociate the sulfur trioxide to sulfur dioxide and oxygen.

The decomposer region is located in the top of the bayonet heat exchanger. It houses the pellets. The pellets are made up silicon carbide which contain approximately 1wt.% of platinum. The pellets are spherical shaped and their packing is simple cubical packing. The diameter of the pellets is 5 mm.

The boundary condition for the outer wall is obtained from the thermocouples. The upper wall is maintained under adiabatic conditions. The value for density of sulfur trioxide mix with water is obtained from commercial software Fluent 6.2.16. The operation is carried out under atmospheric pressure. Pressure outlet is selected as outflow boundary condition.

The calculated geometry is meshed using the mesh generator GAMBIT 2.2.30. The used mesh is quadrilateral mesh. A mesh independent study was done for different nodes and from the results the proper number of cells and nodes (33036 cells, 35374 nodes) are selected for the future meshing. The fluid flow within the investigated geometry is assumed to be laminar because the Reynolds numbers are less than 700.

The governing equations are solved in the Cartesian coordinate system with a control volume finite difference method. A nonstaggered grid storage scheme is adapted to define the discrete control volumes. In this scheme, the same control volume is employed for the integration of all conservation equations, and all variables were stored at the control volume's cell center. The numerical scheme used in this study is a power-law differencing scheme, and the solver used is a segregated solver. The SIMPLE algorithm is used to resolve the coupling between pressure and velocity. The governing equations, which were discrete and nonlinear, are linearized using an implicit technique with respect to set of dependent variables. The algebraic equations are solved iteratively using an additive correction multigrid method with a Gauss-Seidel relaxation procedure.

The flow calculations of decomposition of sulfur trioxide to sulfur dioxide in the presence of catalyst using porous media approach are performed. The pressure drop obtained in the pebble bed region is about 3000 Pa. The temperature in the region increases from the inlet and goes on increasing in the pebble bed region. As the decomposition takes place the mole fraction of SO3 is reduced in the pebble bed region. The mole fractions of SO2 and O2 increases in the decomposition region as the decomposition take place. According to the results the decomposition percentage of sulfur trioxide (SO3) is found to be 19.11%. The percentage is smaller than that with the assumption of constant outer wall temperature (33%). It is because the average temperature in the pebble bed region is smaller than for assumption of constant outer wall temperature. The results are compared with the experimental results obtained in SNL.

2007 AIChE Annual Meeting, November 4-9th, Salt Lake City, Utah

Submitted to Session: **Nuclear Engineering Division**

# NGNP Process Heat Utilization: Evaporative Spiral Heat Exchanger Design

Piyush Sabharwall*,[1], Steven Sherman[1], Vivek Utgikar[2]

1. *Idaho National Laboratory*
2. *University of Idaho*
   *Idaho Falls, Idaho 83415-3860*

## Abstract

The next generation nuclear plants (NGNP) will be configured to provide electricity and process heat for industrial applications such as hydrogen and transportation fuels production, desalination, conversion of non-conventional fossil fuel resources, etc. The heat transfer system is particularly challenging not only due to the elevated temperatures (up to ~ 1300K) and industrial – scale power transport (~ 50 MW), but also due to a potentially large separation distance between the nuclear and industrial plants (100+m) dictated by safety and licensing mandates.

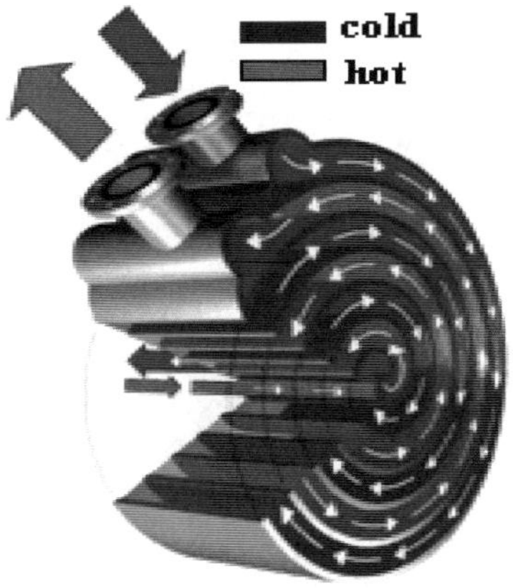


Figure 1: Schematic of a Spiral Heat Exchanger

Spiral heat exchanger (SHE) designs as shown in figure 1, are less susceptible to fouling, and are more compact than conventional shell and tube designs resulting in reduced heat losses. Due to curvature of the tube, a centrifugal force is generated as the fluid flows through the curved tubes. The curved channels create a swirling or spiraling motion within the fluid. This spiraling motion helps keeps the solid present in fluid in suspension, and the heat transfer surfaces are kept clean through a scrubbing action. The flow is generally turbulent, even when flowing at velocities that would otherwise make the flow non-turbulent in straight tubes. Secondary flow induced by the centrifugal force has significant ability to enhance the heat transfer rate. SHE have been extensively used in a wide variety of applications for example, air conditioning and refrigeration systems. Heat transfer characteristics of SHE can be enhanced through manipulation of channel dimension (i.e. by varying the diameter of the pipe) and surface modifications by adding roughness to the pipe.

* Corresponding Author: Tel: 208.526.6494, Email: Piyush.Sabharwall@inl.gov

The SHE could be efficient phase change exchangers. The specific volume change from convective evaporation and condensation is well accommodated by the spiral design. Liquid flows into the eye of the spiral, gains heat, and vapor exits from the spiral 'horn' for transport to the distant condenser at the industrial end user.

In this paper the case study is presented for the evaporative SHE design, where the transfer of thermal energy of the NGNP will be utilized for the boiling of sodium and further this heat will be used in the production of hydrogen. The approach to the development of the model involves writing energy balance equations for the two fluids over a differential element of the SHE. Figure 2 shows the flow paths of the two fluids and clarifies the basis for modeling of the SHE.

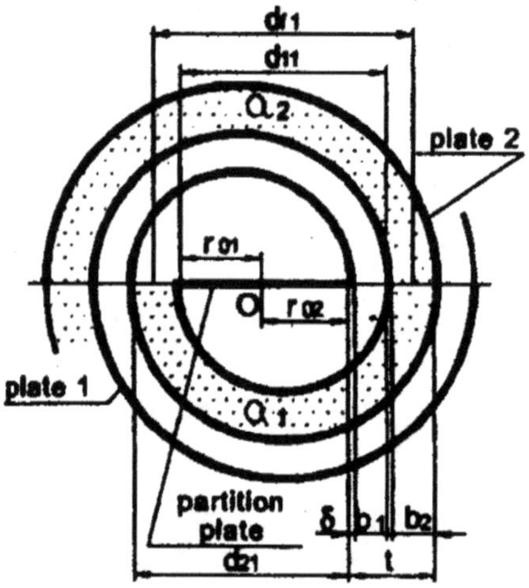


Figure 2: Flow Paths in the SHE

The energy balance equations for the hot and cold fluid are [2]:

$$-C_H \frac{dt_{H,r-\frac{1}{2}}}{d\varphi} = q_{i,j-1} + q_{i,j+1} \qquad (1)$$

$$-C_C \frac{dt_{C,r+\frac{1}{2}}}{d\varphi} = q_{i,j+1} + q_{i+1,j+2} \qquad (2)$$

where $C_H$ and $C_C$ are the heat capacity rates (products of mass flow rates and heat capacities), q are the heat fluxes, t are the dimensionless temperatures, and φ the differential angle over which the energy balance is written. The subscripts i, j and r refer to the spiral positions.

Some of assumptions typically made for developing the mathematical models for the SHE are [6]:

- The flow is fully developed for both the hot and cold streams, i.e. in the eye of the spiral or at center region we have gas (helium) and liquid (sodium), whereas at the periphery we have gas (helium) and vapor (sodium).
- The fluid physical properties are uniform in each phase i.e. they are invariant for liquid sodium and also for sodium vapor, though the properties of liquid sodium are different than that of sodium vapor.

* Corresponding Author: Tel: 208.526.6494, Email: Piyush.Sabharwall@inl.gov

- Radiative heat transfer is negligible
- Heat losses from the outer surface to the surroundings are negligible
- Thermal resistance of the channel wall is negligible

The energy balances shown above will be refined by removing restricting assumptions and developed further to model the performance of the SHEs. Variations and descriptions of SHE designs are also noted in the literature [1,4,5].

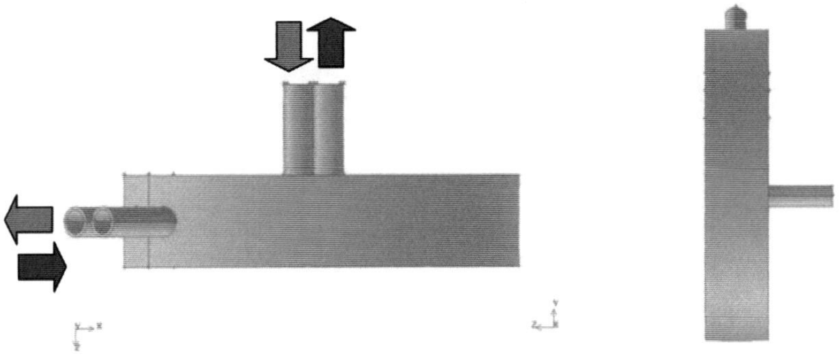

Figure 3. Modeled Geometry of the Spiral Heat Exchanger (SHE)

The inlet and outlet of the hot and cold fluid can be seen from the Figure 3, inlet of the hot fluid will be connected with the next generation nuclear plant and the outlet of the cold fluid (in vaporized state) will be connected to a thermposyphon or a heat pipe for further transport of heat to a hydrogen plant. The direction of the coolant motion in SHE is shown below. The hot fluid enters at the center, as can be seen from the red arrow and flows outward in a spiral path to the outlet on periphery. The cold fluid simultaneously enters the peripheral header, as can be seen from the blue arrow and flow countercurrent to the hot fluid toward the center.

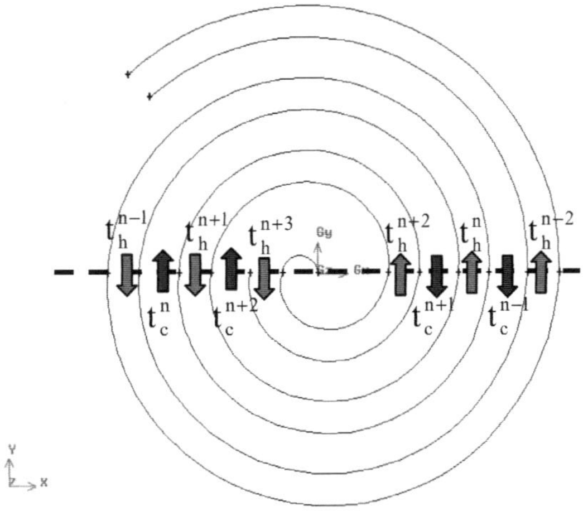

Figure 4. Detail Description of the Fluid Flow Motion and Heat Transfer Between Adjacent elements for the Spiral Heat Exchanger

* Corresponding Author: Tel: 208.526.6494, Email: Piyush.Sabharwall@inl.gov

**Mathematical Formulation (Heat Balance Equation)**

For a SHE with N spirals we will have 2N elements (N for the cooling fluid channel and N for the heating fluid channel). All of these elements, both for cooling and heating channels, should follow a same heat transfer law, though the heat transfer coefficient will vary depending on the interaction between the adjacent element, i.e when the heated He interacts with liquid Na, we have heat transfer between liquid and a gas, further in the SHE spiral the interaction is between He and boiling Na (where the boiling heat transfer coefficient will dominate) and at the periphery the interaction is between He and vaporized Na.

For liquid to liquid heat transfer in the turbulent flow (Re > 5000), we have

$$Nu = 0.023\ \mathrm{Re}^{0.8}\ \mathrm{Pr}^{m}\ \varepsilon_R \quad [7] \tag{3}$$

with m = 0.3 for heating
m = 0.4 for cooling

and $\varepsilon_R = 1 + 10.3\left(\frac{D_h}{R}\right)^3$

Equation 3, has been used to evaluate the convective heat transfer coefficients in a curved tube, but has been found suitable for a curved duct of rectangular cross section as well [7]. Heat transfer will occur between the adjacent elements as can be seen from the figure 4. Thus heat balance equation can be obtained as:

$$\dot{m}_c C_{p,c}[t_c^{(n+1)} - t_c^{(n)}] = U_i A_i [\bar{t}_h^{(n-1)} - \bar{t}_c^{(n)}] + U_j A_j [\bar{t}_h^{(n+1)} - \bar{t}_c^{(n)}] \tag{4}$$

$$\dot{m}_h C_{p,h}[t_h^{(n)} - t_h^{(n+1)}] = U_u A_u [\bar{t}_h^{(n)} - \bar{t}_c^{(n-1)}] + U_v A_v [\bar{t}_h^{(n)} - \bar{t}_c^{(n+1)}] \tag{5}$$

where, the average temperature in the elements $\bar{t}_c^{(n)}$ and $\bar{t}_h^{(n)}$ are the arithmetic mean temperatures:

$$\bar{t}_c^{(n)} = \frac{1}{2}\left[\bar{t}_c^{(n)} + \bar{t}_c^{(n+1)}\right] \tag{6}$$

$$\bar{t}_h^{(n)} = \frac{1}{2}\left[\bar{t}_h^{(n)} + \bar{t}_h^{(n+1)}\right] \tag{7}$$

The overall heat transfer coefficient (U) between every adjacent element can be calculated by:

$$U = \frac{1}{\frac{1}{h_h} + \frac{1}{k} + \frac{1}{h_c}} \tag{8}$$

* Corresponding Author: Tel: 208.526.6494, Email: Piyush.Sabharwall@inl.gov

where, the heat transfer coefficient can be obtained from equation (3) and then be substituted in equation (8). Further, the overall heat transfer coefficient for the entire SHE can be obtained easily by:

$$q = \overline{U} \; A \; \Delta T_m \qquad (9)$$

**Conclusions**

The study carried out in the paper provides valuable insight and options useful in the design and development of NGNP process heat utilization. The flow is generally turbulent in SHE, even when flowing at velocities that would otherwise make the flow non-turbulent in straight tubes. A computer program is being written to carry out this analysis, but one would have to modify the number of spirals N to meet other required condition of outlet temperature of hot and cold fluids. In this analysis, as discussed earlier the fluid properties are assumed to be constant, but in order to model a practical design of SHE with satisfactory engineering accuracy, the properties of the fluid will have to be varied with temperature and also would have to take into account the nonlinearities of the equations, a multi-iterative computation will be necessary.

**Nomenclature**

A Heat Transfer Area ($m^2$)

$C_P$ Heat Capacity (KJ/kg K)

$D_h$ Hydraulic Diameter (m)

h Heat transfer coefficient ($W/m^2K$)

k Thermal Conductivity (W/m K)

$\dot{m}$ Mass flow rate (Kg/sec)

q Heat Transfer Rate (W)

R Duct radius of curvature (m)

$T_m$ Log mean temperature difference (K) (based on overall inlet and outlet temperature for SHE)

t Temperature in ducts (K)

$\bar{t}$ Mean Temperature (K)

U Overall Heat Transfer Coefficient ($W/m^2 K$)

$\overline{U}$ Average Heat Transfer Coefficient for entire SHE ($W/m^2 K$)

v Average longitudinal velocity in the duct (m/sec)

* Corresponding Author: Tel: 208.526.6494, Email: Piyush.Sabharwall@inl.gov

Nu Nusselt number $\left(\frac{h\,D_h}{k}\right)$

Re Reynolds number $\left(\frac{\rho v D_h}{\mu}\right)$

Pr Prandtl number $\left(\frac{\mu\, C_P}{k}\right)$

*Greek Symbols*

$\rho$ Density (Kg/m$^3$)

$\mu$ Dynamic Viscosity (Pa.sec)

$\varepsilon_R$ Bend Factor of curved duct

*Subscripts*

c cold fluid

h hot fluid

i = 2n-2 for even elements
2n-3 for odd elements

j = 2n for even elements
2n-1 for odd elements

u = 2n-3 for even elements
2n-2 for odd elements

v = 2n-1 for even elements
2n for odd elements

*Superscripts*

n element position in computation

* Corresponding Author: Tel: 208.526.6494, Email: Piyush.Sabharwall@inl.gov

**REFRENCES**


1. Alfalaval, available on-line at www.alfalaval.com, January 2006

2. Bes, Th, and Roetzel, W., *Thermal Theory of the Spiral Heat Exchanger*, Int. J. Heat Mass Transfer, 36:765-773, 1993.

3. Davis C.B., Oh, C.H., Barner, R.B., Sherman, S.R., and Wilson, D.F. *Thermal-Hydraulic Analyses of Heat Transfer Fluid Requirements and Characteristics for Coupling a Hydrogen Production Plant to a High-Temperature Nuclear Reactor,* INL/EXT-05-00453, 2005.

4. Deshpande, S., *Spiral Heat Exchangers*, Chem. Eng. World, 28(2):81-83, 1993

5. Dongwu, W., *Geometric Calculations for the Spiral Heat Exchanger*, Chem. Eng. Technol., 26:592-598, 2003.

6. Targett, M.J., Retallick, W.B. and Churchill, S.W., *Solutions in Closed form for a Double-Spiral Heat Exchanger,* Ind. Eng. Chem. Res., 31:658-669, 1992.

7. Nengli, Z., Zhilin, J., Zhenwei, Ni., and Kung, H.C., *A Computational Method For Thermal Design of Spiral Plate Heat* Exchanger, ASME Proceedings of National Heat Transfer, 1: 445-449, 1988


* Corresponding Author: Tel: 208.526.6494, Email: Piyush.Sabharwall@inl.gov

2007 AIChE Annual Meeting, November 4-9th, Salt Lake City, Utah

Submitted to Session: **Nuclear Engineering Division**

# Comparison of Sodium Thermosyphon with Convective Loop

Fred Gunnerson[1], Piyush Sabharwall*[,2], Steven Sherman[2]

1. *University of Idaho*
2. *Idaho National Laboratory*
   *Idaho Falls, Idaho 83415-3860*

## Abstract

The Next Generation Nuclear Plant (NGNP) will most likely produce electricity and process heat, with both being used for hydrogen production. To capture nuclear process heat, and transport it to a distant industrial facility requires a high temperature system of heat exchangers, pumps and/or compressors. The heat transfer system is particularly challenging not only due to the elevated temperatures (up to ~ 1300K) and industrial scale power transport (~ 50 MW), but also due to a potentially large separation distance between the nuclear and industrial plants (100+m) dictated by safety and licensing mandates.

In contrast to single-phased forced convective heat transfer via 'pumping a fluid', a thermosyphon (also called a wickless heat pipe) transfers latent heat through the vaporization / condensing process. The condensate is further returned to the hot end by gravity, i.e. without any requirement of pumps or compressors. Since the latent heat of evaporation is large, a considerable amount of heat can be transported with a minimal temperature difference from end to end.

It is easy to demonstrate that a sodium-filled thermosyphon can transport comparable amounts of thermal energy as a single-phase convective sodium loop, within the same diameter pipe. Sodium thermosyphon and sodium forced convective loops can both deliver comparable rates of heat transfer through a given size pipe. This can be easily demonstrated by considering the ideal rate of convective heat transport through a pipe without losses; modeled in terms of enthalpy by:

$$Q'' = \Delta H \, \rho \, V = \Delta H \, (m' / A) \qquad (1)$$

where $Q''$ is the rate of thermal energy transport per unit flow area (kJ/m$^2$ s), ΔH is the change of the specific enthalpy of the transport fluid (kJ/kg), V is the average flow velocity (m/s), $m'$ is the mass flow rate (kg/s), ρ is density (kg/m$^3$), and A is the cross-sectional flow area (m$^2$).

Two-phase heat transfer by a thermosyphon has the advantage of high enthalpy transport that includes the sensible heat of the liquid, the latent heat of vaporization, and possible vapor superheat. In contrast, single-phase forced convection transports only the sensible heat of the fluid. Additionally, vapor-phase velocities within a thermosyphon can be much greater than single-phase liquid velocities within a forced convective loop.

* Corresponding Author: Tel: 208.526.6494, Email: Piyush.Sabharwall@inl.gov

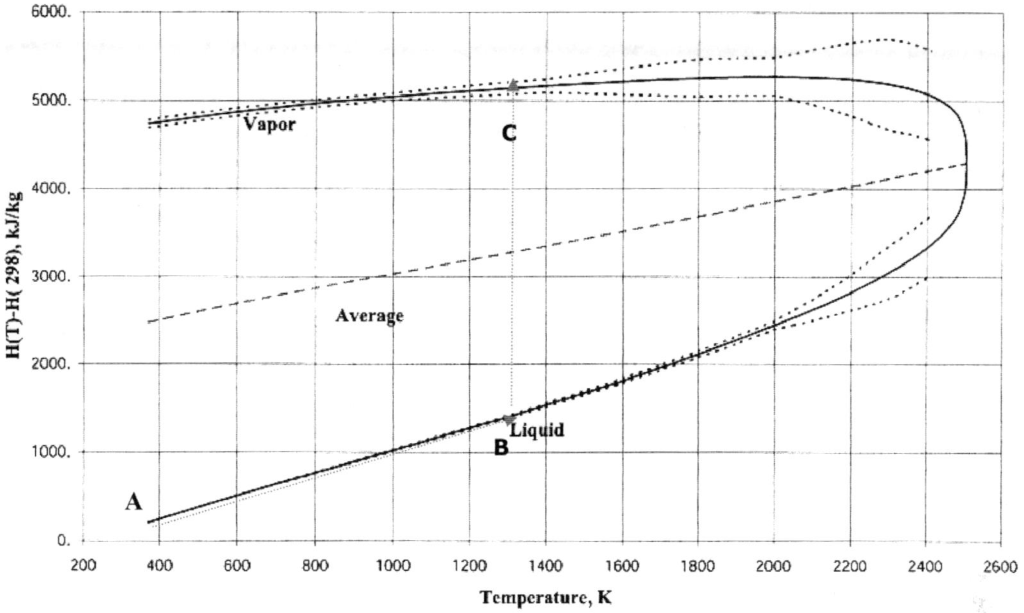


Figure 1 Enthalpy for saturated sodium: liquid and vapor [2]

Figure 1 exemplifies the enthalpy enhancement in heat transfer afforded by a two-phase thermosyphon verses a single-phase convective loop with sodium as the working fluid. The specific enthalpy (ΔH) of saturated liquid and vapor, relative to the solid at 298.15K, are illustrated as a function of temperature [2]. Assuming heat transfer from a high-temperature gas cooled reactor at 1300K (1027°C) to an industrial facility; the maximum single-phase heat transfer is given by the enthalpy gain from points A to B in Figure 1, or approximately 1,200 kJ for each kilogram of sodium. Compare with two-phase heat transfer from points A to B to C where the enthalpy gain is approximately 4,933 kJ per kilogram with no vapor superheat; over 3½ times more heat per kilogram of sodium than the single-phase. The saturation pressure of sodium at 1300K is only 0.326 MPa thus pressure forces are minimized. Vapor flow through a pipe is limited by compressible 'choke flow' when the vapor reaches its sonic velocity. For sodium vapor the sonic velocity is approximately 750m/sec at 1300K [3].

The limiting heat transfer rate for an ideal sodium thermosyphon operating around 1300K can therefore be estimated by equation (1):

$$Q'' = \Delta H \, \rho \, V = (4{,}933)\ \text{kJ/kg}\ (0.8)\text{kg/m}^3(750)\text{m/s} = 2960\ \text{MW/m}^2$$

Similarly, single-phase liquid sodium could transport the same rate of thermal energy with an average flow velocity of about 3.5 m/s, well within the capabilities of advanced liquid metal pumps. This simple analysis for sodium as the working fluid theoretically illustrates that a thermosyphon and a forced convective loop can both deliver comparable rates of heat transfer through a comparable diameter pipe. An inclined thermosyphon, however, has the luxury of controllable heat transfer without the need for high temperature pumping and can deliver the heat at the same approximate temperature as the source.

Ideally, a 15 cm diameter pipe could transport ~ 50 MW of power at 1300 K to a hydrogen production facility some distance away by serving either as a sodium thermosyphon or as part of forced-convective loop. A larger diameter thermosyphon could reduce the vapor velocity below limiting critical choke

* Corresponding Author: Tel: 208.526.6494, Email: Piyush.Sabharwall@inl.gov

velocity. Therefore, the two-phase heat delivered by a sodium thermosyphon is comparable to that delivered by a single-phase forced convection loop.

**REFRENCES**

1. Davis C.B., Oh, C.H., Barner, R.B., Sherman, S.R., and Wilson, D.F. *Thermal-Hydraulic Analyses of Heat Transfer Fluid Requirements and Characteristics for Coupling a Hydrogen Production Plant to a High-Temperature Nuclear Reactor,* INL/EXT-05-00453, 2005.

2. Fink, J.K. and Leibowitz, L., *Thermodynamic and Transport Properties of Sodium Liquid and Vapor*, ANL/RE-95/2, 1995.

3. Vargaftik, et al., *Handbook of Physical Properties of Liquids and Gases*, Hemisphere Publication, Second Edition, 1983.

* Corresponding Author: Tel: 208.526.6494, Email: Piyush.Sabharwall@inl.gov

# 612e Modeling of High Temperature Shell and Tube Heat Exchanger and Decomposer for Hydrogen Production

**Gayatri Kuchi**[1], Valery Ponyavin[1], Yitung Chen[1], Steven Sherman[2], and Anthony E. Hechanova[3]. (1) Mechanical Engineering, University of Nevada, Las Vegas (UNLV), 4505 Maryland Parkway, Las Vegas, NV 89154, (2) Idaho National Laboratory, MFC 774 113 Mail Stop 6122, Idaho Falls, ID 83404, (3) Nuclear Engineering, University of Nevada at Las Vegas, Las Vegas, NV 89119

Hydrogen can be an attractive energy carrier if it can be produced cleanly and in a cost-effective manner. Sulfur-iodine water splitting cycle is a baseline candidate thermo-chemical process for industrial hydrogen production. The cycle consists of the following three chemical reactions which yield the dissociation of water:

$I_2$ (l) + $SO_2$ (g) + $2H_2O$ = 2HI (aq) + $H_2SO_4$ (aq) (1)

2HI (g) = $H_2$ (g) + $I_2$ (g) (2)

$H_2SO_4$ (g) = $H_2O$ (g) + $SO_2$ (g) + $1/2O_2$ (g) (3)

Reaction of the sulfuric acid decomposition (Eq. 3) is the most critical part of the cycle via very high temperature and extremely corrosive environment. The sulfuric acid decomposition is carried out in two steps. In the first step, sulfuric acid is assumed to decompose into water and sulfur trioxide (Eq. 5). The reaction is volumetric type of reaction and requires temperature 450ºC. The second step, sulfur trioxide decomposes to produce oxygen and sulfur dioxide (Eq. 6) The reaction is surface reaction which require catalyst and high temperature (700ºC and higher).

$H_2SO_4$ (g) = $H_2O$(g) + $SO_3$(g) (4)

$SO_3$(g)= $SO_2$(g)+$0.5O_2$(g) (5)

In the literature different types of high temperature heat exchanger and chemical decomposer are used to support the sulfuric acid decomposition reaction. This paper shows numerical simulations of the sulfuric acid decomposition process in shell and tube heat exchanger and chemical decomposer with straight tube configuration. The shell and tube heat exchanger consists of a bundle of tubes enclosed within a cylindrical shell. It is assumed that sulfuric acid is completely decomposed to the sulfur trioxide ($SO_3$) and water ($H_2O$) before the entrance to the decomposer. In the modeled decomposer one fluid (mix of $SO_3$ and $H_2O$ gases) flows through the tubes and second fluid (helium) flows within the space between the tubes and the shell.

The computation geometry of the heat exchanger, used in current study, is created based on the assumption that the inlet and outlet manifolds are properly designed to provide uniform flow distribution for the inner tubes and shell side. Because of the assumption the geometry includes only one tube and part of the shell zone around the tube. The created geometry is two-dimensional axisymmetric geometry.

Helium is made to pass through the shell which provides the required heat for the decomposition of sulfur trioxide. Based on the literature search the dimensions for the computation domain were chosen as follow: the tube is 4.5m in length and 40mm in diameter with an entrance length of 1.5m. The shell is 80mm in diameter. The length of the catalytic bed which is modeled using porous media assumption is 1.5m.

The chemical reaction is modeled using Arrhenius equation. The rate constants for the reaction are computed from are obtained from experiments completed in INL.

A nonstaggered grid storage scheme is adapted to define the discrete control volumes. In this scheme, the same control volume is employed for the integration of all conservation equations, and all variables were stored at the control volume's cell center. The numerical scheme used in this study is a power-law differencing scheme, and the solver used is a segregated solver. The SIMPLE algorithm is used to resolve the coupling between pressure and velocity. The governing equations, which were discrete and nonlinear, are linearized using an implicit technique with respect to set of dependent variables. The algebraic equations are solved iteratively using an additive correction multigrid method with a Gauss-Seidel relaxation procedure.

Because sulfur trioxide decomposition reaction is catalytic surface reaction, the choice of the catalyst is very important to get high decomposition percentage. In the used numerical model the different type of the catalyst are modeled by selection of appropriate activation energy value for the Arrhenius equation. Besides, the concentration of the catalyst in the catalyst carrier influence to the activation energy significantly. For example, for the 1 wt % Pt the activation energy is 32.67 kJ/mol; for the 0.1 wt % Pt the activation energy is 46.24 kJ/mol.

Fluid flow, heat transfer and chemical reaction calculations were done for the shell and tube configuration using porous media approach, with counter and parallel flow arrangements.

The percentage decomposition of sulfur trioxide is found to be 93% for the counter flow arrangement. For the parallel flow arrangement the decomposition percentage of sulfur trioxide is found to be 92%.

The parametric studies for different activation energies have been completed in the research. It was found, that sulfuric trioxide decomposition percentage increases when the activation energy decreases for the both parallel and counter flow cases.

The obtained results can be used for the optimization of the shell and tube high temperature heat exchanger and decomposer design and operating conditions to increase the decomposition performance and minimize pressure drops.

Intermediate Heat Exchanger Dynamic Thermal Response Model

Eugenio Urquiza-Fernández
Per F. Peterson
Ralph Greif

U.C. Berkeley

Sept. 15, 2007

**Abstract**

This report presents UCB progress in developing a comprehensive thermal and fluid dynamics model for the NGNP intermediate heat exchanger (IHX) and other compact heat exchangers. For improved efficiency electricity generation and for nuclear hydrogen applications, an IHX is required to transfer heat from high temperature and high-pressure primary helium coolant to a power plant or hydrogen production process. An intermediate heat transfer loop is used for the purpose. Under these conditions, plate-type heat exchangers with small flow channels, such as the well known Heatric designs, are a major candidates because they can achieve high power densities with small amounts of material, and can be fabricated using a diffusion bonding process so that the entire heat exchanger has the strength of the base material. However, these types of heat exchangers can be susceptible to very large stresses during thermal transients, for example when the flow of one fluid is interrupted abruptly. UCB has proposed a capillary shell and tube IHX configuration that could have lower susceptibility to thermal shock. For all IHX options accurate analysis of global and local thermal stresses are critical to evaluating the heat exchanger reliability and safety.

In order to estimate the stresses in compact heat exchangers a comprehensive thermal and hydraulic model is needed. The model developed here uses an effective porous media (EPM) approach because the evaluation of the detailed global flow with computational fluid dynamics (CFD) as well as finite element methods (FEM) for the mechanical analysis, at the resolution scale of the flow channels involves prohibitive computational time. The EPM fluid dynamics and heat transfer computational code developed at UCB is called the compact heat exchanger thermal and hydraulics (CHEETAH) code. CHEETAH solves for the transient temperature-distribution in the IHX. This temperature distribution can then be imported into a commercial finite element analysis (FEA) code for mechanical stress analysis using the EPM methods developed earlier by UCB for global and local stress analysis [2]. These simulation tools will also allow the designer to optimize the heat exchanger design, to minimize the pressure drop while maximizing the IHX's thermal effectiveness, as well as to optimize the mechanical performance of the IHX particularly as it relates to creep deformation and transient thermal stresses.

## Introduction

The production of hydrogen using high temperature reactors requires intermediate heat exchangers to transfer heat from a primary coolant to an intermediate coolant. The intermediate liquid salt or gas loop acts as a buffer between the nuclear reactor and the hydrogen or chemical plant. This is intrinsically beneficial for overall system safety because by increasing the thermal inertia in the system (especially with LS) the intermediate loop also helps reduce the sensitivity to temperature transients.

An IHX is used to remove heat from the reactor-side of the intermediate loop and a process heat exchanger (PHX) is used to remove the high temperature heat from the intermediate loop. This high quality heat can then be used to produce hydrogen though chemical processes that dissociate water, or to heat steam for high temperature electrolysis. Figure 1 shows a typical Heatric type heat exchanger. Figure 2 shows a schematic of a capillary tube and shell heat exchanger concept developed at UCB intended to reduce the effects of transient thermal stresses [10].

**Figure 1: Photo of a cutaway-model of a typical Heatric heat exchanger showing multiple inlet and outlet manifolds and slices across various plates and flow channels.**

The temperature input into the power plant occurs at the IHX, which makes this temperature critical to the plant's thermodynamic efficiency. Still, the heat exchanger must be able to withstand the rapid temperature changes that can result during a reactor shutdown or interruption of the intermediate coolant flow, as when one of the fluid circulators or pumps suddenly trips off. In scenarios such as these, the thicker sections away for the enhanced heat transfer region (fins and flow channels) will take longer to thermally equilibrate than the thinner heat transfer surfaces. That transition period is critical to the integrity of the IHX because large thermal gradients will occur in areas that also may have internal features that act as stress concentrators. For an offset strip fin HX design,

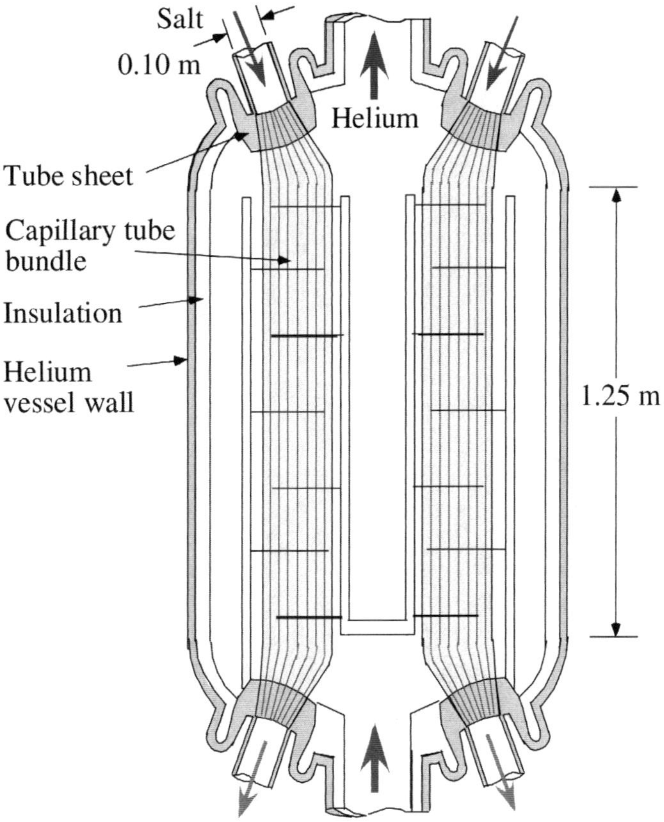


**Figure 1: Capillary tube and shell heat exchanger showing the proposed tube-bundle geometry formed by diffusion bonding of multiple bundles of ~2,500 3.0-mm diameter tubes with hexagonally tapered ends to form inlet and outlet tube sheets [10].**

It is evident from some of the sharp angles near the manifolds in Fig. 1 that the relatively elaborate geometry in the manifold region can lead to a very complex stress distribution over the IHX that will vary in time as the different sections thermally equilibrate during a thermal transient. Again, attempting to model this fluid, temperature and stress distribution directly at the channel level for the entire heat exchanger is computationally prohibitive because of the different length scales that must be resolved. Still, it is critical to be able to predict approximate stresses over the entire IHX (e.g, a global model). This requirement led the current study towards the EPM approach.

Previous work at UC Berkeley [2] developed an alternative method to obtain approximate stresses that greatly reduces computational effort. This method is composed of three steps. First, the heat exchanger is broken down into several regions. Unit cell models are built based on each region that captures all of the most important features of that region. The effective mechanical and thermal properties for each unit cell are then determined using FEM simulations. Second, global stress distributions based on the volume-averaged unit-cell-equivalent model are computed by using the effective mechanical and thermal properties. Third, the values from the stress distributions are then applied to the unit cells to find localized points of high stress.

The fluid dynamics component of the model was first developed for the offset strip fin (OSF) region in only one dimension and with a constant permeability. The model can also apply to other flow channel geometries. Later, spatially varying permeabilities were included as well as areas with no permeability to represent no flow boundaries and the solid portions of the IHX. Analysis for both high-pressure helium and liquid salts has been performed. The small characteristic length (hydraulic diameter) of the region and in the manifolds in addition to the creeping flow of the liquid salt produces a Darcian flow regime in the liquid salt plate. Furthermore, the incompressibility of the liquid salt greatly simplifies the fluid mechanics. This is less so the case on the helium side of the IHX where the flow rates are much higher and where the compressibility of the gas could play a small role in the transient fluid dynamics. Established friction factor correlations can be manipulated to produce an effective permeability (which also serves as a correction factor as the fluid dynamics deviate from the Darcian regime) and Darcy's transport equation again relates the fluid velocity to the pressure drop. This is discussed in more detail in the fluid mechanics section of this report.

The heat transfer part of the problem has been developed progressively from one dimensional models that neglected longitudinal (main flow direction [x]) conduction to two dimensional models that include cross flow in the fluids as well as spatially varying longitudinal and latitudinal (cross flow direction [z]) conduction. Similarly, early models used constant convection coefficients and porosity while the more recent version includes spatially varying convection coefficients and porosity. Eventually, the CHEETAH model may accommodate convection coefficients that vary with the local fluid velocity.

## Media Description

The IHX can be made of one of several high temperature metallic alloys the way current Heatric HX are currently manufactured but UCB has also done extensive work to explore the possibility of making a ceramic IHX through a molding process. While the ceramic IHX would have a lower conductivity and would be less compliant under thermal stresses, it would able to operate at much higher temperatures without creep deformation, increasing plant efficiency and producing heat at temperatures necessary for chemical hydrogen production. These silicon-carbide based ceramic materials are also the primary candidates for use in the process heat exchangers.

In modeling the IHX, the actual material properties can be set and changed with great ease since they are simply represented by a series of constants assigned in the input file of the code. However, the fluidic, thermal and mechanical response of the solid media comes from geometric effects that are central to the model's validity. This is where the EPM model simplifies the problem greatly by creating mass and energy balances at a scale that will both represent the complexity of the thermal hydraulics in the IHX and yet produce simulation results in a time period that allows researchers to improve the IHX design from a fluidic, thermal and mechanical perspective through several design iterations and computational simulations. Calculating representative constants that correspond to effective x-dir conductivity, z-dir conductivity, x-dir permeability, z-dir permeability, porosity, convection coefficients for LS and/or He plates, elastic modulii, shear modulii & Poisson's ratios is very labor intensive because it takes over 20 unique zones with different media properties to fully describe the hydraulic, thermal and mechanical properties of a typical compact heat exchanger. This is done with various scripts within the CHEETAH code.

## Zoning the IHX

The assignment of constants to the property matrices in the CHEETAH code is a cumbersome but essential part of the EPM model. One central purpose of the CHEETAH code is to enable the designer to simulate the performance of the IHX with a given geometry and set of dimensions so that the exchanger design can be improved in an iterative fashion based on the simulation's results. This means that the IHX's dimensions are very much subject to change while the general shape changes little.

With this in mind, the CHEETAH code is designed with a long set of system parameters that make changing the dimensions on the IHX model as easy as changing the values of a few constants that represent those dimensions and system parameters. Then, the CHEETAH code will automatically change the zone boundaries so that the user doesn't have to modify the code with every dimensional variation. Modification of the code may be necessary if any of the dimensions approach the grid size or if the shape of the IHX is modified. A schematic drawing of the IHX's porosity distribution is included in Figure 3 to illustrate some of these zones with different colors.

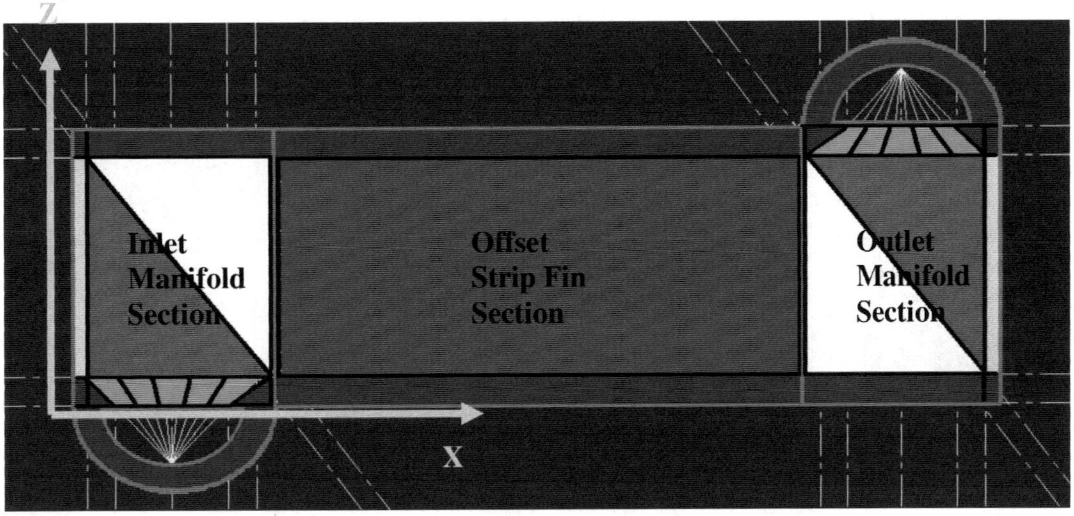


**Figure 3: A schematic of the IHX's porosity distribution by zones**

For example, the zones shown in purple in Figure 3, above, represent a porosity of zero ($\phi$=0) which means that these zones will have a larger thermal diffusion length scale and all else being equal will take longer to equilibrate during thermal transients.

*Diffuser and Reducer Permeability*

The schematic drawing above also is simplified in the diffuser and reducer areas (trapazoids shown in orange). In these zones every control volume has unique effective properties assigned to it depending on the local channel geometry being represented by the EPM. In order to accurately represent the local geometry the assigned EPM uses the angle and spacing of the channels nearest the node in the actual IHX. The CHEETAH code accomplishes this by actually finding the position of each node and the slope of the line connecting the node point to a central fictitious point to which the flow distribution channels align. Using the point and slope information for each node and the line equation from adjacent channels the CHEETAH code finds the normal distance between channels at that node point and uses this to calculate a multiplier for the directional permeability so that the effective x and z-dir permeabilities at that node take into consideration not only the alignment of the channel but also the local channel density. These zones are shown in a uniform color here for simplicity but each is also includes other parameters that allow the CHEETAH code user to manipulate the flow distribution. This will be discussed further in the next section. Schematic drawings similar to Fig. 3 would illustrate the layout of other effective property values such as x-dir conductivity, z-dir conductivity, x-dir permeability, z-dir permeability, porosity, convection coefficients for LS and He, elastic modulii, shear modulii and Poisson's ratios the IHX.

*Adjustable Flow Distribution*

In addition to accommodating the channel orientation and channel density in the nodal permeability assignment the CHEETAH code also allows the designer to manually tune a multiplier coefficient for distinct zones in the manifold sections. The manifolds play a key role in creating uniform flow in the OSF region which in turn is crucial to maximizing the effectiveness of the IHX. The adjustable permeability in these zones allows the user to quickly and directly manipulate the flow distribution in the OSF region as if controlling independent and parallel valves leading into and out of the OSF section. Shown in yellow, orange, pink, red and maroon, these flow regions can be seen in the inlet and outlet manifolds shown below of the liquid plate schematic drawing in Fig. 4.

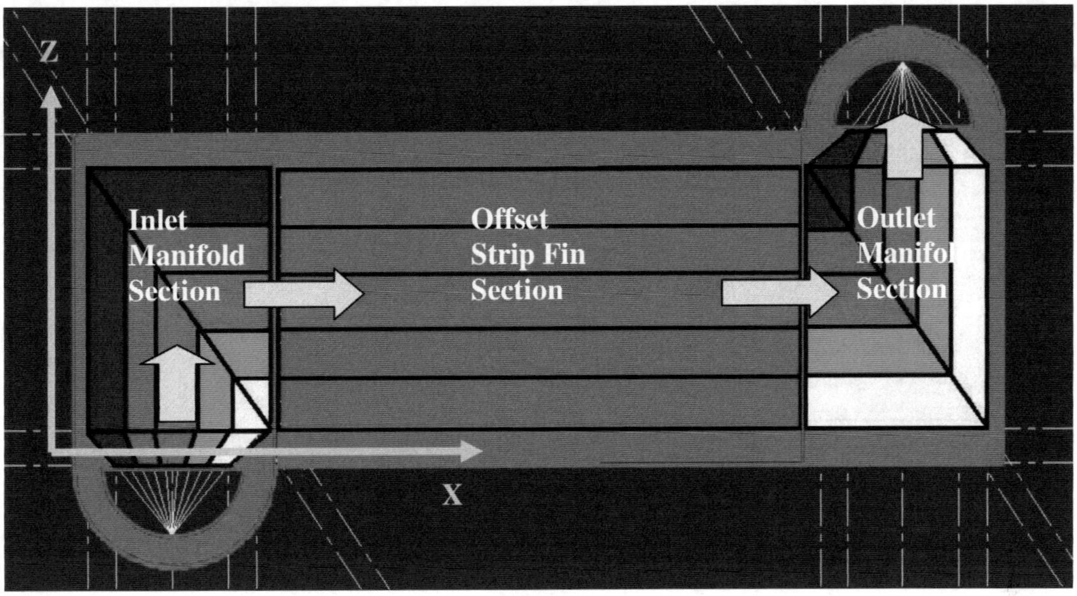


**Figure 4: Independent flow regions in the inlet and outlet manifolds**

The lower flow resistance associated with the shortest flow path in the inlet manifold is already balanced by a higher flow resistance created by the corresponding longer flow path in the outlet manifold so that the total flow resistance is roughly equal regardless of the path followed. However, there are still issues with flow maldistribution in the IHX. Among the factors that affect the flow distribution are the temperature dependant viscosity effects (not yet incorporated into CHEETAH), longer flow paths associated with the diffuser and reducer path as well as the fact that the paths nearest the inner corners of each region are shortest and therefore see higher flow rates than the outside corners in each flow region.

Each manifold (which includes the diffuser and reducer section) is separated into several flow regions. The number of flow regions is controlled by the assignment of a constant (chanmax) in the code. The manifolds can be divided into any number of flow regions but it is important to maintain a region width of at least 3 cells across to avoid discretization errors such as "staircasing", where the coarse mesh creates problems of unintended reductions in flow area and therefore peak velocities that can lead to instabilities by way of excessive CFL (Courant – Friedrichs – Levy = u* t/ x) numbers.

*Specifying the Grid*

The CHEETAH Code uses a rectangular mesh which fits the rectangular geometry of the IHX and with the rectangular matrices produced in MatLab. Originally the code was written with a typical rectangular grid such that each node point lies in the center of the control surface and was identified by four coordinates (x – position, y – position, z – position, iteration number) with the following indexes (i,j,k,n). The y-position ("j" index) in the rectangular grid connotes the phase with index 1 and 3 being the fluids and 2 always representing the solid phase. The CHEETAH Code was later improved by replacing the simple rectangular grid with a more sophisticated staggered grid.

The x and z – position coordinates associated with every node when multiplied by the corresponding spatial discretization ( x & z, respectively), provide a position for the node point in the IHX geometry. That position then led to the assignment of the corresponding local physical properties such as porosity, x & z permeability, x & z conductivity, convection coefficients, pressure and temperature according to the zoning described previously.

*Rectangular Grid*

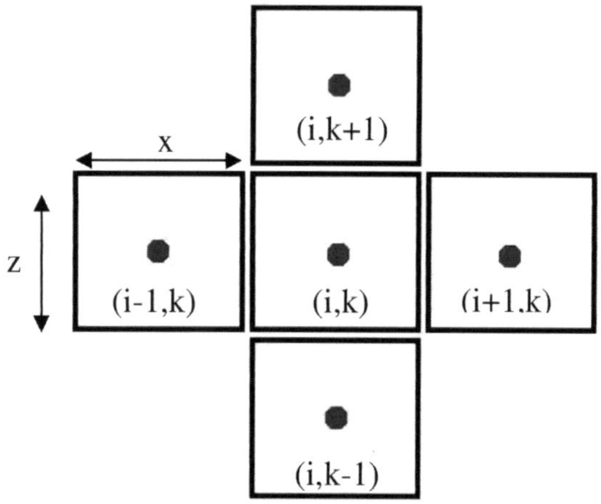


**Figure 5: Rectagular Grid**

In this method of discretization all properties are assigned to the center of the control volume and mass conservation is checked at (i,k) by applying the following equation:

$$u(i-1,k)-u(i+1,k)+w(i,k-1)-w(i,k+1)=0$$

While simple, this type of discretization actually uses a control volume 4 times larger than the actual grid spacing. This is because for continuity the velocities must be summed at the boundary of each cell but since the velocities are defined at the center of each cell the code must calculate velocities from neighboring cells or must find an average velocity between to cells to approximate a boundary velocity. If neighboring cell velocities are used directly then this creates a much coarser grid than intended and significantly diminishes the accuracy of the results. If an average velocity is found then the computation time is increased. Furthermore, this method can also lend itself to "zigzag" or "checkerboard" pressure or temperature distributions since the continuity and momentum equations use information from alternating nodes to check for continuity. Essentially, this means that a wavy pressure distribution will produce a wavy velocity distribution that can satisfy continuity since it is information from alternating and not adjacent nodes that must satisfy continuity [4].

*Staggered Grid*

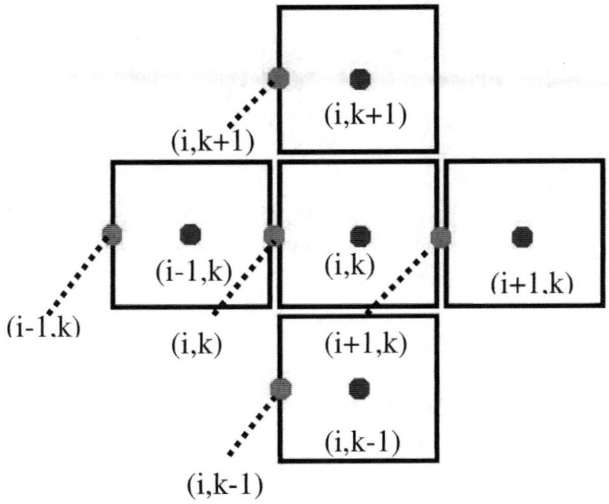


**Figure 6: Staggered Grid in x-dir**

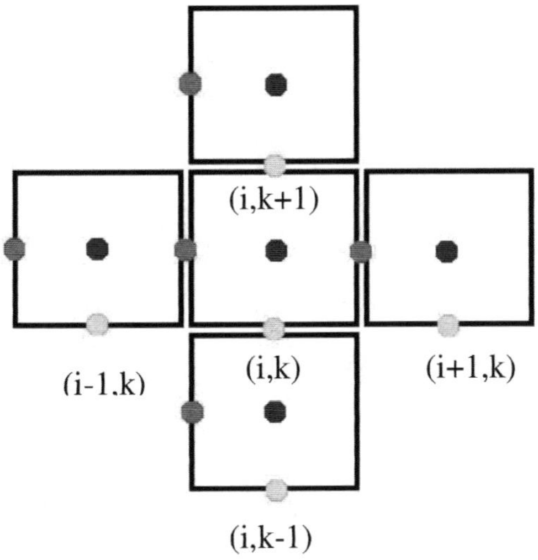


**Figure 7: Staggered Grid in x & z-dir**

An alternative to the rectangular grid is the staggered grid. This method involves using several matrices to describe the physical properties of the control volumes. Those properties that are defined at the center of the control volume (blue nodes) such as temperature, pressure, and porosity are still defined there, as before. However, those properties that are best defined at the left and right boundaries such as x-permeabilities, x-conductivities and x- velocities are defined on a new matrix that is shifted to the left by half grid space (red nodes) so that they lie right on the x boundary.

The same is true for the properties that must be accounted for on the upper and lower boundaries of the control volume. This refers to the z-permeabilities, z-conductivities and z-velocities which are defined on a new matrix that is shifted to the down by half-a-grid space (green nodes) so that they lie right on the x boundary.

By using the staggered grid, the code is able to incorporate a much finer mesh and is able to ensure continuity at every control volume using velocity components from adjacent nodes. Also, the pressure drop between adjacent cells becomes the natural driving force for the velocity that is calculated between them at the face of the control volume of the adjacent cell [4].

### Fluid Dynamics

The fluid mechanics in the IHX are solved for both fluid phases by breaking up the IHX into over 20 flow zones. Each of these zones has unique physical properties that govern the flow for each of the two phases. For the case of an IHX that transfers heat between a liquid and a gas, it is representative to take the case of liquid salt and helium. Figures 8 and 9 show the zoning for these two plates, respectively.

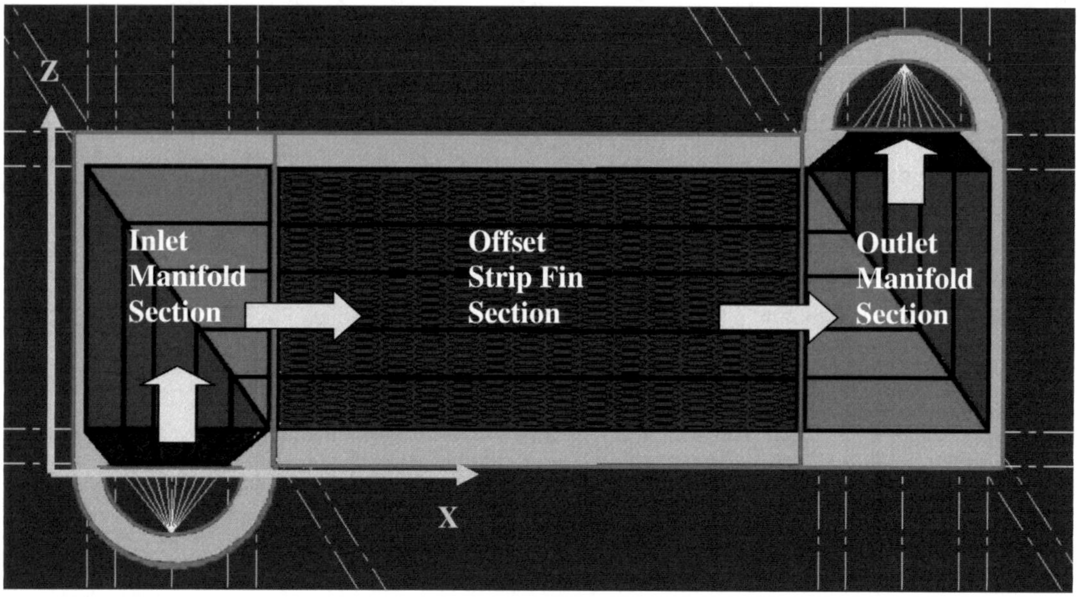


**Figure. 8:** Schematic of the IHX section for a LS plate

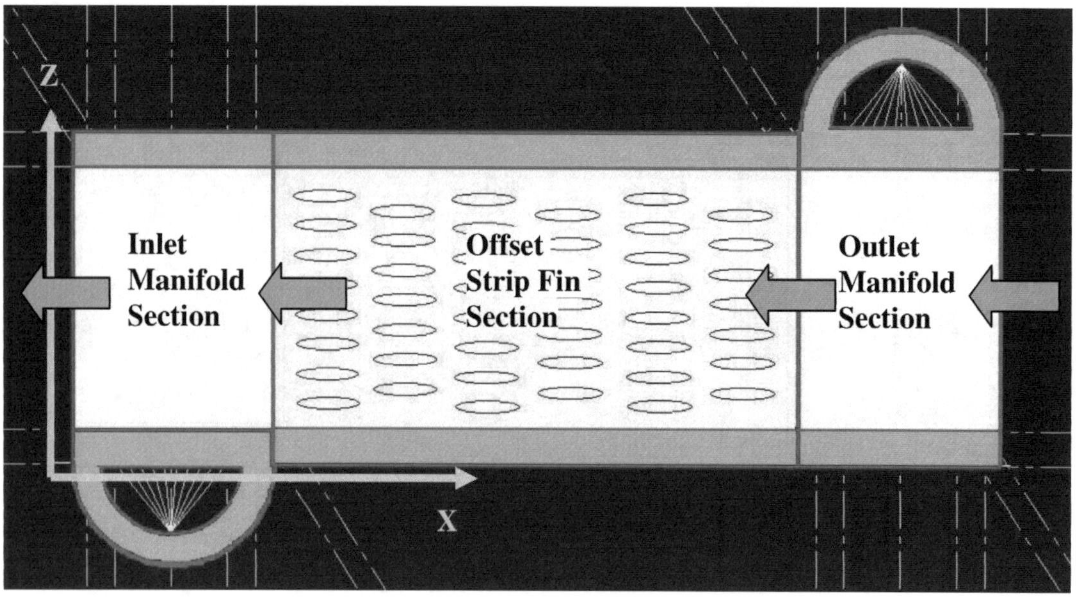


**Figure. 9:** Schematic of the IHX section for a He plate

The IHX is built from a diffusion bonded stack of alternating gas and liquid (or gas) plates as shown in Figure 10. Due to the stacked assembly, the flow areas on the liquid and gas plate overlap all over the IHX and create zones in the solid material with complex thermal properties. The zones in the liquid plate are completely separate form the zones in the gas plate, so the fact that the zones overlap has no effect on the fluid mechanics. The overlap intimately affects the heat transfer, however.

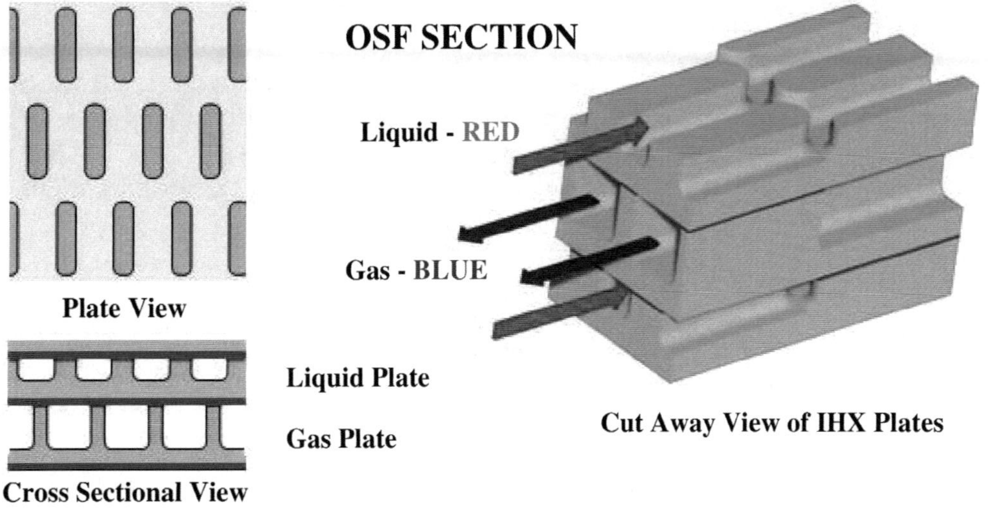


**Figure 10**: Cut away view through the OSF section showing alternating liquid salt (top and bottom arrows) and helium (middle arrows) flow channels. Dark bands at the top of each fin indicate the location of joints between each plate [1].

*Fluid Dynamics Equations*

Beginning with the equation of continuity it is known that mass must be conserved in any representative control volume in the field of flow such that:

$$\frac{\partial u}{\partial x}+\frac{\partial w}{\partial z}=0 \qquad \text{Equation 1}$$

Then, it is known that for very small Reynold's number flows, Darcy's transport equation yields excellent accuracy with a medium dependant permeability. For laminar flows with a Reynold's number larger than 1 the permeability coefficient can be adjusted so that it becomes a function of the Reynold's number. This is done by solving for the Fanning friction factor, $\boldsymbol{f_f}$ and using this to find the effective permeability, $\mathbf{k_x}$ & $\mathbf{k_z}$.

Darcy's transport equation for 1D flow in porous media is

$$u=u_{\text{int}}\cdot\phi_x=-\frac{k_x\rho g}{\mu}\frac{dh}{dx} \quad \text{or} \quad u=-\frac{k_x}{\mu}\frac{d\Phi}{dx} \qquad \text{Equation 2a}$$

$$w=w_{\text{int}}\cdot\phi_z=-\frac{k_z\rho g}{\mu}\frac{dh}{dz} \quad \text{or} \quad w=-\frac{k_z}{\mu}\frac{d\Phi}{dz} \qquad \text{Equation 2b}$$

Here, $\Phi$ is the flow potential [Pa] and u is the Darcy velocity.

The following section will describe how to obtain the effective permeability from correlations for the Fanning friction factor and also will detail some of the advantages in using this approach. In order to find the velocity distribution in the IHX it is necessary to first find the pressure distribution. The following equation shows one way to accomplish this.

The partial derivative of the x component of the fluid velocity with spatially varying permeability **k**, and dynamic viscosity **μ**, gives due to the spatially varying viscosity and permeability:

$$\frac{\partial u}{\partial x} = -\frac{1}{\mu}\frac{\partial k}{\partial x}\frac{\partial \Phi}{\partial x} + \frac{k}{\mu^2}\frac{\partial \mu}{\partial x}\frac{\partial \Phi}{\partial x} - \frac{k}{\mu}\frac{\partial^2 \Phi}{\partial x^2} \qquad \text{Equation 3}$$

Combining equations 1 & 3 gives the following elliptic differential equation:

$$-\frac{1}{\mu}\frac{\partial k_x}{\partial x}\frac{\partial \Phi}{\partial x} + \frac{k_x}{\mu^2}\frac{\partial \mu}{\partial x}\frac{\partial \Phi}{\partial x} - \frac{k_x}{\mu}\frac{\partial^2 \Phi}{\partial x^2} - \frac{1}{\mu}\frac{\partial k_z}{\partial z}\frac{\partial \Phi}{\partial z} + \frac{k_z}{\mu^2}\frac{\partial \mu}{\partial z}\frac{\partial \Phi}{\partial z} - \frac{k_z}{\mu}\frac{\partial^2 \Phi}{\partial z^2} = 0 \quad \text{Equation 4}$$

Initially, this equation was solved only within certain zones with constant x & z permeability, under the assumption of constant fluid properties such as a constant dynamic viscosity, **μ**. For this case the Equation 4 is reduced to the following elliptic equation:

$$\frac{k_x}{\mu}\frac{\partial^2 \Phi}{\partial x^2} + \frac{k_z}{\mu}\frac{\partial^2 \Phi}{\partial z^2} = 0 \qquad \text{Equation 5}$$

This equation was solved with appropriate boundary conditions using finite differencing with Taylor-series approximations for each phase and it provided the steady state pressure fields for each fluid. This pressure field readily gave the velocity field for the areas of constant properties. However, while this set of assumptions worked well within the OSF portion of the IHX due to its constant permeability, the assumptions were clearly invalid in the inlet and outlet manifolds where the geometry changes frequently and abruptly.

In order to estimate the velocity field in these areas the more complicated Equation 4, above, had to be solved. However, when it too was approached using the finite difference scheme with Taylor-series approximations, large instabilities resulted in the zones where the permeability varies from node to node such as in the diffuser, reducer and also at the interface between two zones of different permeability such as between the OSF region and the inlet and outlet manifolds. It seemed the instabilities were coming from the first derivatives of permeability with respect to space, $\frac{\partial k_x}{\partial x}$ and $\frac{\partial k_z}{\partial z}$

*Fluid Dynamics Discretization & Results*

This obstacle was overcome by approaching the problem with finite volume analysis (FVA) or control volume approach. In this approach the conservation equation is applied to a discrete area, volume or grid from the beginning, see Figure 6 below. The velocities come from the pressure relationships given in Darcy's transport equation in Equation 2, above. Unlike the Taylor-series based approaches from before, the FVA does not take the limit by reducing the control volume down to a point. This provides a significantly more stable solution since the conservation equation is asserted on each discrete area or volume in the grid [6].

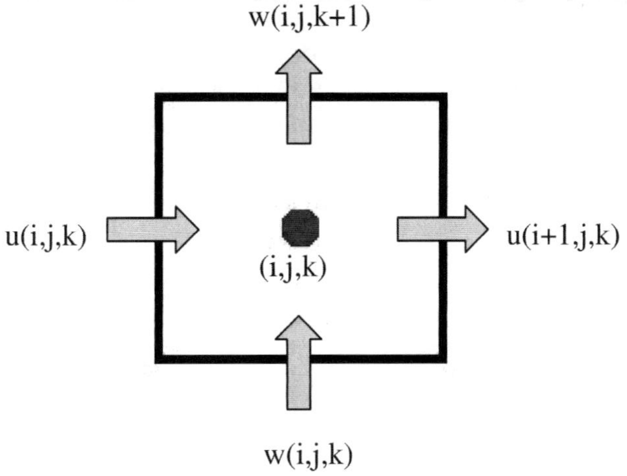


**Figure 11:** Finite volume analysis or control volume approach for fluid dynamics solution.

An iterative FVA technique was used to solve for the pressure distribution in the liquid salt plate. Figure 12 shows the resulting steady state pressure distribution in the IHX.

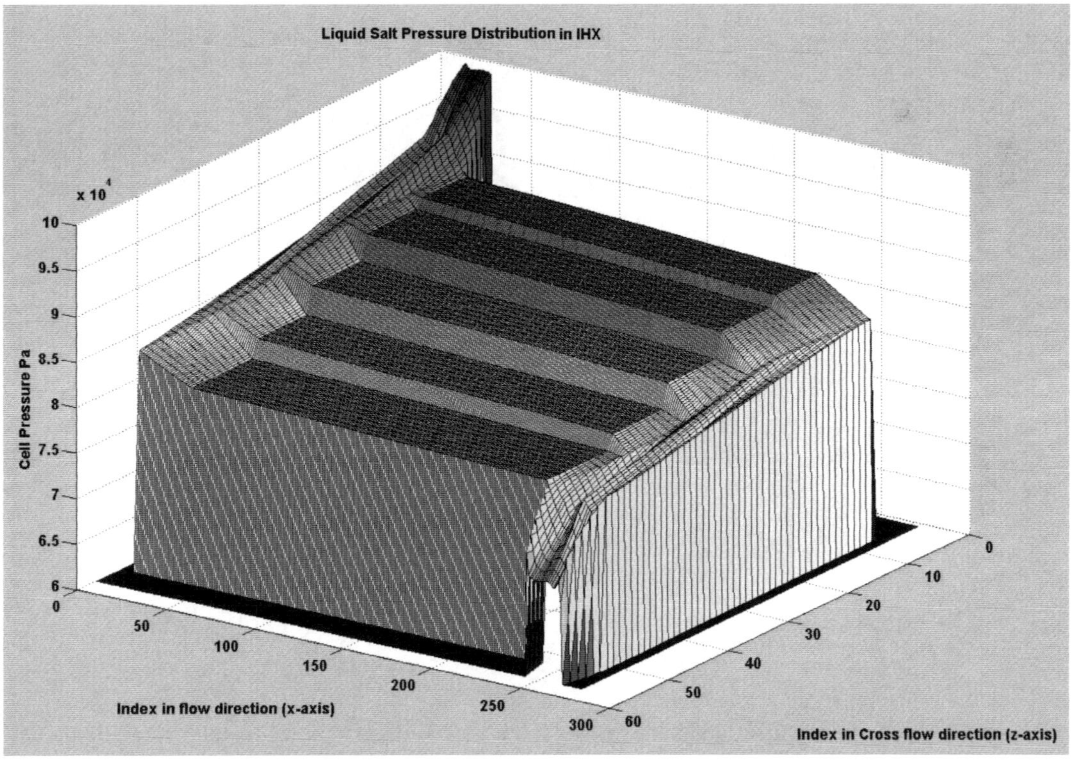

**Figure 12:** Liquid salt plate pressure distribution solved with the finite volume analysis or control volume approach.

Clearly the pressure gradient varies greatly in the diffuser and reducer, as expected. Also the gradient appears flat in the core of the IHX where the permeabilities are constant. Sharp changes in the pressure gradient can be seen where the OSF sections meet the inlet and outlet manifold. This identifies the interface without the instabilities seen with the Taylor-series based finite difference techniques. Outside the flow zones, the pressure gradient is flat and the pressure is set at a lower constant to indicate that this is not a flow region.

Using the pressure distribution solved above and the permeability matrix assigned previously, Equation 2 can give a velocity field in the x and z-direction. Summing the x and z-direction at every mesh provides a non-directional total flow distribution in the liquid plate of the IHX. This flow field can be seen in Figure 13.

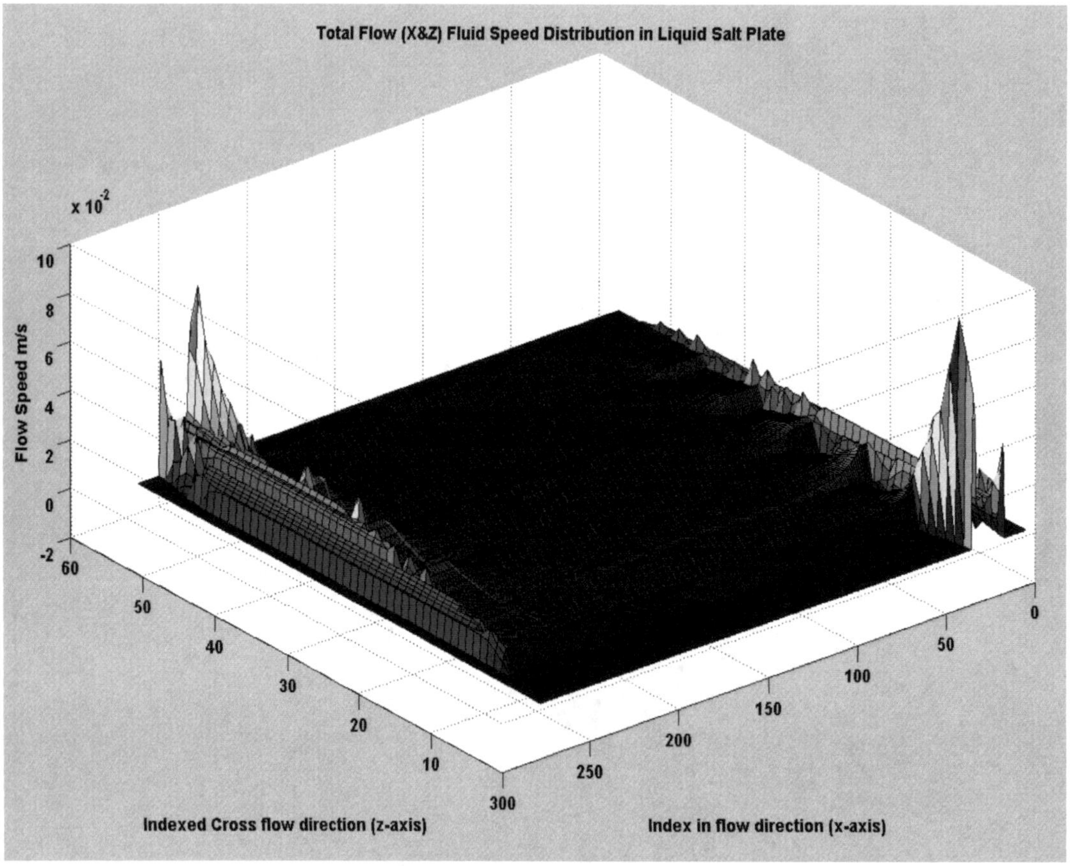


**Figure 13:** Liquid salt plate flow distribution solved with the finite volume analysis.

Notice the increased flow rates in the diffuser and reducer (at the inlet and outlet). As expected in the diffuser (right side of Figure 13) as the cross section increases the flow rate drops sharply and the flow rate accelerates as the cross section is reduced in the nozzle-like reducer (left side of Figure 13). The same effect can be observed with the rise in the flow rate in the z-dir oriented flow channels of the inlet and outlet manifold. Furthermore, at the entrance to the OSF region the liquid can be seen to increase in velocity close to the near-side corner of each flow section, as it flows preferentially through the path of least resistance (in this case the shortest path being just around the nearest the corner).

Following the same procedure for the helium plate Figure 14, below, shows the steady state pressure distribution for the helium plate in the IHX.

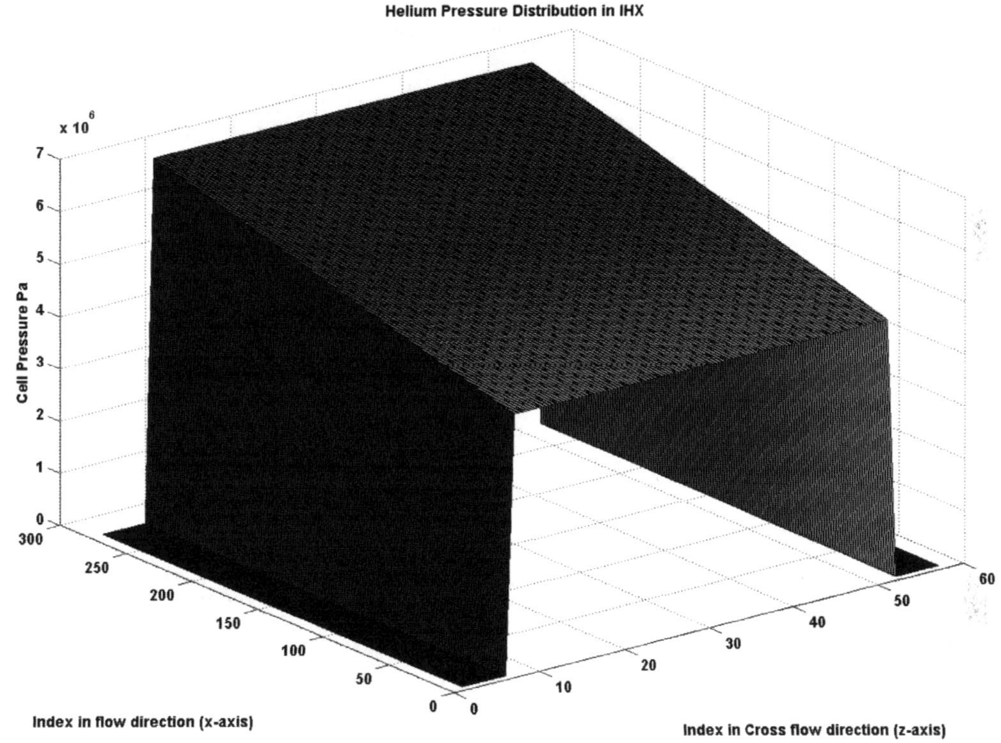


**Figure 14:** Helium plate pressure distribution solved with finite volume analysis.

The simple fluid dynamics of the helium plate are evident in the above pressure distribution where the permeabilities in the helium-side OSF region are constants in both x and z directions.

### Effective Permeability

For this model it is important to show that the flow has reached a hydrodynamically fully developed condition. The offset fin studies of Sparrow et al. and Kelkar and Patankar show that for Graetz number, Gz< 200 the flow has effectively reached a periodic hydrodynamics fully developed condition [7].

$$Gz = \frac{Re * D_h}{x} \qquad \text{Equation 6}$$

For the IHX the hydrodynamic entry length is a fraction of a percent of the length of the core section.

For 1D fully developed laminar flow in a pipe the Navier-Stokes equations of motion reduce to:

$$\bar{u} = \frac{-D^2}{32\mu}\frac{d\Phi}{dx} \quad \text{or} \quad \bar{u}^2 = \bar{u} * \frac{-D^2}{32\mu}\frac{d\Phi}{dx} = -Re\frac{D}{32\rho}\frac{d\Phi}{dx} \qquad \text{Equation 7}$$

Here it can be clearly seen that $\frac{D^2}{32}$ serves as the geometry dependant correction factor similar to the permeability which has the same units.

The Fanning friction factor for laminar flow in a pipe is

$$f_f = \frac{16}{Re} \qquad \text{Equation 8}$$

Combining Equations 7 & 8 yields

$$\bar{u}^2 = -\frac{D}{2f_f\rho}\frac{d\Phi}{dx} \qquad \text{Equation 9}$$

and dividing through by the average fluid velocity gives the following equation for flow in a pipe.

$$\bar{u} = -\frac{D\mu}{2f_f\rho\bar{u}}\frac{1}{\mu}\frac{d\Phi}{dx} \qquad \text{Equation 10}$$

Applying this to the effective porous media which also concerns laminar fully developed flow and applying the hydraulic diameter D becomes $D_h$ and $\bar{u}$ (average velocity in a pipe or analogous to the interstitial velocity in the effective porous medium). Then $u_D$ (Darcy velocity) divided by the directional porosity is equivalent to the interstitial velocity, as follows:

$$\bar{u} = u_{int} = \frac{u_D}{\phi_x} \qquad \text{Equation 11}$$

This in turn allows Equation 10 to be re-written as

$$u_D = -\frac{D_h\mu\phi_x^2}{2f_f\rho u_D}\frac{1}{\mu}\frac{d\Phi}{dx} \qquad \text{Equation 12}$$

This finally determines the relationship between the Fanning friction factor and the effective permeability as:

$$k_x = \frac{D_h \mu \phi_x^{\ 2}}{2 f_f \rho u_D}$$ Equation 13

This result plays a very important role in the implementation of the fluid dynamics of the CHEETAH Code. First, it allows the script to discretize Darcy's transport equation in its simplest form as seen in Equation 2. It is important to note that the permeability is now a function of velocity, but it is equally important to note that the Fanning friction factor is a function of velocity as well through the Reynolds number. At very low flow rates (Re<1) the velocity dependence will be eliminated.

A real advantage of this technique is the fact that the fluid potential distribution is found with the same equation in core flow zones, the boundary conditions and the no-flow areas provided the permeability distribution is correctly specified. This also implies that the best correlations giving the Fanning friction factor for any complex geometry in laminar fully developed flow can be used to generate an effective permeability which will accommodate the flow rates in the range of validity of the Fanning friction factor correlation.

From a programming standpoint this makes the CHEETAH Code very versatile and easy to adjust for different fin arrangements and geometries or dimensions because only the Fanning friction factor correlation in the code must be changed. This is a very significant advantage versus the alternative of discretizing the correlation itself. If the correlation was discretized using the Fanning friction factor directly, while the solution would be the same, this would make changing the code to reflect dimensional or geometric changes very cumbersome because this momentum equation appears all over the code with a few adjustments for the boundaries and zones. In short, the effective porous media (EPM) treatment is very advantageous!

## Heat Transfer

After the fluids module resolves the velocity distribution in the axial (x) and cross-flow (z) directions, the thermal module of the IHX code applies an energy balance using a similar finite volume analysis (FVA) or control volume analysis as discussed previously in the fluid mechanics sections. Here too, the differential equations stemming from the energy equation were initially solved using the FDM with Taylor-series approximations until the CHEETAH code encountered instabilities when modeling the consolidated IHX, which includes many zones with properties that vary spatially and abruptly. In fact, the differential equations below result from the assumption that only temperature varies spatially and that variables such as velocity, density, specific heat, conductivity, convection coefficient, and porosity are all spatially independent. Clearly, this is not the real case, but still the following equations are valid within the independent zones of the IHX where it is reasonable to assume the above mentioned properties as constant and the FDM was used to solve this system of equations before the development of the consolidated CHEETAH code. The equations are also included here because they illustrate the dominant physical phenomena that govern the energy transport in the IHX for each phase.

*Heat Transfer Equations*

Hot Fluid (fh) Equation 14

$$-u_{fh}\rho_{fh}c_{pfh}\frac{\partial T_{fh}}{\partial x}-w_{fh}\rho_{fh}c_{pfh}\frac{dT_{fh}}{dz}+k_{fh}a'_{fhx}\frac{\partial^2 T_{fh}}{\partial x^2}+k_{fh}a'_{fhz}\frac{\partial^2 T_{fh}}{\partial z^2}-h_{fhs}a'_{fhs}(T_{fh}-T_s)=\phi_{fh}\rho_{fh}c_{pfh}\frac{\partial T_{fh}}{\partial t}$$

An important dimensionless number to consider here is the Peclet number, which provides the ratio of the advective energy transport (thermal energy transport due to mass transport) and the conductive thermal energy transport. The Peclet number is defined as:

$$Pe=\frac{u*\Delta x}{\alpha}=\frac{u*\Delta x*\rho*c_p}{k}\quad \text{Equation 15}$$

The Peclet number for the helium flow rate is roughly 7700. This clearly means that the heat is mostly due to mass transport while the conductive contribution is negligible.

Neglecting Conduction in the Hot Fluid the energy balance is reduced to:

$$-u_{fh}\rho_{fh}c_{pfh}\frac{\partial T_{fh}}{\partial x}-w_{fh}\rho_{fh}c_{pfh}\frac{dT_{fh}}{dz}-h_{fhs}a'_{fhs}(T_{fh}-T_s)=\phi_{fh}\rho_{fh}c_{pfh}\frac{\partial T_{fh}}{\partial t}\quad \text{Equation 16}$$

Solid (s):

$$h_{fhs}a'_{fhs}(T_{fh}-T_s)-h_{sfc}a'_{sfc}(T_s-T_{fc})+k_s a'_k\frac{\partial^2 T_s}{\partial x^2}+k_s a'_k\frac{\partial^2 T_s}{\partial z^2}=\phi_s\rho_s c_{ps}\frac{\partial T_s}{\partial t}\quad \text{Equation 17}$$

Cold Fluid (fc) Equation 18

$$-u_{fc}\rho_{fc}c_{pfc}\frac{\partial T_{fc}}{\partial x}-w_{fc}\rho_{fc}c_{pfc}\frac{\partial T_{fc}}{\partial z}+k_{fc}a'_{fcx}\frac{\partial^2 T_{fc}}{\partial x^2}+k_{fc}a'_{fcz}\frac{\partial^2 T_{fc}}{\partial z^2}+h_{sfc}a'_{sfc}(T_s-T_{fc})=\phi_{fc}\rho_{fc}c_{pfc}\frac{\partial T_{fc}}{\partial t}$$

Again, the Peclet number for the liquid salt flow rate is roughly 7500, so the conduction effect is negligible.

Neglecting conduction in the cold fluid the energy balance is reduced to:

$$-u_{fc}\rho_{fc}c_{pfc}\frac{\partial T_{fc}}{\partial x}-w_{fc}\rho_{fc}c_{pfc}\frac{\partial T_{fc}}{\partial z}+h_{sfc}a'_{sfc}(T_s-T_{fc})=\phi_{fc}\rho_{fc}c_{pfc}\frac{\partial T_{fc}}{\partial t}\quad \text{Equation 19}$$

Lastly, it is important to clarify that the CHEETAH code solves the energy equations at every temperature node by summing advective contributions, for the fluid phases, and conductive contributions, for solid phase, at the boundaries of the control volume, x y z, just as was executed for the fluid mechanics portion. This is combined with convective heat transfer between the phases to give the increase or decrease in sensible thermal energy for each control volume which manifests itself as a change in temperature as illustrated in Figures 15 and 16

*Heat Transfer Discretization*

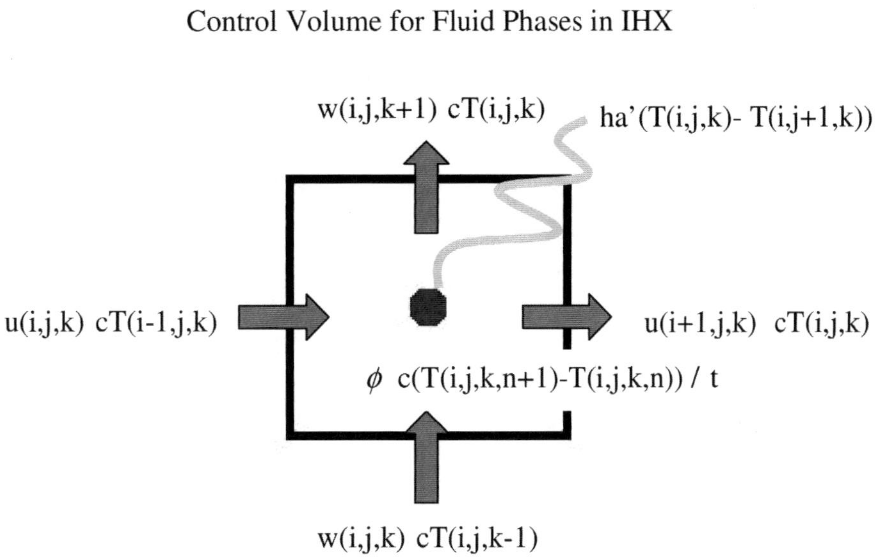


**Figures 15**: Control volume for either the hot or cold fluid as approached via FVA

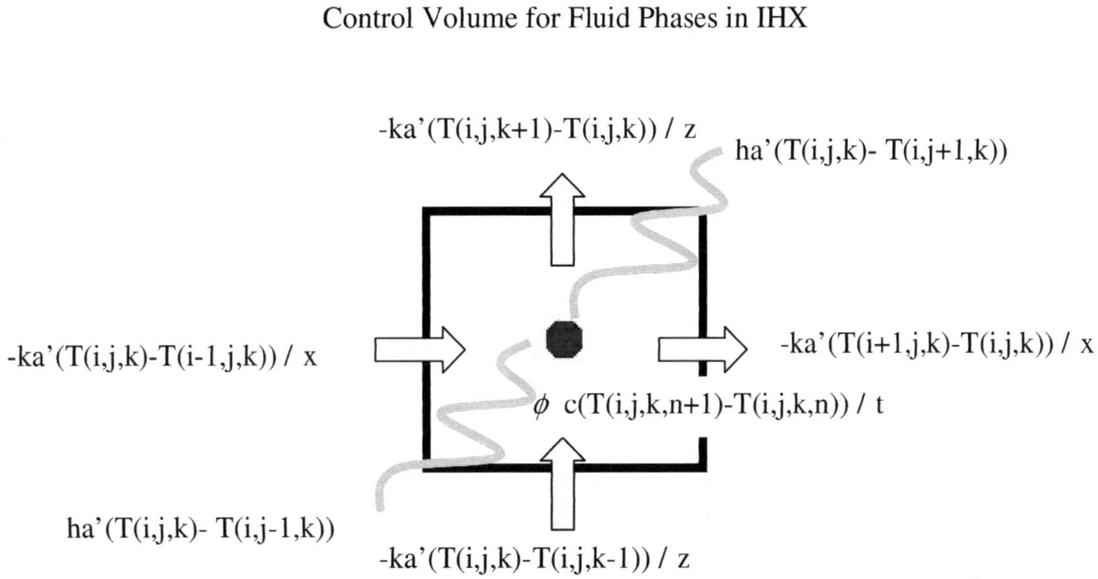

**Figure 16**: Control volume for either the solid phase as approached via FVA

**Preliminary Heat Transfer Results and Conclusions**

The main concern with the IHX is the mechanical performance during rapid thermal transients. The CHEETAH code has been developed to model the transient temperature distribution during such events. In modeling this scenario, it is assumed that the power plant is operating at steady state at full power when a transient event takes place. Among other things, these scenarios may include an almost instantaneous tripping of one of the coolant pumps. From a thermal stress perspective the most likely, relatively severe thermal transient would involve either the primary or intermediate coolant pump tripping while the other continues to operate. The thermal response for this scenario was recently modeled by CHEETAH, and the results for both a liquid salt and helium pump trip are provided in Figure 17.

The bottom center of each graph in Fig. 17 shows the temperature outline from the diffuser in the inlet manifold of the liquid salt plate at the low inlet temperature. Just above it the reducer from the exit manifold of the liquid salt plate can be seen at a much higher outlet temperature. In this case the liquid salt is the cold fluid and it enters the IHX at 900K while helium is the hot fluid which enters the IHX at 1200K. For this reason, the vertical axis of the graphs spans from 850K to 1250K. In this study the OSF section of the IHX is shown to be 1 meter long with an inlet and outlet manifold which are each 60cm long. The IHX is 57cm wide (z-dir) while the diffuser and reducer only stretch 6cm from the opening. While this is a short distance for the diffuser and reducer, it also makes the bounding wall along the length of the OSF region 6 cm thick. This is the region that also appears in purple along the OSF region in Figure 3. This creates a relatively large thermal diffusion length scale, which is evident in observing the delay in thermal response in this region in Figure 17. Even after 50 seconds those bounding walls have not equilibrated with the OSF section and that delay causes very large thermal gradients early in the temperature transients.

This observation comes as a direct result of graphics and animations produced from the CHEETAH code and the information learned from the results that it provides incentive to refocus the attention of the IHX designer towards the most vulnerable areas of the design. Through an iterative process such as this one the IHX designers can quickly optimize the design and update CHEETAH to reflect the new dimensions or shape of the IHX. Eventually, when no obvious problems arise, the temperature distribution can be exported to a mechanical FEA software (likely ANSYS) for detailed thermal/mechanical stress analysis using effective mechanical properties to calculate strains and to then use these strains to back calculate the local stresses as demonstrated in previous work using effective mechanical properties at UCB [1].

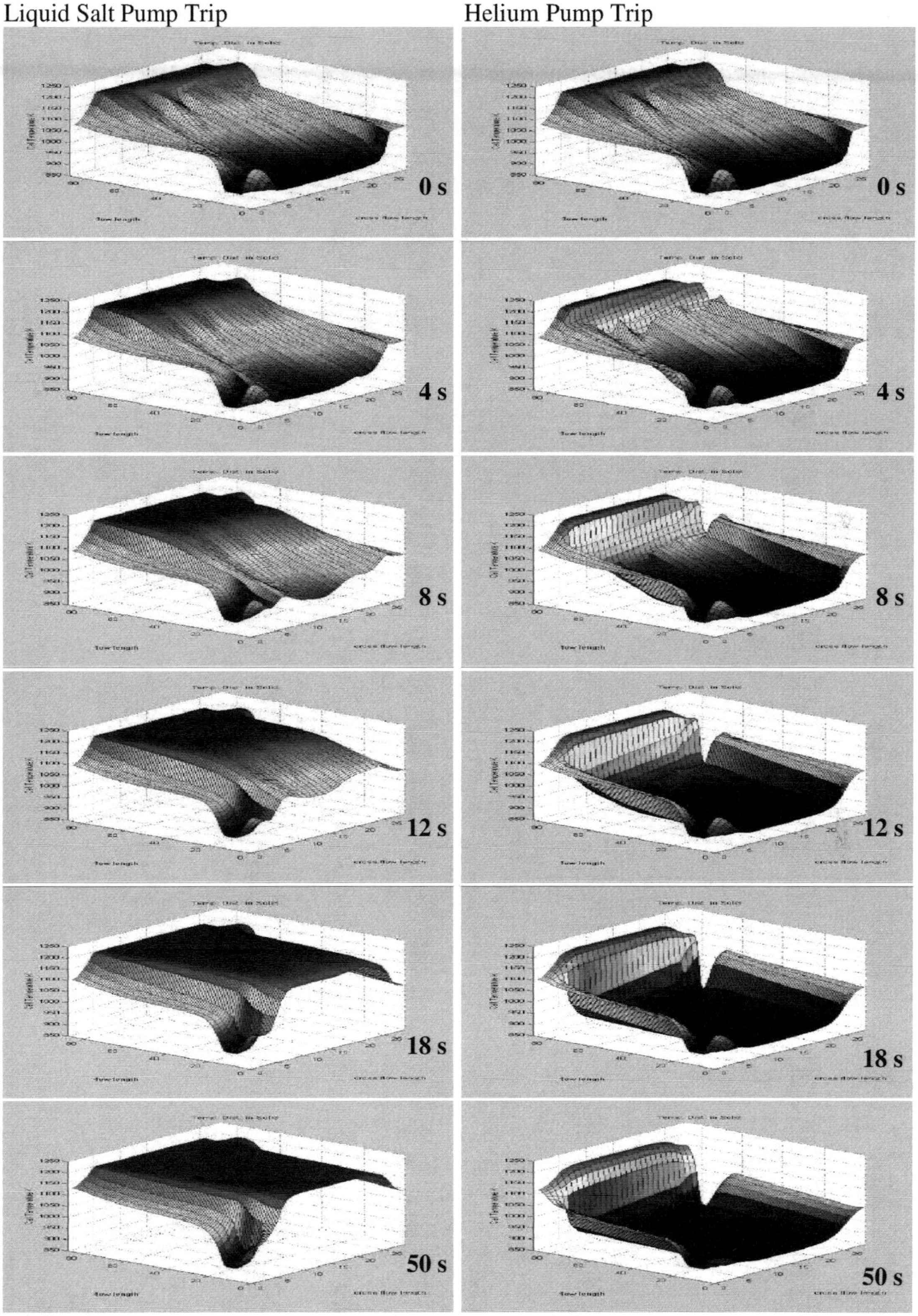


**Figure 17**: Transient temperature response to interruption of flow of liquid salt (left) or helium (right)

**Nomenclature**

– thermal diffusivity [$m^2/s$]
a' – surface area per unit volume [$m^2/m^3$]
c – specific heat at constant pressure [J/kg/K]
x – length of a finite volume (control volume) or grid length
y – height of a finite volume (control volume) or grid height
z – width of a finite volume (control volume) or grid width
$D_h$ – hydraulic diameter [m]
$f_f$ – Fanning friction factor
Gz – Graetz Number
h – convection coefficient [$W/m^2$]
i – x index
j – y index
k – z index
$k_x$ – conductivity in x direction [$m^2$]
$k_z$ – conductivity in z direction [$m^2$]
$k_{px}$ – permeability in x direction [$m^2$]
$k_{pz}$ – permeability in z direction [$m^2$]
μ – dynamic viscosity [Pa*s]
Re – Reynolds Number
– density [kg/ $m^3$]
Pe – Peclet Number
– flow potential [Pa]
$\phi$ – porosity
t – time [s]
T – temperature [K]
$u_D$ – Darcy fluid speed in x direction
$u_{int}$ – interstitial fluid velocity
$\overline{u}$ – averaged fluid speed
u – fluid speed (Darcy) in x direction [m/s]
w – fluid speed (Darcy) in z direction [m/s]
x – primary flow direction [m]
y – direction of plate stacking [m]
z – cross flow direction [m]

**Future Work**

- Calculation of effective thermal and mechanical properties
- Include the welded pipe section on the IHX
    - This could prove to be a critical area
- Reduction of OSF bounding wall thickness to reduce thermal inertia
- Include temperature dependant Fluid Properties
- Include convection coefficients that vary with fluid velocity

**References**


1. W. Huang, P.F. Peterson and H. Zhao, "C/SiC HX Thermal and Mechanical Analysis," Report UCBTH-05-004, September 30, 2005.

2. W. Huang, H. Zhao, and P.F. Peterson, "Multi-scale Stress Analysis for Compact Plate Heat Exchangers," 2006 International Congress on Advances in Nuclear Power Plants (ICAPP '06), Reno, NV, June 4-8, 2006.

3. H. Zhao and P.F. Peterson, "Design and Development Strategy for a First-Generation Commercial Advanced High Temperature Reactor with Metallic Vessel Internals (AHTR-MI)," Report UCBTH-05-005, October 22, 2005

4. Patankar, Suhas V., Numerical heat transfer and fluid flow; McGraw-Hill, New York, NY, 1980.

5. Özı ık, M. Necati., Finite difference methods in heat transfer; CRC Press, Boca Raton, FL, 1994

6. Anderson, Tannehill, Pletcher; Computational fluid mechanics and heat transfer McGraw-Hill: New York, NY, 1984

7. Hu and Herold, "Prandtl number effect on offset fin heat exchanger performance: predictive model for heat transfer and pressure drop," Int. J. Heat Mass Transfer. Vol. 38, No. 6. pp. 1043-1051, 1995

8. Manlik and Bergles, "Heat transfer and pressure Drop Correlations for the Rectangular Offset Strip Fin Compact heat exchanger," Experimental Thermal and Fluid Science 10:171-180; 1995

9. E. Urquiza-Fernández, A. Niquille, and P.F. Peterson, "Intermediate Heat Exchanger Dynamic Thermal Response Model," Report UCBTH-07-004, May 31, 2007

10. Per F. Peterson, "Capillary Tube and Shell Heat Exchanger Design for Helium to Liquid Salt Heat Transfer," Report UCBTH-07-003, U.C. Berkeley, May 7, 2007.

# AUTHOR INDEX

# AUTHOR INDEX

# AUTHOR INDEX

**American Institute of Chemical Engineers**
3 Park Avenue
New York, New York 10016-5991

**ISBN: 978-1-60423-843-3**